Sustainable Agriculture Reviews

Volume 28

Series editor

Eric Lichtfouse, Aix-Marseille Univ, CEREGE, CNRS, IRD, INRA, Coll France, Europole Mediterraneen de l'Arbois, Avenue Louis Philibert, Aix en Provence, France

Sustainable agriculture is a rapidly growing field aiming at producing food and energy in a sustainable way for humans and their children. Sustainable agriculture is a discipline that addresses current issues such as climate change, increasing food and fuel prices, poor-nation starvation, rich-nation obesity, water pollution, soil erosion, fertility loss, pest control, and biodiversity depletion.

Novel, environmentally-friendly solutions are proposed based on integrated knowledge from sciences as diverse as agronomy, soil science, molecular biology, chemistry, toxicology, ecology, economy, and social sciences. Indeed, sustainable agriculture decipher mechanisms of processes that occur from the molecular level to the farming system to the global level at time scales ranging from seconds to centuries. For that, scientists use the system approach that involves studying components and interactions of a whole system to address scientific, economic and social issues. In that respect, sustainable agriculture is not a classical, narrow science. Instead of solving problems using the classical painkiller approach that treats only negative impacts, sustainable agriculture treats problem sources. Because most actual society issues are now intertwined, global, and fast-developing, sustainable agriculture will bring solutions to build a safer world. This book series gathers review articles that analyze current agricultural issues and knowledge, then propose alternative solutions. It will therefore help all scientists, decision-makers, professors, farmers and politicians who wish to build a safe agriculture, energy and food system for future generations.

More information about this series at http://www.springer.com/series/8380

Sabrina Gaba · Barbara Smith
Eric Lichtfouse
Editors

Sustainable Agriculture Reviews 28

Ecology for Agriculture

Editors
Sabrina Gaba
UMR Agroécologie
INRA
Dijon
France

Barbara Smith
Centre for Agroecology, Water and Resilience
Coventry University
Coventry
UK

Eric Lichtfouse
Aix-Marseille Univ, CEREGE, CNRS, IRD, INRA, Coll France
Europole Mediterraneen de l'Arbois
Avenue Louis Philibert, Aix en Provence
France

ISSN 2210-4410 ISSN 2210-4429 (electronic)
Sustainable Agriculture Reviews
ISBN 978-3-030-07988-8 ISBN 978-3-319-90309-5 (eBook)
https://doi.org/10.1007/978-3-319-90309-5

Printed on acid-free paper

This Springer imprint is published by the registered company Springer International Publishing AG part of Springer Nature
The registered company address is: Gewerbestrasse 11, 6330 Cham, Switzerland

Preface

> The true roots of agroecology probably lie in the school of process ecology as typified by Tansley (1935), whose worldview included both biotic entities and their environment.
>
> Dalgaard, Hutchings and Porter https://doi.org/10.1016/S0167-8809(03)00152-X

Food security is and will increasingly be a major world issue in the context of ever-growing population, limitations of land resources and changing climate. Agroecology offers a promising alternative to industrial and pesticide-based crop production. However, agroecology cannot be restricted to the study of ecological processes that underlie the functioning of agroecosystems, and it engages multiple disciplines. Ecology is a science of complexity that provides a panel of theories, concepts and approaches to increase our understanding of farming systems by integrating different levels of life organization at multiple scales of time and space. This book presents reviews that analyse current challenges faced by agriculture from an ecological perspective, through the eye of several disciplines such as

Grass weeds invading maize plots (Fanazo et al., Chap. 4)

Foxglove aphid (Shah et al., Chap. 5)

eco-evolution, ecotoxicology, ecological economics and political ecology. This book is joined initiative of the Agricultural Ecology Group of the British Ecological Society and the Ecologie and Agriculture Group of the Société Française d'Ecologie.

This book presents principles and applications of ecology in agriculture. The first chapter by Gaba et al. reviews ecological concepts that are applicable for agricultural production, with emphasis on the effect of the landscape on biodiversity and ecosystem functions. The use of allelopathy, a kind of biochemical war between species, to control weeds is explained by Aurelio et al. in Chap. 2. Then, Rayl et al. teach us how to manipulate agroecosystems to favour natural pest enemies, a process known as conservation biological control, in Chap. 3. In the same vein, Fanadzo et al. provide in Chap. 4 examples of weeds and pest management using conservation agriculture practices such as cover crops. The ecology of aphids, pests that transmit viruses to tomatoes, is reviewed by Shah et al. in Chap. 5. Francaviglia et al. present the ecosystem services of soil organic carbon, with focus on carbon sequestration and irrigation, in Chap. 6. The effects of conventional and organic fertilizers on soil organic carbon and soil fungi are reviewed by Souza and Freitas in Chap. 7. Deguine et al. reveal successful agroecological control in mango production, with focus on arthropods, in Chap. 8. The last chapter by Keshavarz and Karami presents ecosystem services used to manage drought in agriculture, in the context of climate change.

Dijon, France — Sabrina Gaba
Coventry, UK — Barbara Smith
Aix en Provence, France — Eric Lichtfouse

Contents

Chapter 1
Ecology for Sustainable and Multifunctional Agriculture

Sabrina Gaba, Audrey Alignier, Stéphanie Aviron, Sébastien Barot, Manuel Blouin, Mickaël Hedde, Franck Jabot, Alan Vergnes, Anne Bonis, Sébastien Bonthoux, Bérenger Bourgeois, Vincent Bretagnolle, Rui Catarino, Camille Coux, Antoine Gardarin, Brice Giffard, Antoine Le Gal, Jane Lecomte, Paul Miguet, Séverine Piutti, Adrien Rusch, Marine Zwicke and Denis Couvet

Abstract The Green Revolution and the introduction of chemical fertilizers, synthetic pesticides and high yield crops had enabled to increase food production in the mid and late 20th. The benefits of this agricultural intensification have however reached their limits since yields are no longer increasing for many crops, negative externalities on the environment and human health are now recognized and economic inequality between farmers have increased. Agroecology has been proposed to secure food supply with fewer or lower negative environmental and social

S. Gaba (✉) · M. Blouin · B. Bourgeois
Agroécologie AgroSup Dijon, INRA, Université Bourgogne Franche-Comté, 21000 Dijon, France
e-mail: sabrina.gaba@inra.fr

A. Alignier · S. Aviron
UMR 0980 BAGAP, Agrocampus Ouest, ESA, INRA 35042 Rennes cedex, France

S. Barot
IEES-Paris (CNRS, UPMC, IRD, INRA, UPEC), UPMC, 75252 Paris cedex 05, France

M. Hedde
INRA, UMR 1402 ECOSYS, RD 10, 78026 Versailles Cedex, France

F. Jabot
Irstea, UR LISC, Centre de Clermont-Ferrand, 63178 Aubière, France

A. Vergnes
Université Paul Valery, Montpellier 3, UMR 5175 CEFE, (CNRS, UM UPVM3, EPHE, IRD), 1919 Route de Mende, Montpellier, France

A. Bonis
UMR 6553 ECOBIO, Université Rennes I- CNRS, Campus de Beaulieu, 35042 Rennes Cedex, France

S. Bonthoux
UMR 7324 CITERES, CNRS, INSA Centre Val de Loire, Ecole de la Nature et du Paysage, 41000 Blois, France

S. Gaba et al. (eds.), *Sustainable Agriculture Reviews 28*, Ecology for Agriculture 28,
https://doi.org/10.1007/978-3-319-90309-5_1

impacts than intensive agriculture. Agroecology principles are based on the recognition that biodiversity in agroecosystems can provide more than only food, fibre and timber. Hence, biodiversity and its associated functions, such as pollination, pest control, and mechanisms that maintain or improve soil fertility, may improve production efficiency and sustainability of agroecosystems. Although appealing, promoting ecological-based agricultural production is not straightforward since agroecosystems are socio-ecosystems with complex interactions between the ecological and social systems that act at different spatial and temporal scales. To be operational, agroecology thus requires understanding the relationships between biodiversity, functions and management, as well as to take into account the links between agriculture, ecology and the society. Here we review current knowledge on (i) the effect of landscape context on biodiversity and ecosystem functions and (ii) trophic and non-trophic interactions in ecological networks in agroecosystems. In particular, many insights have been made these two previous decades on (i) the interacting effects of management and landscape characteristics on biodiversity, (ii) the crucial role of plant diversity in delivering multiple services

V. Bretagnolle · R. Catarino · C. Coux
Centre d'Etudes Biologiques de Chizé, UMR7372, CNRS, Université de La Rochelle,
79360 Villiers-en-Bois, France

A. Gardarin
UMR Agronomie, INRA, AgroParisTech Université Paris-Saclay,
78850 Thiverval-Grignon, France

B. Giffard
Bordeaux Sciences Agro, Université de Bordeaux, 33170 Gradignan, France

A. Le Gal · J. Lecomte
Ecologie Systématique Evolution, Université Paris-Sud, CNRS,
AgroParisTech Université Paris-Saclay, 91400 Orsay, France

P. Miguet
INRA, UR 1115, PSH (Plantes et Systèmes de culture Horticoles), 84000 Avignon, France

S. Piutti
UMR 1121 Agronomie et Environnement, INRA, Université de Lorraine,
54500 Vandoeuvre-lès-Nancy, France

A. Rusch
UMR SAVE, INRA, Bordeaux Science Agro, ISVV, Université de Bordeaux,
33883 Villenave d'Ornon, France

M. Zwicke
UPEC, Institute of Ecology and Environmental Sciences of Paris – UMR7618,
61 avenue du Général de Gaulle, 94010 Créteil, France

D. Couvet
UMR CESCO, MNHN-SU-CNRS, 55 rue Buffon, 75005 Paris, France

Present Address:
S. Gaba
USC 1339, Centre d'Etudes Biologiques de Chizé, INRA, Villiers en Bois,
79360 Beauvoir sur Niort, France

and (iii) the variety of ecological belowground mechanisms determining soil fertility in interaction with aboveground processes. However, we also pinpointed the absence of consensus on the effects of landscape heterogeneity on biodiversity and the need for a better mechanistic understanding of the effects of landscape and agricultural variables on farmland food webs and related services. We end by proposing new research avenues to fill knowledge gaps and implement agroecological principles within operational management strategies.

Keywords Agroecology · Ecological intensification · Ecosystem services
Eco-evolutionary dynamics · Biotic interactions · Landscape heterogeneity
Socio-ecological systems

1.1 Introduction

Contemporary agriculture faces conflicting challenges due to the need of increasing or expanding production (i.e. food, feed, bioenergy) while simultaneously reducing negative environmental impacts. The heavy agricultural reliance on synthetic chemical pesticides or fertilizers for crop protection and crop nutrition is leading to soil, air and water pollution (agriculture represents 52 and 84% of global methane and nitrous oxide emissions, Smith et al. 2008; more than 50% of the nitrogen applied to fields is not taken up by crops, Hoang and Allaudin 2011), as well as a dramatic decline of biodiversity (67% of the most common bird species in Europe, i.e. mainly farmland species, (Inger et al. 2014), soil degradation concerning about 40% of cropped areas worldwide (Gomiero et al. 2011)) and the degradations in ecosystem functioning (Tilman et al. 2001; Cardinale et al. 2012). Agroecology principles suggest that strengthening ecosystem functions will improve the production efficiency and sustainability of agroecosystems, while decreasing negative environmental and social impacts (Gliessman 2006; Altieri 1989; Altieri and Rosset 1995; Wezel et al. 2009). One generic term grouping approaches that rely on strengthening ecosystem functions, such as pollination, pest control, and mechanisms that maintain or improve soil fertility, is 'ecological intensification' (Doré et al. 2011; Bommarco et al. 2013; Tittonnel et al. 2016, but see Godfray 2015). Such an approach fits the aim of adopting a sustainable and multifunctional agriculture, i.e. an agriculture that delivers multiple ecosystem services (Fig. 1.1). However, it constitutes a knowledge challenge as it requires to both understand and manage ecosystem functions and also to take into account the relationships between agriculture, ecology and the society.

Agroecosystems are commonly defined as ecological systems that are modified by humans to produce food, fibres or other agricultural products (Conway 1987). They are prime examples of social-ecological systems (Redman et al. 2004; Collins et al. 2007; Mirtl et al. 2013): multiple interactions between farmers, societies and ecological systems are indeed involved in the sociological and ecological dynamics. However, until fairly recently, social and biophysical processes were most often

Fig. 1.1 Example of a technique delivering multiple ecosystem services. This multifunctional cover crop is composed of *Vicia sativa, Trifolium alexandrinum, Phacelia tanacetifolia* and *Avena strigosa* and is designed to enhance soil fertility (nitrogen supply via legumes, nitrogen retention through *A. strigosa*, erosion control and soil organic matter enhancement thanks to biomass production), to support some pollinators (thanks to flowering *P. tanacetifolia* and *T. alexandrinum*) and to maintain natural enemies between successive crops (thanks to legumes providing alternative hosts to aphid predators and *V. sativa* providing extrafloral nectar)

considered separately. For instance, questions regarding agricultural production on one hand, and those regarding social needs and diets on the other hand, were treated apart. Hence one avenue to improve sustainability in agriculture is to treat agriculture ecological impacts with the same attention than question of optimal food production. This requires to adopt an ecological perspective with interactions and networks as core concepts. Research is currently dealing with many issues, from ecological point of view, such as: *How can greenhouse gas emissions be minimized? How can the impacts on biodiversity be reduced? Where and how should biofuels be produced to avoid or limit impacts on biodiversity? How can we solve the land-sharing/land-sparing debate (Green* et al. *2005) regarding biodiversity conservation*? *How to design efficient biodiversity based agricultural systems to ensure the availability of natural resources (water, fossil resources, phosphorus…)? How production types and biodiversity interact with social issues? How can we alleviate poverty and hunger through innovative food production systems (Griggs* et al. *2013), as well as appraise the new diet challenges of developed countries?* All these burning questions require to be addressed together and to solve the nexus between provisioning goods, climate, social context and biodiversity (Tomish et al. 2011).

A better understanding of the interactions within and between the ecological and social templates, and processes underlying them will help to improve the analysis of farming system and public policies (Cumming et al. 2013). Yet, both the ecological and the social templates have their own and peculiar characteristics, that must be accounted for. For example, arable fields are dominated by one single plant species (the crop), and both the abiotic and biotic environment are modified to increase biomass production by human practices (Swift et al. 2004), which thereafter affects nutrients and ecological processes (e.g. competition for resources). The conventional practices tend to reduce the magnitude of ecologically-driving mechanisms beneficial for crop production: for instance, pesticides may may reduce tri-trophic interactions between pest and their predator or parasitoid by killing non-targeted potentially beneficial organisms (Potts et al. 2010; Pelosi et al. 2014); losses of soil organic matter and tillage practices tend to reduce the abundances of soil fauna and microorganisms (Kladivko 2001; Roger-Estrade et al. 2010) and thereafter their beneficial effect on soil fertility.

Biodiversity is one of the mostly affected dimension of ecosystem due to intensively managed agroecosystems: in croplands, the plant biodiversity is strongly biased towards short-lived disturbance-tolerant plant strategies. Together, tillage impedes the development of a structured soil profile with organic-rich layer at the soil surface. As a consequence of selection of new crop varieties through intensive breeding technics for fast growth rates in nutrient and water rich environment, crop plants have evolved from resource-conservation towards resource-acquisition traits in comparison to wild species (Tribouillois et al. 2015; Delgado-Baquerizo et al. 2016; Milla et al. 2015). This contributes to nutrients' leaching from agroecosystems (Gardner and Drinkwater 2009). Rather to make the most of ecological processes, the current practices thus limit ecological interactions and keep them as neutral as possible to reduce their imponderable effects on crop production.

Ecological and socioeconomic processes act at different spatial scales, since field or farm scales are rarely ecologically meaningful (Cummings et al. 2013). Agroecosystems are complex, in particular because they are driven by spatially nested decision-making that range from farmer decisions at local scale (e.g. field) to societal management and political decisions at regional and national scales. Given that complexity, it is perhaps not surprising that in several cases ecological laboratory studies do not reflect the results obtained in long-term field studies. This is the case for the study of the impact of genetically modified (GM) crops on natural enemies (Lövei, Andow, and Arpaia (2009); Box 1) and biological control (Frank van Veen et al. 2006). Taken together, these arguments suggest that studying the relationships between agricultural practices and ecological processes (i.e. biotic interactions related to pest control, pollination, biogeochemical cycles and soil fertility) at nested scales (field, farm, landscape) is mandatory to develop sustainable and multifunctional agriculture.

Box 1: Technology, Agro-Ecological Engineering, and Socio-Cultural Mismatch: The Case of Genetically Modified Crops

Technologies can reshape interactions between humans and ecosystems, namely between agro-ecosystems, the agri-food system and the overall socio-ecological system. Since the green revolution, modern food production has become highly dependent on agricultural technological advances (Altieri and Nicholls 2012). Despite its numerous claimed benefits and widespread adoption (Lu et al. 2012; Klümper and Qaim 2014), no other agriculture technological advance has been as controversial as the development of GM crops (Stone 2010). There is still intense discussion in the research community on whether the use of this technology in agriculture may contribute to a sustainable agriculture reaching the world nutritional demand (Ervin et al. 2011; Godfray et al. 2010). The arguable environmental uncertainty of GM crops allied with the feasibility (or even ethicality) of food monopolization, and the enormous economic interests at stake for the biotechnology industry make this topic rather complex (Glover 2010). Besides the conceivable ecological risks directly caused by the employment of GM crops (Dale et al. 2002) and the dispersion of its contents (Piñeyro-Nelson et al. 2009), which may take several years to manifest (Catarino et al. 2015), other questions and challenges have arose.

Biotechnology companies and some academic proponents claim that GM crops are a crucial scientific step forward in order to meet food security demands (Tester and Langridge 2010; Qaim and Kouser 2013), however some evidence dispute these assertions. Research and political priorities, and the consequent employment of new plant strains usually occur with little knowledge on the intricacies of their impact on the complex socio and agri-food systems of small-scale farmers (Glover 2010; Altieri and Rosset 2002). A key example is the case of the Golden Rice (for details see Stein et al. 2006 and Paine et al. 2005), more than a decade after its development, is still not available (Whitty et al. 2013). Instead, in developing countries, two plants dominate the GM market, *Bt* Cotton and *Bt* Maize (James 2014). Since the intellectual property rights system implemented in many countries promote a restricted number of private companies with an excessive dominance (Rao and Dev 2009; Russell 2008), it has been argued that strong adoption of GM crops in developing countries, such as *Bt* maize in South Africa, may actually result from a lack of choice rather than being a direct benefit of the technology (Witt et al. 2006), or as Gouse et al. (2005) claim "a technological triumph but institutional failure".

In addition, evaluating the suitability of this technology has mainly focused on immediate ecological and economic impact (Fischer et al. 2015). There is a clear lack of knowledge regarding the actual social impacts of GM crops introduction, particularly within smaller-scale and resource-poor farmers (Fischer et al. 2015; Stone 2011). Still, it is clear that the amalgamation of these factors create a technological regime and a lock-in situation that delays the development of alternative agriculture solutions (Vanloqueren and Baret 2009; Dumont et al. 2016) and limited food sovereignty (Jansen 2015). Thus, the sustainability of an agriculture innovation, including biotechnology, is dependent on the relationship between economic performance while addressing key social, ecological and political challenges facing the adopting farmers (Ervin et al. 2011). The latest gene editing techniques, including CRISPR-Cas 9 method, relaunch this debate and highlight the importance to focus on broad issues on sustainability rather than on technologies (Abbott 2015).

References cited in Box 1

Altieri M A, Nicholls CI (2012) Agroecology scaling up for food sovereignty and resiliency. In: Sustainable agriculture reviews. Springer, pp. 1–29

Altieri MA, Rosset P (1999) Ten reasons why biotechnology will not ensure food security, protect the environment, or reduce poverty in the developing world. AgBioForum **2**:155–162.

Catarino R, Ceddia G, Areal FJ Park J (2015) The impact of secondary pests on Bacillus thuringiensis (Bt) crops. Plant Biotechnol J **13**:601–612.

Dale PJ, Clarke B, Fontes EMG (2002) Potential for the environmental impact of transgenic crops. Nat Biotech **20**:567–574.

Dumont AM, Vanloqueren G, Stassart PM, Baret PV (2016) Clarifying the socioeconomic dimensions of agroecology: between principles and practices. Agroecol Sustain Food Syst **40**:24–47.

Ervin DE, Glenna LL, Jussaume RA (2011) The theory and practice of genetically engineered crops and agricultural sustainability. Sustainability **3**:847–874.

Fischer K, Ekener-Petersen E, Rydhmer L, Björnberg K (2015) Social impacts of GM crops in agriculture: a systematic literature review. Sustainability **7**:8598–8620.

Glover, D (2010). The corporate shaping of GM crops as a technology for the poor. J Peasant Stud **37**:67–90.

Godfray HCJ, Beddington JR, Crute IR, Haddad L, Lawrence D, Muir JF, Pretty J, Robinson S, Thomas SM, Toulmin C (2010) Food security: the challenge of feeding 9 billion people. Science **327**:812–818.

Gouse M, Kirsten J, Shankar B, Thirtle C (2005). Bt cotton in KwaZulu Natal: Technological triumph but institutional failure. AgBiotechNet **7**:1–7.

James C (2014). Global status of commercialised biotech/GM crops: 2014, ISAAA Brief No. 49. International service for the acquisition of agri-biotech applications. 978-1-892456-59-1, Ithaca, NY.

Jansen K (2015) The debate on food sovereignty theory: agrarian capitalism, dispossession and agroecology. J Peasant Stud **42**:213–232.
Klümper W, Qaim M (2014) A meta-analysis of the impacts of genetically modified crops. PLoS ONE **9**:e111629.
Lu Y, Wu K, Jiang Y, Guo Y, Desneux N (2012) Widespread adoption of Bt cotton and insecticide decrease promotes biocontrol services. Nature **487**:362–365.
Paine JA, Shipton CA, Chaggar S, Howells RM, Kennedy MJ, Vernon G, Wright SY, Hinchliffe E, Adams JL, Silverstone AL (2005). Improving the nutritional value of Golden Rice through increased pro-vitamin A content. Nat Biotechnol **23**:482–487.
Piñeyro-Nelson A, Van Heerwaarden J, Perales HR, Serratos-Hernández JA, Rangel A, Hufford MB, Gepts P, Garay-Arroyo A, Rivera-Bustamante R, Álvarez-Buylla ER (2009) Transgenes in Mexican maize: molecular evidence and methodological considerations for GMO detection in landrace populations. Mol Ecol **18**:750–761.
Qaim M, Kouser S (2013) Genetically modified crops and food security. PloS ONE **8**:e64879.
Rao NC, Dev SM (2009) Biotechnology and pro-poor agricultural development. Econ Polit Wkly 56–64.
Russell AW (2008) GMOs and their contexts: a comparison of potential and actual performance of GM crops in a local agricultural setting. Geoforum **39**:213–222.
Stein AJ, Sachdev HPS, Qaim M (2006) Potential impact and cost-effectiveness of Golden Rice. Nat Biotechnol **24**:1200–1201.
Stone GD (2010) The anthropology of genetically modified crops. Annu Rev Anthropol **39**:381–400.
Stone GD (2011). Field versus farm in Warangal: Bt cotton, higher yields, and larger questions. World Dev **39**:387–398.
Tester M, Langridge P (2010) Breeding technologies to increase crop production in a changing world. Science **327**:818–822.
Vanloqueren G, Baret PV (2009) How agricultural research systems shape a technological regime that develops genetic engineering but locks out agroecological innovations **38**:971–983.
Whitty CJM, Jones M, Tollervey A, Wheeler T (2013). Biotechnology: Africa and Asia need a rational debate on GM crops. Nature **497**:31–33.
Witt H, Patel R, Schnurr M (2006). Can the poor help GM crops? Technology, representation & cotton in the Makhathini flats, South Africa. Rev Afr Polit Econ **33**:497–513.

Here, we review ecological theories and concepts, that may be useful to understand and enhance biodiversity and ecosystem functions in agroecosystems. We first discuss the specific characteristic of agroecosystems as social-ecological systems in order to highlight the need to study ecological processes in interaction with management and human decisions, while taking into account the socio-economic context. We then present several contributions of ecological sciences on (i) the effect of landscape on biodiversity and ecosystem functions and (ii) biotic interactions in ecological networks in agroecosystems. Finally, we discuss relevant perspectives to fill current knowledge gaps to implement agroecological principles in agriculture and to go from theories to practices.

1.2 Agroecosystems are Social-Ecological Systems at Work

Dynamics of social-ecological systems depend on interactions and feedbacks between environmental and social processes (Ström 2007). Feedbacks result from human actions on one side, and from amenities and ecosystem services, environmental constraints, stochastic events or vulnerabilities on the other side. Various socio-ecosystem models (e.g. DPSIR, MEFA, HES…) emphasize different interactions, or feed-backs (Binder et al. 2013). To understand these feed-backs in the case of agriculture, different systems may be considered, agroecosystems (Loucks 1977), agri-food system (Busch and Bain 2004), and the overall socio-ecological system, emphasizing different entities, processes (Fig. 1.2). Public policies, markets and technologies determine relationships between agroecosystem and the agri-food system (Fig. 1.2). These two systems further interact with the overall socio-ecological system, through global change and society dynamics. Considering these three systems and their interactions, is necessary to analyze the nexus between food-price-, energy, available land and sustainable development goals (Obersteiner et al. 2016), or at a finer scale the relation between biodiversity conservation and poverty traps (Barrett et al. 2011).

Given their environmental impact, the way public policies are scrutinized and evaluated by the different stakeholders, is a major feedback mechanism. Agricultural policies are technically quite complex, involving at least four strata of decision-makers, from voters to politicians, administration and managers, related through a principal-client relationship (Wolfson, 2014). As a result, social choice to change agricultural modes of production faces many complexities, uncertainties, and rigidities. Indeed, social and environmental consequences of decisions, involve path-dependence and lock-in processes, particularly between technologies and social organizations (Vanloqueren and Baret 2009), accounting for difficulties to decide technical changes, even though detrimental environmental effects of the present techniques have been shown.

Beyond public policies, social processes having major environmental effects involve human demography, life-styles, including urbanization and, more specifically in regards to agriculture, types of food distribution, consumption (Seto and Ramankutty 2016) and diets overall (Bonhommeau et al. 2013), but also related

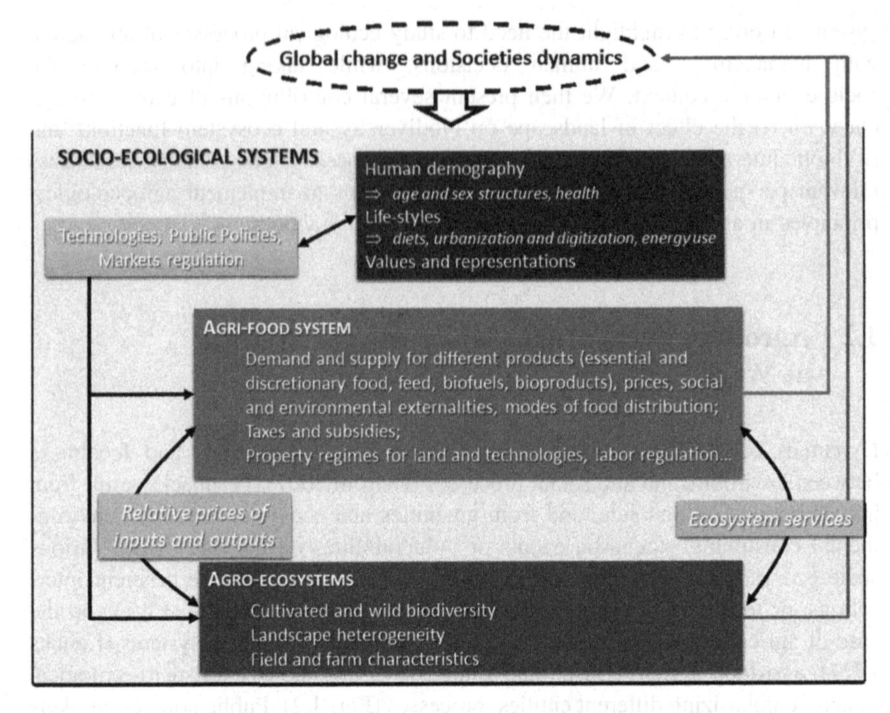

Fig. 1.2 Interaction between Agro-ecosystems, Agri-food systems and Socio-Ecosystems (adapted from Hubeau et al. 2016)

institutions (Kessler and Sperling 2016). That concerns social norms, through representations and preferences relative to diets, for example preferences for discretionary food (sensu Hadjikakou 2017). These processes determine the relationship between supply and demand, through the agri-food system, relating different kinds of producers and consumers, affecting the dynamics of local, regional and global agroecosystems. In this regard, understanding and integration of environmental impacts of diets by consumers is a major mechanism determining the relationship between societies and agro-ecosystems, promoting some types of agricultural production such as conventional, agro-ecological or organic farming, at the expense of others. For example, changes in social norms require knowledge on the relationships between the local effects on food preference induced by the global agricultural markets (Lenzen et al. 2012) and the dietary information according to nutritional requirements (deFries et al. 2015). Such information depends on life-cycle analyses (LCA, e.g. Kareiva et al. 2015; Schouten and Bitzer 2015) that can estimate the impact of market including economic incentives, such as taxes and subsidies, on agroecosystem dynamics. Then, the effect of incentives such as public policies, designed beyond the national levels and mediated through international treaties, can be evaluated on local agro-ecosystems (Friedmann 2016). Rules or guidelines may specify a desired environmental state or limit to alterations of the environment by human activities. Competition between different standards, could thus become a major determinant of the dynamics of agro-ecosystems, in the context of rigid public policies (see above). Such standards were developed first by non-governmental organisations (NGOs), in close collaboration with northern retailer actors, based at first on environmental criteria. Southern countries production actors now propose competing standards, putting more emphasis on socio-economic criteria (Schouten and Bitzer 2015), potentially leading to a different kind of agro-environmental changes. In other words, through its input on the making of environmental standards of food products, ecology could have a major impact on the dynamics of agro-ecosystems.

1.3 Reconciling Production and Biodiversity Using the Concept of Ecosystem Services

Ecosystem services concept formalizes the dependency of human societies to ecosystem functioning, between social-ecological and agro-ecosystems (Fig. 1.2). From this, ecosystem services have an operational value for rethinking the links between ecological processes and functions and expected agriculture-related services. As such, ES embraces all complexity and interactions involved and present promising avenues for addressing the sustainable production challenges, more generally to consider sustainable livelihoods.

This concept emerged during the 70's and the 80's in the scientific literature, but grew faster since 1997 and the seminal publications of Daily et al. (1997) and Costanza et al. (1997). The Millennium Ecosystem Assessment (MEA 2005) ratified a definition of ecosystem services (ES) actually proposed by Daily et al. (1997). The concept has, since then, been used as a framework in numerous initiatives and international platforms such as IPBES (Intergovernmental Science-Policy Platform on Biodiversity and Ecosystem Assessment) or SGA-Network (Sub-Global Assessment Network, operated by the United National Environment Program, UNEP) (Tancoigne et al. 2014). As being part of a socio-ecological system (SES) framework (e.g. Collins et al. 2007), ecosystem service concept emphasizes the interdependency between economic systems and ecosystems. It also offers a common framework to initiate debates between the different stakeholders, allowing operational ways of thinking for collective design and assessment of management options. For agriculture issues, the evaluation of ecosystem services requires considering, regulating and cultural services jointly to provisioning services (Bateman et al. 2013). The analyses of bundles of services relying on processes acting at different spatial scales require landscape-scale investigations (see for an example Nelson et al. 2009).

1.4 Landscape Scale, Key Scale for Agroecology

Landscape is a level of organization of ecological systems that is characterized by its heterogeneity and its dynamics that are partly driven by human activities (Burel and Baudry 2003). Agricultural landscapes are spatially heterogeneous because of the variety of cultivated land-cover types that are distributed in a complex spatial pattern and interspersed with semi-natural and/or uncultivated habitats like woodlands, hedgerows, field margins or permanent grasslands. Farmers' decision rules about cropping systems led to the highly variable of landscape mosaic in time with a diversity of crop types, organized in inter-annual sequences and with within-year management practices (Vasseur et al. 2013). In agricultural landscapes, farming activities generally operate at the field scale but their type and intensity strongly depend on processes acting at larger scales such as the farm such as type of agriculture or availability of agricultural material, the territory such as agricultural cooperatives and agri-food market, and administrative scales relevant for policy making such as national or European levels. Biodiversity patterns and their associated ecosystem functions occur at several spatial scales from some few mm^2 (e.g. soil micro-organisms) to worldwide (e.g. carbon cycle). Accordingly, ecological processes act at a variety of spatial and temporal scales, and they generate patterns at scales that may differ from that at which processes act (Levin 1992). Such nested patterns in the ecosystems drivers bring complexity that need to be taken into account for the management of biodiversity and ecosystems functions. There are therefore mismatches between the scales of ecological processes and the scale of management (Pelosi et al. 2010).

1.4.1 Absence of Consensus About the Effects of Landscape Heterogeneity on Biodiversity and Ecosystem Services

Landscape heterogeneity, defined as the composition (diversity, quality and surface of habitats) and configuration (spatial arrangement of habitats) of a landscape (Fahrig et al. 2011), has been recognized as a key driver of biodiversity and ecological processes in most agro-ecological studies (Benton et al. 2003; Bianchi et al. 2006). Landscape heterogeneity influences a variety of ecological responses, including animal movement (reviewed in Fahrig 2007), population persistence (Fraterrigo et al. 2009), species diversity (Benton et al. 2003), species interactions (Polis et al. 2004), and ecosystem functions (Lovett et al. 2005). In relation with the island biogeography theory (MacArthur and Wilson 1967), studies investigating the effects of landscape heterogeneity on ecological processes have traditionally focused on the role of semi-natural habitats viewed as embedded in a hostile agricultural matrix (Fig. 1.3).

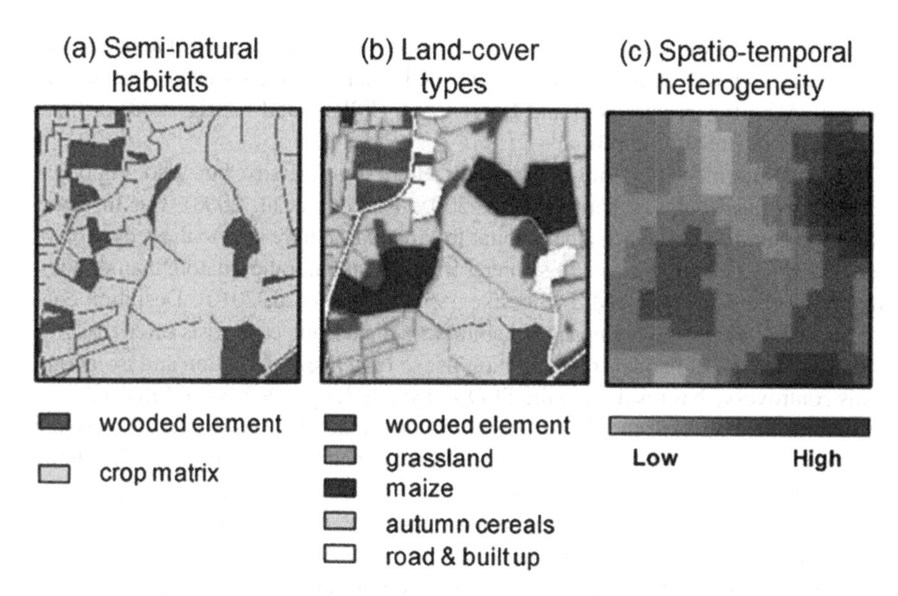

Fig. 1.3 Different representations of spatio-temporal heterogeneity, adapted from Vasseur et al. (2013): **a** spatial heterogeneity related to semi-natural habitats, **b** spatial heterogeneity related to land-cover types, and **c** spatio-temporal heterogeneity related to shift intensity in the relative crop composition of crop successions over years

Meanwhile, studies have measured landscape heterogeneity, also called "landscape complexity", as the amount or surface area of semi-natural habitats in agricultural landscapes (Benton et al. 2003). They have highlighted its role in maintaining farmland biodiversity (Baudry et al. 2000; Tscharntke et al. 2005) and enhancing ecosystem functions of economic importance such as pest predation and parasitism (Bianchi et al. 2006; Rusch et al. 2011). Indeed, semi-natural habitats provide resources (e.g. food, nesting places, shelters) for many taxa, and are often considered as "sources" of pest natural enemies in the landscape (Landis et al. 2000; Médiène et al. 2011). However, several empirical evidence also demonstrated that semi-natural habitats might fail to enhance biological control of crop pests in various context (Tscharntke et al. 2016). Generalist predators, such as aphidophagous coccinellids, may also spillover from crops to semi-natural habitats, as they exploit resources (i.e. aphid resources, overwintering sites) in both habitats (Tscharntke et al. 2005). Other studies have underlined the detrimental effect of spatial isolation of semi-natural habitats on the diversity and abundance of many taxa such as invertebrates, plants or birds (Steffan-Dewenter and Tscharntke 1999; Tewksbury et al. 2002; Petit et al. 2004; Bailey et al. 2010). Indeed, spatial isolation of semi-natural habitats alters the physical continuity of resources. The abundance and richness of species inhabiting semi-natural habitats varied with the success of finding a patch, which decreases with isolation (Goodwin and Fahrig 2002). Thus, decreasing isolation by increasing spatial connectivity[1] of semi-natural habitats with ecological corridors is a way to promote biodiversity and associated functions as demonstrated for insects (Petit and Burel 1998; Holland and Fahrig 2000) or birds (Hinsley and Bellamy 2000).

Habitat fragmentation might not only affect biodiversity but also important ecosystem functions (Tscharntke et al. 2005; Ricketts et al. 2008). For instance, there are empirical evidences that habitat fragmentation may lead to the reduction of pest control as a consequence of stronger impacts on natural predators than on their herbivore preys (Kruess and Tscharntke 1994; Bailey et al. 2010). Despite a consensus about the negative impact of habitat loss, landscape ecologists often disagree about the impact of habitat fragmentation per se (patch size reduction and isolation). This controversy has resulted in the SLOSS (Single Large Or Several Small) debate regarding how species should be conserved in fragmented landscape, i.e. through the promotion of "Single Large" or "Several Small" habitat patches (Diamond 1975; May 1975). It has been reinforced by the difficulty to quantify the relative effects of both aspects of fragmentation that are often strongly correlated in non-manipulative studies (Fahrig et al. 2011).

Semi-natural habitats play a key role in agricultural landscape. For instance, pollinators, crop pests and their natural enemies use alternatively semi-natural habitats (e.g. overwintering in hedgerows or forest edges) and crop fields to complement or supplement their resources during their life cycle for example. feeding and breeding in crop fields (Kromp 1999; Westphal et al. 2003; Rand et al. 2006; Macfadyen and Muller 2013). Other species may also interact with the whole agricultural mosaic whilst simply moving between semi-natural habitats (Vos et al. 2007). The growing awareness that the "matrix matters" for ecological processes (Ricketts 2001; Jules and Shahani 2003; Kindlmann and Burel 2008) has resulted in growing consideration of the heterogeneity of the agricultural mosaic itself. Characterizing the mosaic is not straightforward because of the strong correlation between landscape composition and configuration (Box 2). Fahrig et al. (2011) has proposed a framework to decorrelate these features (Box 2), and its use led to contradictory results about the effects of crop heterogeneity on biodiversity. For instance, Fahrig et al. (2015) have found a higher effect of crop configurational heterogeneity on multi-taxa diversity while Duflot et al. (2016) showed that carabid diversity was more affected by crop compositional heterogeneity and Hiron et al. (2015) did not find any effect of crop heterogeneity on bird diversity. The effects of crop heterogeneity thus appear highly species and case study dependent, which emphasizes the need for further researches and alternative approaches.

---

[1]Landscape connectivity is defined as the ability of landscapes to facilitates or impedes the movement of organisms (Taylor et al. 1993)

Box 2: Methodological Issues to Investigate the Effect of the Spatial and Temporal Heterogeneity of Agricultural Landscapes

Landscape heterogeneity, defined as the composition (diversity, quality and surface of habitats) and configuration (spatial arrangement of habitats) of a landscape (Fahrig et al. 2011), is a fundamental concept in landscape ecology (Wiens 2002; Fig. 1.5a). Distinguishing the relative effects of these two components is a challenge to identify on which aspects management measures should focus. Recently, several authors have proposed a conceptual framework towards a more functional view of landscape heterogeneity for farmland biodiversity, no more based upon the amount of semi-natural habitats, but considering the agricultural mosaic as composed of cultivated habitat patches with varying quality for species (Fahrig et al. 2011). This representation of functional heterogeneity is derived from a map of cover types that are characterized according to species requirements, and not according to the perception by the human observer (or remote-sensing device). Fahrig et al. (2011) have also proposed a pseudo-experimental design to disentangle metrics of landscape compositional heterogeneity (e.g. richness or Shannon diversity of cover types) and configurational heterogeneity (e.g. mean patch size, edge density) in mosaics of crops (Fig. 1.4).

From a methodological point of view, and comparatively to spatial heterogeneity, efforts are still needed to account for landscape temporal

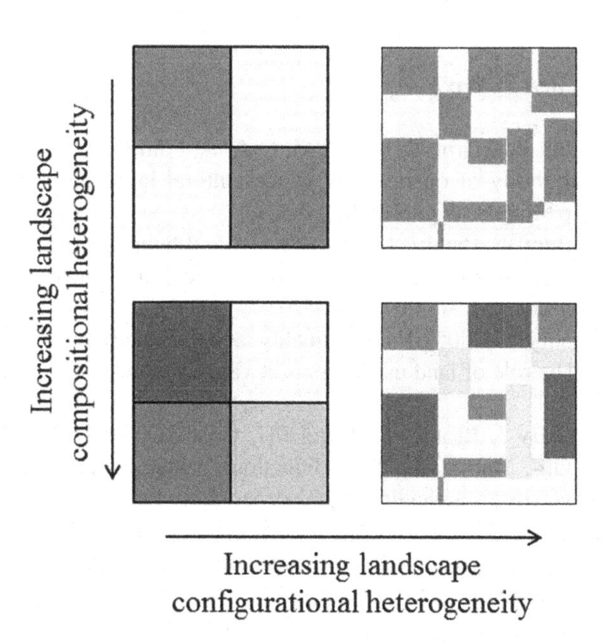

Fig. 1.4 Theoretical definition of spatial heterogeneity into its two components i.e. landscape composition and landscape configuration (adapted from Fahrig et al. 2011)

heterogeneity. Bertrand et al. (2016) proposed four general metrics to account for temporal heterogeneity of cropped areas across a short period of time. However, the authors underlined that the relevance and meaning of these metrics are strongly dependent on the cropping system under evaluation and should be studied in conjunction with other landscape factors. Indeed, in a simulation work, Baudry et al. (2003) showed that landscape changes over long time were determined by changes in the farming systems and associated changes in cropping systems.

Such dynamics of agricultural landscapes may also determine temporal variability of connectivity. Usually, measures of connectivity consider only one state of the landscape that can be past (Petit and Burel 1998) or most of the time current (Tischendorf and Fahrig 2000). Studies that relate the temporal variability of connectivity with actual agricultural systems and crop rotations are rare (but see Baudry et al. 2003, Vasseur et al. 2013). Burel and Baudry (2005) also showed high variability of connectivity from year to year in a given landscape, due to the variation in area of the crops, but also on their spatial organization. Such measure of connectivity based on dynamic structural patterns of landscapes offers the possibility to more closely link biological and landscape processes and thus, to assess the ecological outcomes of various landscape scenarios.

References cited in Box2

Baudry J, Burel F, Aviron S, Martin M, Ouin A, Pain G, Thenail C (2003). Temporal variability of connectivity in agricultural landscapes: do farming activities help? Landsc Ecol 18:303–314.

Bertrand C, Burel F, Baudry J (2016) Spatial and temporal heterogeneity of the crop mosaic influences carabid beetles in agricultural landscapes. Landsc Ecol **31**:451–466.

Burel F, Baudry J (2005) Habitat quality and connectivity in agricultural landscapes: The role of land use systems at various scales in time. Ecol Indic **5**:305–313.

Fahrig L, Baudry J, Brotons L, Burel FG, Crist TO, Fuller RJ, Sirami C, Siriwardena GM, Martin J-L (2011) Functional landscape heterogeneity and animal biodiversity in agricultural landscapes. Ecol Lett **14**:101–112.

Petit S, Burel F (1998) Effects of landscape dynamics on the metapopulation of a ground beetle (Coleoptera, Carabidae) in a hedgerow network. Agric Ecosyst Environ **69**:243–252.

Vasseur C, Joannon A, Aviron S, Burel F, Meynard J-M, Baudry J (2013). The cropping systems mosaic: How does the hidden heterogeneity of

agricultural landscapes drive arthropod populations? Agric Ecosyst Environ **166**:3–14.
Wiens JA (2002) Central concepts and issues of landscape ecology. In: KJ Gutzwiller (ed) Applying landscape ecology in biological conservation. Springer, New York, pp 3–21.

1.4.2 Towards an Explicit Account of Agricultural Practices in the Characterization of Farmland

The diversity of farming practices in fields such as plowing, direct seeding or different levels of pesticide use, and their landscape-level organization bring additional heterogeneity. Such "hidden heterogeneity" (Vasseur et al. 2013) may be as important, or even more relevant to consider, than the diversity of crop types, in driving biodiversity in agricultural landscapes. At a given time, the agricultural mosaic can indeed be viewed as a mosaic of cropped and ephemeral habitats of varying quality for species in terms of food resources, reproduction sites or, shelters. The quality of cropped habitats depends on crop type and phenology, and on disturbances induced by agricultural practices (Vasseur et al. 2013, Fig. 1.3). This mosaic of cropping systems is therefore likely to drive the source-sinks dynamics of species between crop fields, as demonstrated for pests (Carrière et al. 2004) or between crop fields and adjacent semi-natural habitats in the case of predatory insects (Carrière et al. 2009). In addition, the variable amount of suitable resources i.e. flowering resources in the agricultural mosaic has been shown to influence pollinators, that exhibit either concentrated or diluted patterns when flowering resources are rare, or on the contrary, when resources are largely distributed in agricultural landscapes (e.g. large areas as oilseed rape) (Holzschuh et al. 2011; Le Féon et al. 2013; Requier et al. 2015).

The agricultural mosaic is characterized by variations of resource localization and accessibility (i.e. landscape connectivity) for species (Burel et al. 2013). The connectivity of resource patches is expected to be crucial for species survival but few studies have addressed this issue and attempted to integrate it in the study of ecological processes (Baudry et al. 2003; Burel and Baudry 2005). All these studies have mainly focused on the variability in resource availability and quality related to crop type and phenology. However, the effects of landscape heterogeneity induced by agricultural practices have been less investigated. The few studies addressing this issue analyzed the effects of the amount of organic vs. conventional farming at the landscape scale. They have generally found a positive influence of large surfaces of organic farming in landscapes on the diversity of plants, butterflies, pollinators, and some groups of natural enemies and pest arthropods (Holzschuh et al.

2008; Rundlöf et al. 2008; Gabriel et al. 2010; Gosme et al. 2012; Henckel et al. 2015). Other studies have however failed to confirm the positive effect of organic farming at the landscape scale on communities of natural enemies (Puech et al. 2015). Several authors have underlined the need to go beyond the simple dichotomy "organic *versus* conventional" and to account for the diversity of farming practices at local and landscape scales (Vasseur et al. 2013; Puech et al. 2014). One of the key challenges to go further is to solve the difficulty of characterizing and mapping farming practices at the landscape scale (Vasseur et al. 2013).

1.4.3 Taking into Account the Temporal Variability of Agroecosystems

Agricultural landscapes are highly dynamic at various temporal scales. Temporal changes occurred from fine scale to long-temporal scales. Within-year variations are related to crop phenology and to the successive agricultural operations during the cropping season such as ploughing, sowing, fertilization application or pesticides sprays. Over decades, changes may intervene that affect the size and the shape of cropping areas and of semi-natural or extensively farmed areas (Baudry et al. 2003). Studies that have used diachronic data, mostly focused on long-term land use changes and their effects on various taxonomical groups such as plants (Lindborg and Eriksson 2004; Ernoult et al. 2006), vertebrates (Metzger et al. 2009) and invertebrates (Petit and Burel 1998; Hanski and Ovaskainen 2002). However, only few of these studies explicitly investigated impacts of landscape changes on populations dynamics (but see Wimberly 2006; Bommarco et al. 2014; Baselga et al. 2015), most probably because of the rarity of long-term monitoring data covering several years at the landscape scale. Similarly to space, no consensus has been found when investigating the effect of temporal dynamic of landscape on population or communities (e.g. for different results about bird communities, see Sirami et al. 2010; Wretenberg et al. 2010; Bonthoux et al. 2013). In particular, changes over short periods due to crop succession have been poorly investigated. At the field level, some studies have considered the impact of crop successions on invertebrates (e.g. for Carabidae, Marrec et al. 2015; Dunbar et al. 2016). At the landscape scale, temporal heterogeneity of the crop mosaic has mainly been assessed by changes in the proportions of specific crop types over time. For instance, high diversity in crop succession, with one year of grassland, positively affected solitary bee richness (Thies et al. 2008; Le Féon et al. 2013).

To sum up, few studies have accounted for the whole cropping system at a landscape scale and the effects of the multi-year temporal heterogeneity of crop mosaics on biodiversity are still largely unknown (but see Baudry et al. 2003; Vasseur et al. 2013; Bertrand et al. 2016, Fig. 1.3). This suggests that the effects of landscape heterogeneity should be assessed simultaneously in space and time and for several organisms rather than being extrapolated from static maps (Wimberly 2006).

1.5 Ecological Networks, Productivity and Biological Regulation

One pillar of agroecology is to take advantage of biotic interactions to ensure productivity and pest management instead of relying on chemical products (Shennan 2008; Médiène et al. 2011; Kremen and Miles 2012). Biotic interactions have been studied in various ecological subfields: community ecology has primarily focused on horizontal interactions between individuals of a same trophic level, while trophic ecology has primarily focused on vertical interactions between different trophic levels (Duffy et al. 2007; Fig. 1.4). An emerging line of research in network ecology focuses on interactions per se rather than through the lens of their impact on ecosystem dynamics (Tylianakis et al. 2008). The findings from these various subfields can be useful for agroecology, since they provide theoretical frameworks to interpret empirical observations (Vandermeer 1992).

1.5.1 Horizontal Diversity and Biotic Interactions

Hundreds of ecological studies have demonstrated that multispecies assemblages of plants are more productive and temporally stable than monocultures (Tilman et al. 2014). Two general mechanisms may explain these effects. First, complementarity and positive interactions between species increase production, and may even lead to transgressive overyielding i.e. some mixtures of species may have a higher production than the best monoculture. Second, species rich communities are more likely to contain the species that are more productive in local conditions during a given year. If these productive species compensate for the less productive species this can lead to overyielding through a sampling effect. Some study, however, suggested that such positive effect of diversity may be conditioned by soil fertility, that affects both functional traits and production ability of the most competitive species in the assemblage (Chanteloup and Bonis 2013). Some results also suggest that mixtures of cultivars, i.e. field genetically diverse crops, lead to the same types of benefice as species rich communities through the same ecological mechanisms (Barot et al. 2017).

One agronomic counterpart of these sample effect is a yield benefits and the higher efficiency in resource use in intercropping systems (Vandermeer 1992). The challenge for agroecology is accordingly to design multiple cropping systems that can combine several species or cultivars simultaneously in the same area or sequentially in the crop sequence (Gaba et al. 2015), and that provide food but also others ecosystem services. For instance multiple cropping systems may generate low levels of interspecific competition between crop species, or even lead to facilitative interactions, for instance through nutrient cycling as in agroforestry systems (Auclair and Dupraz 1997) or through an increased availability of minerals (Hinsinger et al. 2011). Beyond yield, multiple cropping systems may also regulate pests by preventing their growth, reproduction or dispersal as well as by enhancing natural enemies' efficiency. For instance, resource dilution of a host plant in the plant mixture can reduce both pest dispersal and reproduction by making the pest less efficient in locating and colonizing its host (Ratnadass et al. 2012). Push-pull strategies can also help controlling pest through the use of "push" plants which restrain pest settlement on crops and "pull" plants which attract them to neighboring plants (Cook et al. 2007). Finally, multiple cropping systems such as intercropping plants may control weeds by directly competing for resources with these wild plants (Liebman and Dyck 1993; Trenbath 1993).

Increasing horizontal diversity may also improve water and soil quality. Plant diversity is one of the most important drivers of belowground processes and increasing plant diversity in multiple cropping systems acts directly on soil fertility by increasing soil organic matter and promoting N_2 fixation by legumes, reducing soil erosion and the associated loss of nutrients (Dabney 1998). Indeed, multiple cropping systems influence faunal, microbial and soil organic matter dynamics through the diversity of root architecture, the quantity and quality of rhizodeposits, and the quality of plant litter. A transition from a monoculture to a diversified crop succession was shown to significantly increase microbial biomass carbon with a rapid saturation threshold due to a strong effect of cover crops (Mc Daniel et al. 2014). This overyielding of microbial biomass was observed as soon as one crop species was added in a crop sequence contrary to grasslands where the threshold is generally reached with six or eight plant species (Zak et al. 2003; Guenay et al. 2013). In addition, cereal-legume intercrops lead to a reduction of soil mineral nitrogen after harvest compared with pea sole crops (Pelzer et al. 2012), thus mitigating nitrate losses by drainage, as with cover crops and relay intercropping (Di and Cameron 2002), and hence preserving the quality of ground and drinking water.

1.5.2 Vertical Diversity and Biotic Interactions

Trophic ecology (sensu Lindeman 1942) is another body of ecological science that may be of interest for agroecology. Most of current researches focuses on pairwise trophic interactions, including the benefits brought by mutualist consumers like pollinators (Potts et al. 2010), by parasitoids (Godfray 1994; Langer and Hance 2004; Zaller et al. 2009) or by predators (Blubaugh et al. 2016; Kromp 2016; Fig. 1.6). A more recent research focus aims at understanding trophic interactions within ecological networks (Bascompte and Jordano 2007). For instance several studies investigate the potential of generalist predators for biological control (Symondson et al. 2002), the global positive relationship between natural enemy diversity and herbivore suppression (Letourneau et al. 2009; Griffin et al. 2013), or the effects of community evenness and functional diversity of natural enemy communities on biological regulation efficiency (Schmitz 2007; Crowder et al. 2010). Similarly to biotic interactions at the same trophic level (horizontal

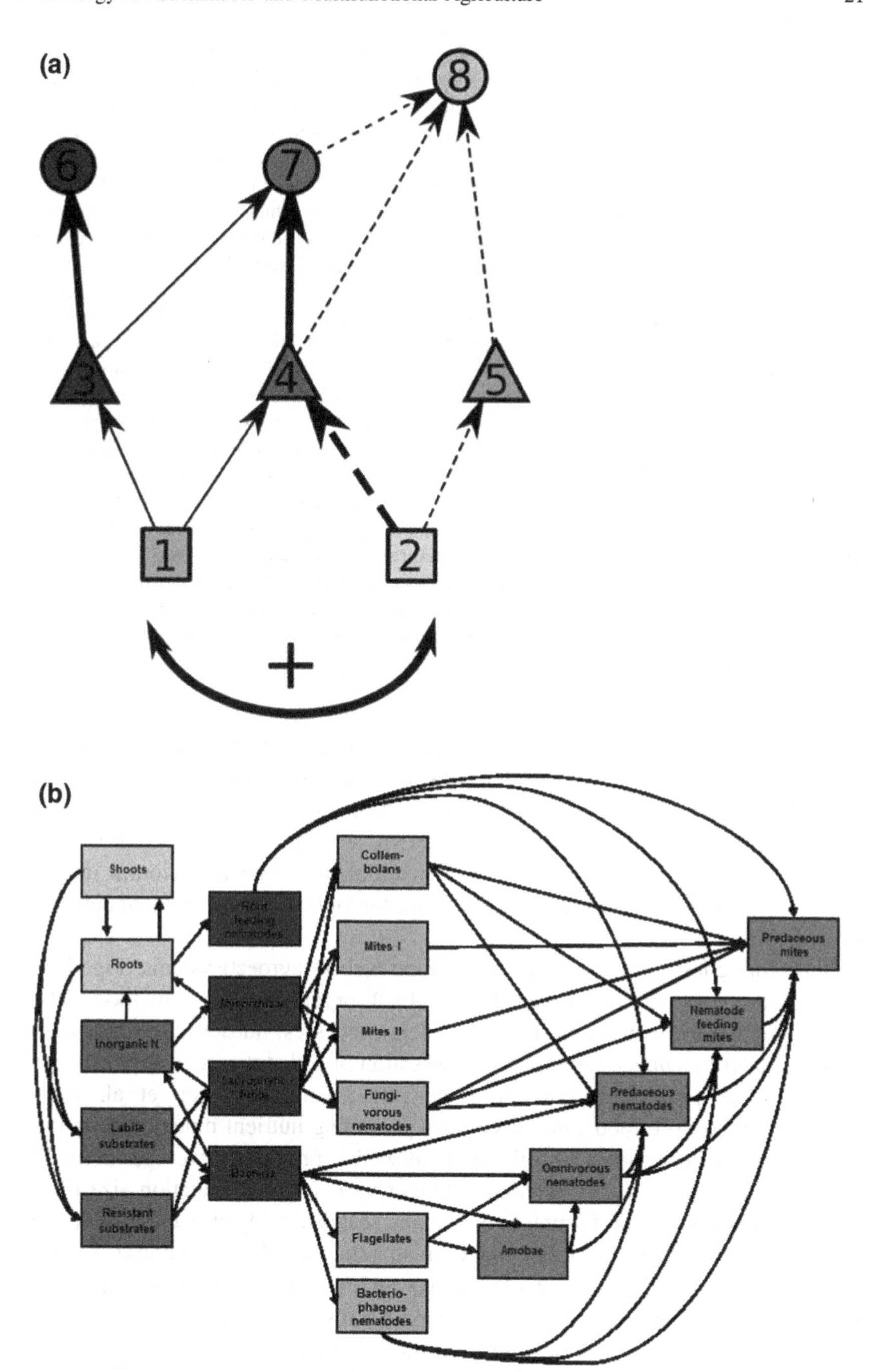
(a)
8
6
7
3
4
5
1
2
+
(b)
Shoots
Roots
Inorganic N
Labile substrates
Resistant substrates
Collem-bolans
Mites I
Mites II
Fungi-vorous nematodes
Flagellates
Bacterio-phagous nematodes
Amobae
Omnivorous nematodes
Predaceous nematodes
Nematode feeding mites
Predaceous mites

◂**Fig. 1.5 a Example of a possible interaction network in agroecosystems**. Plant hosts are symbolised by squares; triangles represent pest species and circles could symbolize either predator or parasitoid species (but we will continue with parasitoids for clarity). All arrows represent a feeding interaction between two given species, with the arrow head pointing in the direction of the biomass flow and arrow width showing the intensity of the interaction (e.g. frequency). If we consider only the species involved in interactions represented by solid arrows, then we face a system with only one plant species in isolation (crop, square 1), which is consumed by two pest species (3 and 4). In turn, one of these pests is consumed by a specialist parasitoid (6), whereas the other parasitoid (7) consumes both pest species. If we now add a second plant species (2) in the system known to have beneficial effects on the first crop species (e.g. a nitrogen-fixing species), then the second set of interactions represented by dashed arrows could occur: this second plant species is heavily consumed pest species 4, which could relieve the first crop from some of the pest pressure. This second plant could also have its own specialized pest (5). Another parasitoid species (8) also arrives in the system, feeding both on the specialized pest species (5) and the pest common to both plants (4). Finally, the link between parasitoids 8 and 7 demonstrates intraguild predation (or in this case, hyperparasitism), which confers species 8 a higher trophic position than the other species of its guild and which would potentially disserve the pest suppression process. Altogether, this figure illustrates an example of the possible benefits of increased biodiversity at all trophic levels for biotic control, and how these benefits may also, however, come with costs. **b Representation of detrital food web in shortgrass grassland**. Fungal-feeding mites are separated into two groups (I and II) to distinguish the slow-growing cryptosfigmatids from faster-growing taxa. Flows omitted from the figure for the sake of clarity include transfers from every organism to the substrate pools (death) and transfers from every animal to the substrate pools (defecation) and to inorganic N (ammonification) (Hunt et al. 1987)

diversity), the efficiency of functionally-rich networks in herbivore control can be due to a variety of ecological mechanisms including niche partitioning (Cardinale et al. 2003), sampling effect (Loreau et al. 2001) and facilitation (Gravel et al. 2016). However, it remains largely unknown how agricultural practices could promote such functionally-rich networks, without favouring behavioural interference and intraguild predation that can disrupt biological control (Rosenheim et al. 1995; Schmitz 2007).

Soil biota constitutes one of the main food web in agroecosystems (Fig. 1.5b), based on the consumption of plant litter and soil organic matter. This food web is very influential for many important ecosystem processes: nutrient cycling and soil organic matter dynamics, setting and conservation of soil structure, water fluxes and plant biomass production (Blouin et al. 2013; van Groenigen et al. 2014). Modelling detritivore food web thus allows predicting nutrient mineralization rates that are a major determinant of soil fertility as well as providing leverages to modify mineralization rates through direct or indirect variation in population size of the different soil taxa. Soil biota spreads over many scales of body size and is divided in microorganisms (fungi, bacteria), microfauna (*e.g.* amoeba, nematoda), mesoinvertebrates (e.g. collembola, enchytraeids) and macroinvertebrates (e.g. earthworms) (Lavelle et al. 2001). It also encompasses roots and vertebrates (e.g. moles, shrews). Soil biota gathers a heap of taxa, that can be translated into a diversity of shapes, stadia, ways of life or diets, that impact soil functioning from textural (arrangement of mineral particles) to structural scale (organization of aggregates, pores and horizons). The modification of nutrient availability (i.e. N or P) for plants

Fig. 1.6 Semi-natural habitats (e.g. field margins and hedges, top left) provide shelters and trophic resources for numerous components of the biodiversity in agroecosystems (e.g. hoverflies, top right). In particular, these habitats support predators and parasitoids contributing to crop herbivore suppression (hoverfly larva and aphid mummies, bottom left). To quantify the level of biological regulation, sentinel preys can be used (bottom right, ladybird predating aphids fixed on a predation card) to estimate the potential of herbivore consumption represented by predators in field conditions

is generally reported as the major mechanism explaining the direct role of soil organisms on plant growth (for earthworms, van Groenigen et al. 2014; protists, Forde 2002; collembolans, Kaneda and Kaneko 2011; or microbes, van der Heijden et al. 2008). This mechanism can provide 10–40% of the plant requirements (Parmelee and Crossley 1988; James 1991). The soil food web contains both crop pests (e.g. wireworms, pathogenic fungi) and beneficial organisms (e.g. generalist predators, parasitoids, symbiotic fungi, plant growth promoting rhizobacteria). However the characteristics of these organisms vary with their feeding requirements as well as with agricultural practices (e.g. soil burrowing, particle relocalization).

1.5.3 Integrating Trophic and Non-tropic Interactions

Although studies on ecological interaction networks have mainly focused on trophic relationships, there is an increasing interest in co-occurrence networks analyses with no a priori on the underlying interactions. Interactions can be either physical modifications of the environment with consequences on resource availability or signal molecule effects. Although the effects of organisms on abiotic environment have been recognized for decades, unifying concepts such as "ecosystem engineers", or "functional domains", are relatively recent (Fig. 1.7a). While ecosystem engineers are organisms which physically modify their environment with consequences on the availability of resources for other organisms (Jones et al. 1994), "functional domains" are defined as the sum of structures produced by a population or community of engineers (Lavelle et al. 2002). Both are particularly relevant in soil ecology where soil structure directly affects the creation, maintenance and destruction of soil organism habitats. Similarly, several soil functions such as water regulation, are deeply influenced by ecosystem engineers such as earthworms (reviewed in Blouin et al. 2013). Taking into account the effect of management practices on these organisms is thus necessary to ensure agroecosystem sustainability. Soil organisms can also interact with their environment through exchanges of information among organisms, which mainly rely on biochemical molecules. As a consequence, many soil organisms have developed the ability to detect "signal molecules", even at very low concentration in the environment which can alter the morphology, the metabolism or the behaviour of these organisms (Zhuang et al. 2013). Signal molecules can be hormones such as auxins, cytokinins, ethylene, jasmonic acid, or salicylic acid, but also other kinds of molecules (Ping and Boland 2004). In soil networks, interactions between soil organisms and plants may therefore, not necessary, require physical contacts between the soil organism and the root. As trophic interactions, they may positively or negatively impact plant development and/or immunity (Fig. 1.7b). This ability to interact with plants, responsible for the entry of energy into soils, is a vital adaptation for soil organisms (Puga-Freitas and Blouin 2015).

1.5.4 Climate and Agricultural Practices Deeply Modify Ecological Networks

Ecological networks are constituted of multiple interactions whose type (positive, neutral or negative) and strength are highly dependent on environmental properties (Tylianakis et al. 2008; Médiène et al. 2011). Understanding how exogenous factors such as climate or agricultural practices affect the network structure and consequences for ecosystem functioning, e.g. production, is however not straightforward, and this is even truer in heterogeneous and changing environments such as agroecosystems. For instance, Tylianakis et al. (2008) highlighted that climate change or habitat modifications "disproportionately affect" predator-prey interactions. They also showed that these responses differ between generalist and specialist predators, with specialists being more negatively affected by habitat simplification than generalists, as showed for aphids in Western Europe (Rand and Tscharntke 2007). Exogenous factors can also have differential effects depending if they are acting alone or in combination. For example, Hoover and Newman (2004) showed that the interaction between atmospheric CO_2 and climate may have lower negative effect on prey-predator interactions than considered separately. Recently, Romo and Tylianakis (2013) highlighted that positive effects of temperature elevation and drought on aphid host-parasitoid interactions observed when studied separately, may 'hide" deleterious effects on the efficiency of host regulation when temperature and drought are combined. The introduction of disturbances in food webs can also affect trophic network dynamics. For instance, a modelling approach revealed that manipulating low matter transfer rates had a strong influence on the trophic network dynamic, while disturbing the major path of material flow had a weak influence (Paine 1980). More recent models proposed to take into account the seasonal dynamics of trophic groups, to study the temporal coupling between mineralization by soil organisms and mineral uptake by plants. Despite increasing evidence of the effects of exogenous factors on food web structure and ecosystem functioning, we still lack of a good mechanistic understanding of the relative and interactive effects of these environmental variables.

1.6 Perspectives

1.6.1 Novel Tools to go Further in the Understanding of Ecological Network Dynamics

Considering trophic relationships with a whole food web perspective further enables to understand indirect effects propagating through food webs that would otherwise be overlooked through the lens of pairwise interactions. Indirect effects such as apparent competition have indeed potentially major implications for biological control (Chailleux et al. 2014) and the design of efficient nutrient cycling strategies within agroecosystems (Wardle et al. 2004). Integrated food web studies come at a cost, however, since achieving sufficient sampling effort to describe complete food webs remains challenging (Chacoff et al. 2012; Jordano et al. 2016). Excitingly, novel technologies offer great promises to tackle such challenges (detailed in Box 3). Another challenging task is to develop innovative modelling strategies to assess the functioning of complex food webs or even of coupled multi-type networks (Fontaine et al. 2011; Georgelin and Loeuille 2014). For instance, there is a strong indeterminacy of how indirect effects propagate through imperfectly documented food webs (Novak et al. 2011). Though, the field of ecological network modelling is a very active area of research (Tixier et al. 2013), such that modelling solutions to these challenges might be at reach. In particular, Bayesian propagation of uncertainties in ecological networks can be sufficient to obtain robust qualitative predictions (Jabot and Bascompte 2012). Nevertheless, one on the main challenges in ecological network studies is to move from a comprehensive description of networks to the understanding of the effect of external factors (climate, agricultural practices, etc.…) on the multiple interactions of different types (positive, neutral or negative) and strength in order to promote multifunctional landscapes.

Box 3: Approaches to Characterize Biological Regulation

Agroecosystems shelter complex interaction networks both above and below ground (Fig. 1.3.). In that context, quantifying biological regulation through trophic linkages is a difficult task (Traugott et al. 2013), especially for the many small and cryptic invertebrates that composed communities (Symondson et al. 2002) and that could be effective biocontrol agents. In most studies, predation rates are estimated through the consumption of preys disposed in the field for a determined amount of time (Fig. 1.6). For host-parasitoid interactions, hosts can be conserved until the emergence of adult parasitoids, which allows a good approximation of biological regulation. Notwithstanding, cryptic interactions such as multiparasitism, hyperparasitism, or when immune host system is well adapted (e.g., encapsulation) often remain challenging (Traugott et al. 2013). In a large majority of studies, the level of biological regulation exerted by generalist predators is estimated by explaining the rates of consumption of preys/hosts by predator activity-density using correlative approaches. In such cases, no correlation is usually observed, mainly because of the high variability inherent of agroecosystems. This highlights the limits of these indirect approaches and the lack of precise knowledge on species roles in food webs (Östman 2004). Thus, other approaches have been developed to provide a more mechanistic understanding of "*who eats whom?*".

First, the use of functional traits to inform food web structure through interaction rules offers great promises to tackle food web complexity with operational simplifications (Jordano et al. 2003; Allesina et al. 2008). Functional traits are defined here as morphological, physiological, phenological and behavioural features (sensu Pey et al. 2014) and may give more mechanistic understanding of predation, pollination or parasitism. For example Rusch et al. (2015) highlighted that functional diversity explained a greater part of variation in predation rates that taxonomic diversity or activity-density. However this approach is still correlative and mainly relies on the availability and the quality of data on processes, communities and traits.

Second, the democratization of DNA-based methods such as PCR diagnostic approach or next-generation sequencing (NGS) is an opportunity to better identify trophic linkage and particularly for varied diet organisms like generalist predators (reviewed by Pompanon et al. 2012; Traugott et al. 2013; Vacher et al. 2016) and pollinators (Pornon et al. 2016). The identification of alimentary items through the molecular analysis of gut content has already given important insights such as the role of carabids in the regulation of *Ceratitis capitata* in citrus orchards (Monzó et al. 2011). It also allowed detecting intraguild predation between generalist predators that could have major consequences on biological regulation efficiency and strategies (Gomez-Polo et al. 2015). However, this approach of molecular ecology is still facing many methodological such as building reliable database of DNA sequences with fine taxonomic resolution, going beyond the semi-quantitative approaches of trophic interactions or distinguishing between direct and secondary predation.

Third, the analysis of movements can help to determine the links within a food web. As most generalist predators actively hunt for preys, biological regulation is driven by specific movement behaviors of both predators and their prey. Movement behavior includes habitat use, paths, home range size, foraging behavior or activity patterns (Daniel Kissling et al. 2014). Passive tags (no battery) like Radio Frequency identification (RFID) have been used to precise the effects of pesticides on bees (Henry et al. 2012) but are still scarcely used for potential bioagents such as carabids or spiders.can be used on Curculionidae in banana plantations (Vinatier et al. 2010) or in strawberry crops (Pope et al. 2015). The significant decrease in both device size and cost of radio telemetry techniques in less than a decade opens tremendous opportunities to study predator movement behaviors.

References cited in Box3

Daniel Kissling W, Pattemore DE, Hagen M (2014) Challenges and prospects in the telemetry of insects. Biol Rev 89:511–530.

Gomez-Polo P, Alomar O, Castañé C, Aznar-Fernández T, Lundgren JG, Piñol J, Agustí N (2016). Understanding trophic interactions of Orius spp. (Hemiptera: Anthocoridae) in lettuce crops by molecular methods. Pest Manag Sci **72**:272–279.

Henry M, Beguin M, Requier F, Rollin O, Odoux J-F, Aupinel P Aptel J, Tchamitchian S, Decourtye A (2012) A common pesticide decreases foraging success and survival in honey bees. Science **336**(6079): 348–350.

Monzó C, Sabater-Muñoz B, Urbaneja A, Castañera P (2011) The ground beetle Pseudophonus rufipes revealed as predator of Ceratitis capitata in citrus orchards. Biol Control **56**: 17–21.

Pompanon F, Deagle BE, Symondson WO, Brown DS, Jarman SN, Taberlet P (2012) Who is eating what: diet assessment using next generation sequencing. Mol Ecol **21**:1931–1950.

Pornon A, Escaravage N, Burrus M, Holota H, Khimoun A, Mariette J, Vidal M (2016) Using metabarcoding to reveal and quantify plant-pollinator interactions. Sci Rep **6**:27282.

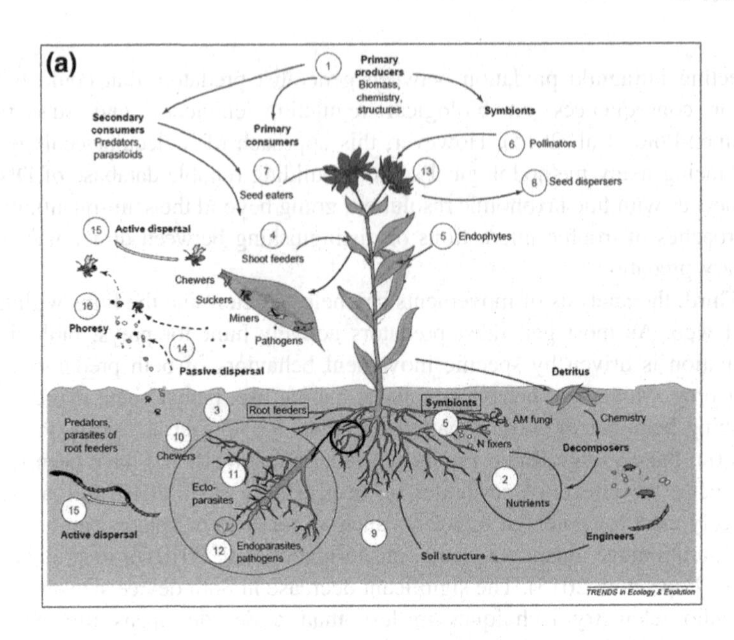
(a)
Primary producers
Biomass, chemistry, structures
Secondary consumers
Predators, parasitoids
Primary consumers
Symbionts
Pollinators
Seed eaters
Seed dispersers
Active dispersal
Shoot feeders
Endophytes
Chewers
Suckers
Miners
Pathogens
Phoresy
Passive dispersal
Detritus
Root feeders
Symbionts
AM fungi
N fixers
Chemistry
Decomposers
Predators, parasites of root feeders
Chewers
Ecto-parasites
Nutrients
Active dispersal
Endoparasites, pathogens
Engineers
Soil structure
TRENDS in Ecology & Evolution

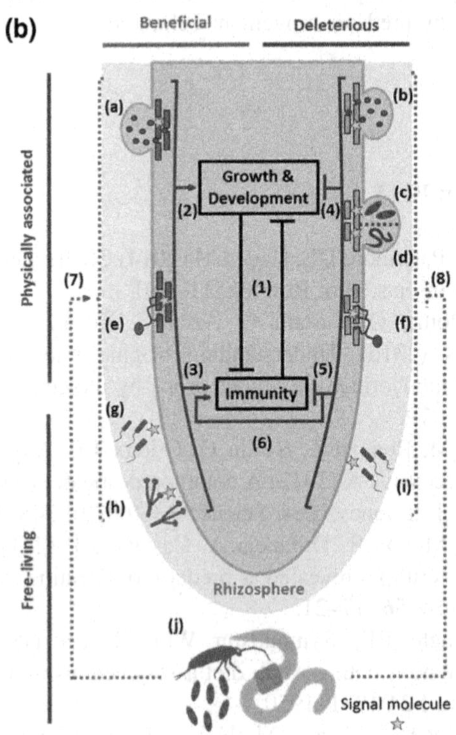
(b)
Beneficial
Deleterious
Physically associated
Free-living
(a)
(b)
(c)
(d)
(e)
(f)
(g)
(h)
(i)
(j)
Growth & Development
Immunity
(1)
(2)
(3)
(4)
(5)
(6)
(7)
(8)
Rhizosphere
Signal molecule

◂**Fig. 1.7 a Interdependency of aboveground and belowground biodiversity**. Aboveground plant community biomass and chemical and structural composition (1) drive the abundance and diversity of aboveground higher trophic levels, although these aboveground plant characteristics depend upon the net activity of soil functional groups, such as decomposers and symbionts (5), which make nutrients available (2), and on aboveground and belowground herbivores and pathogens (3,4), which reduce plant growth [17]. Heterotrophic organisms that interact with plants affect plant metabolism, potentially altering litter, shoot and root biomass production, distribution and chemical composition by feeding on roots (3) or shoots (4) or living symbiotically in shoots, leaves or roots (5). In the longer term, pollinators (6) as well as seed eaters (7) and seed dispersers (8) affect the persistence of the plant species and, thus, the specialist organisms associated with it. Soil organisms are constrained in their mobility and, as a result, organisms interacting with a single plant root system are subsets of the total species pool present in the direct surrounding soil (9). Depending on their size and mobility, these organisms occupy microhabitats of different sizes and might have different effects on plant growth. Although active roots have high turnover rates and are distributed throughout the soil, root herbivores and pathogens (3) can account for this 'unstable food' source by being relatively mobile generalist feeders (10, 11), similar to many aboveground chewing insects and free-living suckers, by adapting a specialized endoparasitic plant association (12) or by having an aboveground life phase enabling targeted active dispersal (15). Aboveground plant structures might be easier to find than are roots, and although the availability of more-specific aboveground plant tissues [e.g. buds, flowers, fruits or seeds (13)] is often brief, these can still affect the aboveground diversity of plant-associated organisms owing to the large active range sizes of aboveground organisms. Large aboveground and belowground organisms might disperse actively in a directional way (15), by flying, walking, crawling or borrowing, whereas smaller organisms (or small structures of larger organisms, such as seeds) disperse more randomly via passive dispersal (14) by air, water or via phoresy (16) (i.e. using other organisms as transport vectors). Abbreviations: AM fungi, arbuscular mycorrhizal fungi; N-fixers, nitrogen-fixing microorganisms. From de Deyn and Van der Putten (2005). **b Ecological interactions between soil organisms and plants mediated through signal molecules**. Soil organisms can be distinguished according to their physical association with plants: (a–f) root physically associated microorganisms/microfauna, (g–i) free-living microorganisms and (j) free-living micro-, meso- and macro-fauna; (a) Rhizobium, (b) Agrobacterium, (c) protozoa, (d) nematodes, (e) AM, ECM and endophytic fungi, (f) pathogenic fungi, (g) PGPR, (h) PGPF, (i) DRB and (j) micro-, meso-, macro-fauna. Arrows lines indicate a promotion, bar-headed lines an inhibition and dotted arrow lines an indirect effect of interactions mediated by signal molecules produced by soil organisms. (1) Those affecting plant growth can be detrimental to plant defence and vice et versa, due to the trade-off between these two processes. (2) Beneficial organisms promote development, induce formation of specialized organs or (3) elicit ISR. (4) Deleterious ones inhibit development, induce formation of aberrant organs, (5) hijack plant defence or (6) upon restricted infection induce SAR. We suggest that (7) free-living micro-, meso- and macro-fauna impact is mediated through the activation of beneficial microorganisms and (8) an inhibition of deleterious ones. Note that described interactions can occur within the whole root system and not exclusively at the root tip level, and induce local as well as systemic responses (Puga-Freitas and Blouin 2015)

Rusch A, Birkhofer K, Bommarco R, Smith HG, Ekbom B (2015). Predator body sizes and habitat preferences predict predation rates in an agroecosystem. Basic Appl Ecol **16**:250–259.

Traugott M, Kamenova S, Ruess L, Seeber J, Plantegenest M (2013) Empirically characterising trophic networks: what emerging DNA-based methods, stable isotope and fatty acid analyses can offer. Adv Ecol Res **49**:177–224.

Vacher C, Tamaddoni-Nezhad A, Kamenova S, Peyrard N, Moalic Y, Sabbadin R, Schwaller L, Chiquet J, Smith MA, Vallance J (2016). Chapter one-learning ecological networks from next-generation sequencing data. Adv Ecol Res **54**:1–39.
Vinatier F, Chailleux A, Duyck PF, Salmon F, Lescourret F, Tixier P (2010) Radiotelemetry unravels movements of a walking insect species in heterogeneous environments. Anim Behav **80**:221–229.

Among others, one important challenge is the multi-functionality of soils: soils should sustain primary and secondary production, provide sustainably mineral nutrients to crops, resist erosion, and regulate water fluxes and climate. This latter ecosystem function has raised high attention with the international "4 per 1000" initiative (http://4p1000.org) put on the political agenda since the COP21 held in Paris in 2015. Soils through their content in organic matter contain three times more of carbon than the atmosphere (2400 Gt carbon vs 800 Gt carbon; Derrien et al. 2016). Small changes in organic matter content can act as a carbon sink or source within a decade. A decrease in atmospheric carbon by 3.5–4 Gt/year would limit the rise in temperature by +1.5/2 °C by 2100, a threshold beyond which climate change would have unpredictable (and likely disastrous) effects (IPCC 2013). One of the most credible ways to meet this goal is to increase the concentration of organic matter in soil by 0.4% (4 per 1000) per year in the 30 first centimeters of soils at the planetary scale (Balesdent and Arrouays 1999; Paustian et al. 2016). Though some strategies can already be imagined to reach this goal (Dignac et al. 2017) research is still needed. In particular, belowground trophic and non-trophic interactions should be jointly considered to increase carbon storage within agricultural soils.

1.6.2 Landscape Issues for the Delivery of Multiple of Functions in Agroecosystems

There are increasing evidences that landscape characteristics play a strong role in maintaining biodiversity and ecosystem functions (Rundlöf and Smith 2006; Henckel et al. 2015). This offers the opportunity to manage agroecosystems at a higher scale than the field, and mainly leads to the "land sparing- land sharing" debate (Green et al. 2005; Fischer et al. 2014). Land sparing strategies rely on seting aside agricultural land for wildlife whilst intensifying agricultural production on the rest of the land. By contrast, land sharing strategies focus on the reduction of production intensity throughout the land to maintain farmland biodiversity ("wildlife friendly farming", Fig. 1.6; Green et al. 2005), whilst less amount of land is set aside for conservation in comparison with land sparing strategies. However, these two strategies are the extremes of a continuum and are not mutually exclusive and, a combination of both natural reserves and wildlife farming practices have been shown to be efficient to conserve biodiversity (Fischer et al. 2008; Scariot 2013). While this debate has been fruitful in conservation biology, the varieties of species (with different habitat requirements, and behaviours), management strategies (from field to the territory) and agricultural landscapes, require studies devoted to the agri-environmental contexts. Current knowledge either supports land sparing or land sharing strategies, depending on the organism or the ecological process considered. For instance, semi-natural habitats in agricultural landscapes are crucial for specialist species (like forest species) as well as for flagship species (e.g. wild bees, or little bustards). Friendly farming practices such as organic farming of agri-environmental schemes, may also be useful for maintaining biodiversity in agricultural landscape throughout various ecological processes (e.g. resource complementation/ supplementation, source-sink, concentration/dilution, Fig. 1.6) (Fahrig et al. 2011; Vasseur et al. 2013). Future ecological research should focus on further assessing strategies of landscape organization to determine those who can provide multifunctional agricultural mosaic according to the socio-economic and environmental contexts.

1.6.3 Towards an Eco-Evolutionary Perspective of Agroecology

Organisms are able to adapt to rapid environmental changes and to respond to strong selection pressure. Selection is an important facet of agriculture. First, crop selection has already been applied in agriculture toward pest resistant or high-yielding cultivars (Thrall et al. 2011; Denison et al. 2013). Given that the selected genotypes (crop and livestock) interact in complex ecological networks, at different trophic levels and through different types of interactions, evolutionary processes may affect ecological dynamics in multiple ways (Loeuille et al. 2013) by altering (i) important attributes of ecological communities (such as the connectance or the number of trophic levels), (ii) the structure of interaction networks (especially those with mutualistic interactions), or (iii) ecosystem functioning such as productivity or nutrient cycling in essential ways for the agricultural activity. Second, rapid natural selection of resistance to pesticides has been very often shown in various herbivores, pathogens and weeds targeted by pesticides (Alyokhin et al. 2015). Similarly, rapid evolutionary changes related to Red Queen dynamics have been demonstrated in antagonistic interactions such as host-parasite (Decaestecker et al. 2007) or prey-parasitoid or predator relationships (Diehl et al. 2013). Ecological intensification principles put forward, the substitution of strong chemical selective pressure by biotic selective pressure. Pest adaptation to their biological control is therefore highly likely, and may have collateral effects in this extreme case (Gaba et al. 2014). Third, there is an evolutionary potential of feedbacks across soil ecosystem, particularly given likely changes in soil function such as water regulation, and soil organisms such as earthworms (Blouin et al. 2013). Such environment and evolutionary feedbacks due to the change in selective constraints has led to the emergence of concepts such as the extended phenotype (Dawkins 1982) or the niche construction (Odling-Smee et al. 1996). Considering eco-evolutionary feedbacks and their potential counter-intuitive effects is therefore critical for the development of a more sustainable agriculture. Through experimental approaches, community evolution models could be valuable to better understand and predict the evolutionary consequences of management strategies on ecological networks over a wide range of agricultural context (Loeuille et al. 2013).

1.6.4 Is the Ecosystem Service Concept Relevant for Designing Sustainable Multifunctional Agriculture?

The ecosystem service concept has spread rapidly these last decades even in agricultural sciences (see Sect. 1.3) and allows exploring the relationships between ES, which could be synergies or trade-offs, despite most studies still focus on a single service (generally biomass production). Considering bundles of ecosystem services is essential to reach multifunctional agricultural and may enhance organization of agricultural landscapes in order to bond a social optimum (Couvet et al. 2016). Still, ecosystem service research is facing severe challenges which could limit, in some respects, its operability (Birkhofer et al. 2015). First, as ecosystem service concept has been initially designed to deal with "natural" ecosystems, its implementation to managed ecosystems required some adaptations (Barot et al. 2017b). Second, it is often hard to evaluate the relationships between indicators or proxies of functions whereas strong uncertainties persist on ecological processes that underlie these services. Third, the lack of theoretical framework limits the exploration of the links between services (Lescourret et al. 2015). Fourth, although land use allocation is important for agricultural production, but also for emissions and greenhouse gases sequestration, open-access recreational visits, urban green space and wild species diversity (Bateman et al. 2013), land use decisions generally ignore the valuation of ecosystem services at the landscape level. Finally, most studies still focus on small spatial and temporal scales, while environmental research should address larger scales. In addition, the ecosystem service concept has been differentially embraced according to epistemic culture of scientists coming from different academic fields. For instance, the agricultural academic field elaborated on the notion of "disservices" or "negative externalities" which are now mainstreamed into the agricultural literature (Tancoigne et al. 2014). Such negative externalities include for instance land use changes which affect natural habitats, or overgrazing that result in erosion and initiates desertification. In parallel, the concept of "environmental service" has been created to address the fact that stakeholders can be responsible for the quality of an ecosystem service and might be paid for such service (Engel et al. 2008). Debates on terminology may not only reflect the various perspectives across academic fields, but may further influence public policies. Different paradigms that are based on diverse motivations and values may also differ between stakeholders. Besides regulating and provisioning services, sociocultural services (e.g. the place attachment or the aesthetics of the landscape) or ethical considerations also contribute to human well-being. Finally, the temporal and evolutionary perspective (e.g. human-biodiversity interaction) is almost absent in the ecosystem system concept, though Faith et al. (2010) recently proposed the "evosystemic" service approach to include "the capacity for future evolutionary change and the continued discovery of useful products in the vast biodiversity storehouse that has resulted from evolution in the past". In the same vein, Sarrazin and Lecomte (2016) advocate for "evocentric approach" in order to go beyond the ecosystem service concept and ensure evolutionary freedom. It should result in a better resilience for the organisms and systems that provide ecosystem service and for all organisms involved in their ecological networks. This approach "fosters a long term, sustainable interaction that promotes both the persistence of the nature and the well-being of humans".

All these criticisms on the concept of ecosystem services express concerns on restrictive (one or few stakeholders) and static (both in time and space) views of the value of ecosystem services and the relationships between biodiversity and ecosystem services, respectively. However, we believe that a proper recognition of the benefits provided by biodiversity can increase the weight given to biodiversity per se and its functions in decision making for agriculture management. While agroecosystems have been mainly managed for provisioning services, we must also pay attention to possible overestimation of regulation processes that would lead to a utility perception of nature. With all these concerns in mind, ecosystem services remain a useful concept to address current challenges in agriculture mainly because they formalize the dependency of human societies on ecosystem functioning, between social-ecological and agro-ecosystem, which is key in socio-ecosystems.

1.7 Conclusion

In this review, we draw a picture of several relevant contributions of ecological sciences to the understanding of the spatio-temporal dynamic of biodiversity and ecosystem functions in agroecosystems. This new insights should help to design innovative farming systems at various temporal and spatial scales (from local practices during a crop cycle to the long term management of agricultural landscapes). Many insights in agroecosystem functioning have been obtained this last decade. However, we still lack a robust theoretical framework to support ecological intensification of food production systems and to translate knowledge and understanding into operational management strategies. We also need to go further in the understanding of interaction between ecological and social-economical processes and to develop approaches allowing scientist to build practices of ecological intensification together with stakeholders and policy makers. This calls for further interdisciplinary researches to investigate the importance of social-ecological processes, at various temporal (ecological and evolutionary time scale) and spatial (from small to global) scales.

Acknowledgements This study is an initiative of the "Ecology and Agriculture" group of the French Ecological Society (Société Française d'Ecologie, Sfe).

References

Altieri MA, Rosset PM (1995) Agroecology and the conversion of large-scale conventional systems to sustainable management. Int J Env Stud **50:**165–185.

Altieri MA (1999) Applying agroecology to enhance productivity of peasant farming systems in Latin America. Env Dev Sustain **1**:197–217.

Altieri MA, Rosset PM (2002) Ten reasons why biotechnology will not ensure food security, protect the environment, or reduce poverty in the developing world. In: Ethical issues in biotechnology. Rowman and Littlefield, Lanham, MD, pp. 175–82.

Alyokhin A, Mota-Sanchez D, Baker M, Snyder WE, Menasha S, Whalon M, Dively G, Moarsi WF (2015) The Red Queen in a potato field: integrated pest management versus chemical dependency in Colorado potato beetle control. Pest Manag Sci **71**:343–356.

Auclair D, Dupraz C (eds) (1997) Agroforestry for sustainable land-use: fundamental research and modelling with emphasis on temperate and Mediterranean applications. In: Forestry sciences, 60.Workshop, Montpellier, FRA (1997-06-23-1997-06-29). Dordrecht, NLD: Kluwer Academic Publishers.

Bailey D, Schmidt-Entling MH, Eberhart P, Herrmann JD, Hofer G, Kormann U, & Herzog F. 2010. Effects of habitat amount and isolation on biodiversity in fragmented traditional orchards. Journal of Applied Ecology 47:1003-1013.

Balesdent J, Arrouays D (1999) An estimate of the net annual carbon storage in French soils induced by land use change from 1900 to 1999 (note présentée par Jérôme Balesdent). Comptes rendus-académie d'Agriculture de France, **85**(6):265–277

Barot S, Yé L, Blouin M, Frascaria N (2017) Ecosystem services must tackle anthropized ecosystems and ecological engineering. Ecol Eng **99**:486–495.

Barot S, Allard V, Cantarel A, Enjalbert J, Gauffreteau A, Goldringer I, Lata J-C, Le Roux X, Niboyet A, Porcher E (2017) Designing mixtures of varieties for multifunctional agriculture with the help of ecology. A review. Agron Sustain Dev (in press).

Barrett CB, Travis AJ, Dasgupta P (2011) On biodiversity conservation and poverty traps. Proc Natl Acad Sci **108**: 13907–13912.

Bascompte J, Jordano P (2007) Plant-animal mutualistic networks: the architecture of biodiversity. Annu Rev Ecol Evol Syst **38**: 567–593.

Baselga A, Bonthoux S, Balent G (2015) Temporal beta diversity of bird assemblages in agricultural landscapes: land cover change vs. stochastic processes. PLoS ONE **10**:e0127913.
Bateman IJ et al (2013) Bringing ecosystem services into economic decision-making: land use in the United Kingdom. Science **341**: 45–50.
Baudry J, Burel F, Thenail C, Le Cœur D (2000) A holistic landscape ecological study of the interactions between farming activities and ecological patterns in Brittany, France. Landsc Urban Plan **50**:119–128.
Baudry J, Burel F, Aviron S, Martin M, Ouin A, Pain G, Thenail C (2003) Temporal variability of connectivity in agricultural landscapes: do farming activities help? Landsc Ecol **18**:303–314.
Bengtsson J, Ahnström J, Weibull AC (2005) The effects of organic agriculture on biodiversity and abundance: a meta-analysis. J Appl Ecol **42**:261–269.
Benton TG, Vickery JA, Wilson JD (2003) Farmland biodiversity: is habitat heterogeneity the key? Trends Ecol Evol **18**:182–188.
Bertrand C, Burel F, Baudry J (2016) Spatial and temporal heterogeneity of the crop mosaic influences carabid beetles in agricultural landscapes. Landsc Ecol **31**:451–466.
Bianchi FJJA, Booij CHJ, Tscharntke T (2006) Sustainable pest regulation in agricultural landscapes: a review on landscape composition, biodiversity and natural pest control. Proc R Soc B Biol Sci **273**:1715–1727.
Binder CR et al (2013) Comparison of frameworks for analyzing social-ecological systems. Ecol Soc **18**:26.
Birkhofer K et al (2015) Ecosystem services—current challenges and opportunities for ecological research. Front Ecol Evol **2**:87.
Blouin M, Hodson ME, Delgado EA, Baker G, Brussaard L, Butt KR, Dai J, Dendooven L, Pérès G, Tondoh JE, Cluzeau D, Brun J-J (2013) A review of earthworm impact on soil function and ecosystem services. Eur J Soil Sci **64**:161–182.
Blubaugh CK, Hagler JR, Machtley SA, Kaplan I (2016) Cover crops increase foraging activity of omnivorous predators in seed patches and facilitate weed biological control. Agric Ecosyst Environ **231**:264–270
Bommarco R, Kleijn D, Potts SG (2013) Ecological intensification: harnessing ecosystem services for food security. Trends Ecol Evol **28**:230–238.
Bommarco R, Lindborg R, Marini L, Öckinger E (2014) Extinction debt for plants and flower-visiting insects in landscapes with contrasting land use history. Divers Distrib **20**(5):591–599.
Bonhommeau S et al (2013) Eating up the world's food web and the human trophic level. Proc Natl Acad Sci **110**:20617–20620.
Bonthoux S, Barnagaud JY, Goulard M, Balent G (2013) Contrasting spatial and temporal responses of bird communities to landscape changes. Oecologia **172**:563–574.
Burel F, Baudry J (2003) Landscape ecology: concepts, methods, and applications. Science Publishers. p 378, ISBN:9781578082148

Burel F, Baudry J (2005) Habitat quality and connectivity in agricultural landscapes: the role of land use systems at various scales in time. Ecol Indic **5**:305–313.
Burel F, Lavigne C, Marshall E (2013) Landscape ecology and biodiversity in agricultural landscapes. Agric Ecosyst Environ 1–2.
Busch L, Bain C (2004) New! Improved? The transformation of the global agrifood system. Rural Soc **69**:321–346.
Cardinale BJ, Harvey CT, Gross K, Ives AR (2003) Biodiversity and biocontrol: emergent impacts of a multi-enemy assemblage on pest suppression and crop yield in an agroecosystem. Ecol Lett **6**(9):857–865.
Cardinale BJ, Duffy JE, Gonzalez A, Hooper DU, Perrings C, Venail P, Narwani A, Mace GM, Tilman D, Wardle DA (2012) Biodiversity loss and its impact on humanity. Nature **486**:59–67.
Carrière Y, Ellers-Kirk C, Cattaneo MG, Yafuso CM, Antilla L, Huang CY, Rahman CM, Orr BJ, Marsh SE (2009) Landscape effects of transgenic cotton on non-target ants and beetles. Basic Appl Ecol **10**:597–606.
Carrière Y, Sisterson MS, Tabashnik BE (2004). Resistance management for sustainable use of Bacillus thuringiensis crops in integrated pest management. In Horowitz AR, Ishaaya I (eds) Insect pest management: field and protected crops. Springer, Berlin, Heidelberg, pp 65–95
Chacoff NP, Vazquez DP, Lomascolo SB, Stevani EL, Dorado J, Padron B (2012) Evaluating sampling completeness in a desert plant–pollinator network. J Anim Ecol **81**:190–200.
Chailleux A, Mohl EK, Teixeira Alves M, Messelink GJ, Desneux N (2014) Natural enemy-mediated indirect interactions among prey species: potential for enhancing biocontrol services in agroecosystems. Pest Manag Sci **70**:1769–1779.
Chanteloup P, Bonis A (2013) Functional diversity in root and above-ground traits in a fertile grassland shows a detrimental effect on productivity. Basic Appl Ecol **4**:208–2016
Collins SL, Swinton SM, Anderson CW et al (2007) Integrated science for society and the environment: a strategic research initiative. In: Miscellaneous Publication of the LTER Network, http://www.lternet.edu.
Conway G (1987) The properties of agroecosystems. Agric Syst **24**:95–117
Cook SM, Khan Z, Pickett JA (2007). The use of push-pull strategies in integrated pest management. Annu Rev Entomol **52**(52):375–400.
Costanza R et al (1997) The value of the world's ecosystem services and natural capital. Nature **387**:253.
Couvet D et al (2016) Services écosystémiques: des compromis aux synergies. In: Roche P et al (eds) Valeurs de la biodiversité et services écosystémiques: Perspectives interdisciplinaires. Editions Quae, 2016.
Crowder DW, Northfield TD, Strand MR, Snyder WE (2010). Organic agriculture promotes evenness and natural pest control. Nature **466**:109–112.
Cumming GS, Olsson P, Chapin FSI et al (2013) Resilience, experimentation, and scale mismatches in social-ecological landscapes Landsc Ecol **28**:1139–1150.
Dabney SM (1998) Cover crop impacts on watershed hydrology. J Soil Water Cons **53**(3):207–213.

Daily G (1997) Nature's services: societal dependence on natural ecosystems. Island Press.
Dawkins R (1982) The extended phenotype. Oxford: Oxford University Press, ISBN 0-19-288051-9.
de Snoo GR (1999) Unsprayed field margins: effects on environment, biodiversity and agricultural practice. Landsc Urban Plan **46**:151–160.
DeFries R et al (2015) Metrics for land-scarce agriculture. Science **349**:238–240.
Decaestecker E, Gaba S, Raeymaekers JAM, Stoks R, Van Kerckhoven L, Ebert D, De Meester L (2007) Host-parasite 'Red Queen' dynamics archived in pond sediment. Nature **450** 870-U16.
Denison RF, Kiers ET, West SA (2003). Darwinian agriculture: when can humans find solutions beyond the reach of natural selection? Quart Rev Biol **78**:145–168.
Derrien D, Dignac MF, Basile-Doelsch I, Barot S, Cécillon L, Chenu C, Barré, P (2016) Stocker du Carbone dans les sols: Quels mécanismes, quelles pratiques agricoles, quels indicateurs ? Étude et Gestion des Sols **23**:193–223
Di HJ, Cameron KC (2002). Nitrate leaching in temperate agroecosystems: sources, factors and mitigating strategies. Nutr Cycl Agroecosystems **64**(3):237–256.
Diamond JM (1975) The island dilemma: lessons of modern biogeographic studies for the design of natural reserves. Biol Conserv **7**:129–146.
Diehl E, Sereda E, Wolters V, Birkhofer K (2013). Effects of predator specialization, host plant and climate on biological control of aphids by natural enemies: a meta-analysis. J Appl Ecol **50**:262–270.
Doré T, Makowski D, Malezieux E et al (2011) Facing up to the paradigm of ecological intensification in agronomy: Revisiting methods, concepts and knowledge. Eur J Agron **34**:197–210.
Duffy JE, Cardinale BJ, France KE, McIntyre PB, Thébault E, Loreau M (2007) The functional role of biodiversity in ecosystems: incorporating trophic complexity. Ecol Lett **10**:522–538.
Duflot R, Ernoult A, Burel F, Aviron S (2016). Landscape level processes driving carabid crop assemblage in dynamic farmlands. Popul Ecol **58**:265–275.
Dunbar MW, Gassmann AJ, O'Neal ME (2016). Impacts of rotation schemes on ground-dwelling beneficial arthropods. Environ Entomol, https://doi.org/10.1093/ee/nvw104
Engel S, Pagiola S, Wunder S (2008) Designing payments for environmental services in theory and practice: An overview of the issues. Ecolog Econ **65**(4):663–674
Ernoult A, Tremauville Y, Cellier D, Margerie P, Langlois E, Alard D (2006) Potential landscape drivers of biodiversity components in a flood plain: Past or present patterns? Biol Conserv **127**:1–17.
Fahrig L (2007) Non-optimal animal movement in human-altered landscapes. Funct Ecol **21**:1003–1015.
Fahrig L, Baudry J, Brotons L, Burel FG, Crist TO, Fuller RJ, Sirami C, Siriwardena GM, Martin J-L (2011) Functional landscape heterogeneity and animal biodiversity in agricultural landscapes. Ecol Lett **14**:101–112.

Fahrig L, Girard J, Duro D, Pasher J, Smith A, Javorek S, King D, Lindsay KF, Mitchell S, Tischendorf L (2015). Farmlands with smaller crop fields have higher within-field biodiversity. Agric Ecosyst Environ **200**:219–234.
Faith DP, Magallón S, Hendry AP, Conti E, Yahara T, Donoghue MJ (2010) Evosystem services: an evolutionary perspective on the links between biodiversity and human well-being. Curr Opin Environ Sustain **2**:66–74
Fontaine C, Guimarães PR, Kéfi S, Loeuille N, Memmott J, van Der Putten WH, van Veen FJ, Thebault E (2011) The ecological and evolutionary implications of merging different types of networks. Ecol Lett **14**:1170–1181.
Forde B (2002) Local and long range signalling pathways regulating plant response to nitrate. Ann Rev Plant Biol **53**:203–224.
Frank van Veen FJ, Morris RJ, Godfray HCJ (2006) Apparent competition, quantitative food webs, and the structure of phytophagous insect communities. Annu Rev Entomol **51**:187–208.
Fraterrigo JM, Pearson SM, Turner MG (2009) Joint effects of habitat configuration and temporal stochasticity on population dynamics. Landsc Ecol **24**:863–877.
Friedmann H (2016) Towards a natural history of food getting. Soc Rural, https://doi.org/10.1111/soru.12144
Gaba S, Bretagnolle F, Rigaud T, Philippot L (2014) Managing biotic interactions for ecological intensification of agroecosystems. Front Ecol Evol **2**, https://doi.org/10.3389/fevo.2014.00029.
Gaba S, Lescourret F, Boudsocq S, Enjalbert J, Hinsinger P, Journet E-P, Navas M-L, Wery J, Louarn G, Malézieux E, Pelzer E, Prudent M, Ozier-Lafontaine H (2015) Biodiversity as a driver in multispecies cropping systems: new strategies for providing multiple ecosystem services. Agron Sustain Dev **35**(2):607–623
Gabriel D, Sait SM, Hodgson JA, Schmutz U, Kunin WE, Benton TG (2010) Scale matters: the impact of organic farming on biodiversity at different spatial scales. Ecol Lett **13**:858–869.
Gardner JB, Drinkwater LE (2009) The fate of nitrogen in grain cropping systems: a meta-analysis of 15 N field experiments. Ecol Appl **19**:2167–2184.
Georgelin E, Loeuille N (2014) Dynamics of coupled mutualistic and antagonistic interactions, and their implications for ecosystem management. J Theor Biol **346**:67–74.
Gliessman SR (2006) Agroecology: the ecology of sustainable food systems. CRC Press, pp 408.
Godfray HC J (1994) Parasitoids: behavioral and evolutionary ecology. Princeton University Press.
Godfray HCJ (2015) The debate over sustainable intensification Food Secur **7**:199–208.
Goodwin BJ, Fahrig L (2002) How does landscape structure influence landscape connectivity? Oikos **99**:552–570.
Gosme M, de Villemandy M, Bazot M, Jeuffroy MH (2012) Local and neighbourhood effects of organic and conventional wheat management on aphids, weeds, and foliar diseases. Agric Ecosyst Environ **161**:121–129.

Griffin JN, Byrnes JE, Cardinale BJ (2013) Effects of predator richness on prey suppression: a meta-analysis. Ecology **94**:2180–2187.

Griggs D, Stafford-Smith M, Gaffney O, Rockstrom J, Ohman MC, Shyamsundar P, Steffen W, Glaser G, Kanie N, Noble I (2013) Policy: sustainable development goals for people and planet. Nature **495** (7441). (Nature Publishing Group, a division of Macmillan Publishers Limited. All Rights Reserved: 305–7).

Guenay Y, Ebeling A, Steinauer K, Weisser WW, Eisenhauer N (2013) Transgressive overyielding of soil microbial biomass in a grassland plant diversity gradient. Soil Biol Biochem **60**:122–124.

Gravel D, Albouy C, Thuiller W (2016) The meaning of functional trait composition of food webs for ecosystem functioning. Phil Trans R Soc B **371**: 20150268.

Grime JP (2001) Plant strategies, vegetation processes, and ecosystem properties. Wiley, New York, USA, pp 417.

Hadjikakou M (2017) Trimming the excess: environmental impacts of discretionary food consumption in Australia. Ecol Econ **131**:119–128.

Hanski I, Ovaskainen O (2002). Extinction debt at extinction threshold. Conserv Biol **16**:666–673.

Henckel L, Börger L, Meiss H, Gaba S, Bretagnolle V (2015). Organic fields sustain weed metacommunity dynamics in farmland landscapes. Proc R Soc Lond B **282**(1808) 20150002

Hinsinger P, Betencourt É, Bernard L, Brauman A, Plassard C, Shen J, Tang X, Zhang F (2011) P for Two, sharing a scarce resource: soil phosphorus acquisition in the rhizosphere of intercropped species. Plant Physiol **156**(3): 1078–1086.

Hinsley SA, Bellamy PE (2000). The influence of hedge structure, management and landscape context on the value of hedgerows to birds: A review. J Environ Manag **60**:33–49.

Hiron M, Berg Å, Eggers S, Berggren Å, Josefsson J, Pärt T (2015). The relationship of bird diversity to crop and non-crop heterogeneity in agricultural landscapes. Landsc Ecol 1–13.

Hoang VN, Alauddin M (2010) Assessing the eco-environmental performance of agricultural production in OECD countries: the use of nitrogen flows and balance. Nutr Cycle Agroecosystem **87**:353–368

Holland J, Fahrig L (2000) Effect of woody borders on insect density and diversity in crop fields: a landscape-scale analysis. Agric Ecosyst Environ **78**:115–122.

Holzschuh A, Steffan-Dewenter I, Tscharntke T (2008) Agricultural landscapes with organic crops support higher pollinator diversity. Oikos **117**:354–361.

Holzschuh A, Dormann CF, Tscharntke T, Steffan-Dewenter I (2011) Expansion of mass-flowering crops leads to transient pollinator dilution and reduced wild plant pollination. Proc R Soc B **278**:3444–3451

Hoover JK, Newman JA (2004) Tritrophic interactions in the context of climate change: a model of grasses, cereal aphids and their parasitoids. Global Change Biol **10**:1197–1208.

Jabot F, Bascompte J (2012) Bitrophic interactions shape biodiversity in space. Proc Natl Acad Sci **109**:4521–4526.

James C (2014) Global status of commercialised biotech/GM Crops: 2014, ISAAA Brief No. 49. International Service for the Acquisition of Agri-Biotech Applications. Ithaca, NY: 978-1-892456-59-1.
Jordano P (2016) Sampling networks of ecological interactions. Funct Ecol **30**:1883–1893.
Jules ES, Shahani P (2003). A broader ecological context to habitat fragmentation: Why matrix habitat is more important than we thought. J Veg Sci **14**:459–464.
Kaneda S, Kaneko N (2011) Influence of Collembola on nitrogen mineralization varies with soil moisture content. Soil Sci Plant Nutr **57**:40–49.
Kareiva PM, McNally BW, McCormick S, Miller T, Ruckelshaus M (2015) Improving global environmental management with standard corporate reporting. Proc Natl Acad Sci **112**:7375–7382
Kessler J, Sperling D (2016). Tracking US biofuel innovation through patents. Energy Policy **98**:97–107.
Kindlmann P, Burel F. 2008. Connectivity measures: a review. Landsc Ecol **23**:879–890.
Kladivko EJ (2001) Tillage systems and soil ecology. Soil Tillage Res **61**(1): 61–76.
Kremen C, Miles A (2012) Ecosystem services in biologically diversified versus conventional farming systems: benefits, externalities, and trade-offs. Ecol Soc **17**:40.
Kromp B (1999). Carabid beetles in sustainable agriculture: a review on pest control efficacy, cultivation impacts and enhancement. Agric Ecosyst Environ **74**:187–228.
Kruess A, Tscharntke T (1994). Habitat fragmentation, species loss, and biological-control. Science **264**:1581–1584.
Landis DA, Wratten SD, Gurr SM (2000). Habitat management to conserve natural enemies of arthropod pests in agriculture. Annu Rev Entomol **45**:175–201.
Langer A, Hance T (2004) Enhancing parasitism of wheat aphids through apparent competition: a tool for biological control. Agric Ecosyst Environ **102**: 205–212.
Lavelle P, Spain AV (2001) Soil ecology. Kluwer Academic Publishers, Dordrecht.
Le Féon V, Burel F, Chifflet R, Henry M, Ricroch A, Vaissière BE, Baudry J (2013) Solitary bee abundance and species richness in dynamic agricultural landscapes. Agric Ecosyst Environ **166**:94–101.
Letourneau DK, Jedlicka JA, Bothwell SG, Moreno CR (2009) Effects of natural enemy biodiversity on the suppression of arthropod herbivores in terrestrial ecosystems. Annu Rev Ecol Evol Syst **40**:573–592.
Lescourret F, Magda D, Richard G, Adam-Blondon AF, Bardy M, Baudry J, Doussan I, Dumont B, Lefèvre F, Litrico I, Martin-Clouaire R, Montuelle B, Pellerin S, Plantegenest M, Tancoigne E, Thomas A, Guyomard H, Soussana J-F (2015) A social–ecological approach to managing multiple agro-ecosystem services. Curr Opin Environ Sustain **14**:68–75.
Levin SA (1992). The problem of pattern and scale in ecology: the Robert H. MacArthur award lecture. Ecology **73**:1943–1967.

Liebman M, Dyck E (1993) Crop rotation and intercropping strategies for weed management. Ecol Appl **3**:92–122.
Lindborg R, Eriksson O (2004) Historical landscape connectivity affects present plant species diversity. Ecology **85**:1840–1845.
Lindeman RL (1942) The trophic-dynamic aspect of ecology. Ecology **23**:399–417.
Loeuille N, Barot S, Georgelin E, Kylafis G, Lavigne C (2013). Eco-evolutionary dynamics of agricultural networks: Implications for sustainable management. Adv Ecol Res **49**:339–435.
Loreau M, Naeem S, Inchausti P, Bengtsson J, Grime JP, Hector A, Hooper DU, Huston MA, Raffaelli D, Schmid B, Tilman D, Wardle DA (2001). Biodiversity and ecosystem functioning: current knowledge and future challenges. Science **294**:804–808.
Loucks O (1977) Emergence of research on agro-ecosystems. Annu Rev Ecol Syst **8**:173–192
Lövei GL, Andow D, Arpaia S (2009). Transgenic insecticidal crops and natural enemies: a detailed review of laboratory studies. Environ Entomol **38**:293–306, https://doi.org/10.1603/022.038.0201.
MacArthur R, Wilson E (1967) The theory of island biogeography. Princeton University Press, Princeton, New Jersey.
Macfadyen S, Muller W (2013) Edges in agricultural landscapes: species interactions and movement of natural enemies. PLoS ONE **8**:e59659.
Marrec R, Badenhausser I, Bretagnolle V, Börger L, Roncoroni M, Guillon N, Gauffre B (2015) Crop succession and habitat preferences drive the distribution and abundance of carabid beetles in an agricultural landscape. Agric Ecosyst Environ **199**:282–289.
May RM (1975) Islands biogeography and the design of wildlife preserves. Nature **254**:177–178.
Mc Daniel MD, Tiemann LK, Grandy AS (2014) Does agricultural crop diversity enhance soil microbial biomass and organic matter dynamics? A meta-analysis. Ecol Appl **24**:560–570.
Médiène S, Valantin-Morison M, Sarthou J-P, de Tourdonnet S, Gosme M, Bertrand M, Roger-Estrade J, Aubertot JN, Rusch A, Motisi N (2011) Agroecosystem management and biotic interactions: a review. Agron Sustain Dev **31**:491–514.
Metzger JP, Martensen AC, Dixo M, Bernacci LC, Ribeiro MC, Teixeira AMG, Pardini R (2009) Time-lag in biological responses to landscape changes in a highly dynamic Atlantic forest region. Biol Conserv **142**:1166–1177.
Millenium Ecosystem Assessment (2005) Ecosystems and human well-being. Washington, DC.
Mirtl M, Orenstein DE, Wildenberg M, Peterseil J, Frenzel M (2013) Development of LTSER platforms in LTER-Europe: challenges and experiences in implementing place-based long-term socio-ecological research in selected regions. In: Long term socio-ecological research. Springer Netherlands, pp 409–442

Nelson E et al (2009) Modelling multiple ecosystem services, biodiversity conservation, commodity production, and trade-offs at landscape scales. Front Ecol Environ **7**: 4–11.

Nielsen UN, Wall DH, Six J (2015) Soil biodiversity and the environment. Annu Rev Environ Resour **40**:1–28

Novak M, Wootton JT, Doak DF, Emmerson M, Estes JA, Tinker MT (2011). Predicting community responses to perturbations in the face of imperfect knowledge and network complexity. Ecology **92**:836–846.

Obersteiner M et al (2016) Assessing the land resource–food price nexus of the Sustainable Development Goals. Sci Adv **2**(9): e1501499.

Odling-Smee FJ, Laland KN, Feldman MW (1996). Niche construction. Am Nat **147**:641–648.

Ostrom E (2007) A diagnostic approach for going beyond panaceas. Proc Natl Acad Sci **104**:15181–15187.

Paine RT (1980) Food webs: linkage, interaction strength and community infrastructure. J Anim Ecol **49**(3): 667–685.

Parmele, RW, Crossley Jr DA (1988). Earthworm production and role in the nitrogen cycle of a no tillage agroecosystem on the Georgia piedmont. Pedobiologia **32**:353–361.

Parsa S, Ccanto R, Rosenheim JA (2011). Resource concentration dilutes a key pest in indigenous potato agriculture. Ecol Appl **21**:539–546.

Paustian L, Babcock B, Hatfield JL, Lal R, McCarl BA, McLaughlin S, Rosenzweig C (2016) Agricultural mitigation of greenhouse gases: science and policy options. In: 2001 conference proceedings, first national conference on carbon sequestration. Washington, DC: Conference on Carbon Sequestration.

Pelosi C, Goulard M, Balent G (2010). The spatial scale mismatch between ecological processes and agricultural management: do difficulties come from underlying theoretical frameworks? Agric Ecosyst Environ **139**: 455–462

Pelosi C, Barot S, Capowiez Y, Hedde M, Vandenbulcke F (2014) Pesticides and earthworms. A review. Agron Sustain Dev **34**:199–228.

Pelzer E, Bazot M, Makowski D, Corre-Hellou G, Naudin C, Al Rifai M, Baranger E, Bedoussac L, Biarnes V, Boucheny P, Carrouee B, Dorvillez D, Foissy D, Gaillard B, Guichard L, Mansard MC, Omon B, Prieur L, Yvergniaux M, Justes E, Jeuffroy MH (2012) Pea-wheat intercrops in low-input conditions combine high economic performances and low environmental impacts. Eur J Agron **40**:39–53.

Petit S, Burel F (1998) Effects of landscape dynamics on the metapopulation of a ground beetle (Coleoptera, Carabidae) in a hedgerow network. Agric Ecosyst Environ **69**:243–252.

Petit S, Griffiths L, Smart SS, Smith GM, Stuart RC, Wright SM (2004) Effects of area and isolation of woodland patches on herbaceous plant species richness across Great Britain. Landsc Ecol **19**:463–471.

Phalan B, Balmford A, Green RE et al (2011). Minimising the harm to biodiversity of producing more food globally. Food Policy **36** Suppl 1: S62–S71.

Polis GA, Power ME, Huxel GR (2004). Food webs at the landscape level. University of Chicago Press.
Potts SG, Biesmeijer JC, Kremen C, Neumann P, Schweiger O, Kunin WE (2010) Global pollinator declines: trends, impacts and drivers. Trends in Ecol Evol **25**:345–353.
Puech C, Baudry J, Joannon A, Poggi S, Aviron S (2014) Organic vs. conventional farming dichotomy: does it make sense for natural enemies? Agric Ecosyst Environ **194**:48–57.
Puech C, Poggi S, Baudry J, Aviron S (2015). Do farming practices affect natural enemies at the landscape scale? Landsc Ecol **30**:125–140.
Purtauf T, Dauber J, Wolters V (2004) Carabid communities in the spatio-temporal mosaic of a rural landscape. Landsc Urban Plan **67**:185–193.
Rand TA, Tscharntke T. 2007. Contrasting effects of natural habitat loss on generalist and specialist aphid natural enemies. Oikos 116:1353-1362.
Rand TA, Tylianakis JM, Tscharntke T (2006). Spillover edge effects: the dispersal of agriculturally subsidized insect natural enemies into adjacent natural habitats. Ecol Lett **9**:603–614.
Ratnadass A, Fernandes P, Avelino J, Habib R (2012). Plant species diversity for sustainable management of crop pests and diseases in agroecosystems: a review. Agron Sustain Dev **32**: 273–303.
Redman CL, Grove JM, Kuby LH (2004) Integrating social science into the long-term ecological research (LTER) network: Social dimensions of ecological change and ecological dimensions of social change. Ecosystems **7**:161–171.
Ricketts TH (2001) The matrix matters: Effective isolation in fragmented landscapes. Am Nat **158**:87–99.
Ricketts TH, Regetz J, Steffan-Dewenter I, Cunningham SA, Kremen C, Bogdanski A, Gemmill-Herren B, Greenleaf SS, Klein AM, Mayfield MM, Morandin LA, Ochieng A, Viana BF (2008) Landscape effects on crop pollination services: are there general patterns? Ecol Lett **11**:499–515.
Roger-Estrade J, Anger C, Bertrand M, Richard G (2010) Tillage and soil ecology: partners for sustainable agriculture. Soil Tillage Res **111**(1): 33–40.
Romo CM, Tylianakis JM (2013). Elevated temperature and drought interact to reduce parasitoid effectiveness in suppressing hosts. PLoS ONE **8**:e58136.
Rosenheim JA, Kaya HK, Ehler LE, Marois JJ, Jaffee BA (1995) Intraguild predation among biological-control agents: theory and evidence. Biol Control **5**:303–335.
Rusch A, Valantin-Morison M, Sarthou J-P, Roger-Estrade J (2011) Multi-scale effects of landscape complexity and crop management on pollen beetle parasitism rate. Landsc Ecol **26**:473–486.
Sarrazin F, Lecomte J (2016) Evolution in the Anthropocene. Science **351**:922–923.
Schmitz OJ (2007) Predator diversity and trophic interactions. Ecology **88**:2415–2426.
Schouten G, Bitzer V (2015) The emergence of Southern standards in agricultural value chains: a new trend in sustainability governance? Ecol Econ **120**:175–184.

Seto KC, Ramankutty N (2016) Hidden linkages between urbanization and food systems. Science **352**:943–945.
Shelton AM, Badenes-Perez FR (2006) Concepts and applications of trap cropping in pest management. Annu Rev Entomol **51**:285–308.
Shennan C (2008) Biotic interactions, ecological knowledge and agriculture. Philos Trans R Soc B Biol Sci **363**:717–739.
Sirami C, Nespoulous A, Cheylan JP, Marty P, Hvenegaard GT, Geniez P, Schatz B, Martin J-L (2010). Long-term anthropogenic and ecological dynamics of a Mediterranean landscape: Impacts on multiple taxa. Landsc Urban Plan **96**:214–223.
Smith P, Martino D, Cai Z et al (2008) Greenhouse gas mitigation in agriculture. Philosophical Trans R Society London B Biological Sciences 363:789–813.
Swift MJ, Izac AMN, van Noordwijk M (2004) Biodiversity and ecosystem services in agricultural landscapes - are we asking the right questions? Agric Ecosyst Environ **104**: 113–134.
Symondson W, Sunderland K, Greenstone M (2002) Can generalist predators be effective biocontrol agents? Annu Rev Entomol **47**:561–594.
Tancoigne et al. (2014) The place of agricultural sciences in the literature on ecosystem services. Ecosyst Serv **10**:35–48.
Tewksbury JJ, Levey DJ, Haddad NM, Sargent S, Orrock JL, Weldon A, Danielson BJ, Brinkerhoff J, Damschen EI, Townsend P (2002) Corridors affect plants, animals, and their interactions in fragmented landscapes. Proc Natl Acad Sci **99**:12923–12926.
Thies C, Roschewitz I, Tscharntke T (2005) The landscape context of cereal aphid–parasitoid interactions. Proc R Soc B Biol Sci **272**:203–210.
Thies C, Steffan-Dewenter I, Tscharntke T (2003) Effects of landscape context on herbivory and parasitism at different spatial scales. Oikos **101**:18–25.
Thies C, Steffan-Dewenter I, Tscharntke T (2008). Interannual landscape changes influence plant–herbivore–parasitoid interactions. Agric Ecosyst Environ **125**:266–268.
Thrall PH, Oakeshott JG, Fitt G, Southerton S, Burdon JJ, Sheppard A, Russell RJ, Zalucki M, Heino M, Ford Denison R (2011) Evolution in agriculture: the application of evolutionary approaches to the management of biotic interactions in agro-ecosystems. Evol Appl **4**:200–215.
Tilman D, Fargione J, Wolff B, D'Antonio C, Dobson A, Howarth R, Schindler D, Schlesinger WH, Simberloff D, Swackhamer D (2001) Forecasting agriculturally driven global environmental change. Science **292**:281–284.
Tilman D, Isbell F, Cowles JM (2014) Biodiversity and ecosystem functioning. Annu Rev Ecol Evol Syst **45**(1):471.
Tilman D, May RM, Lehman CL, Nowak MA. (1994). Habitat destruction and the extinction debt. Nature 371:65–66.
Tittonell P, Klerkx L, Baudron F, Félix GF, Ruggia A, van Apeldoorn D, Dogliotti S, Mapfumo P, Rossing PAH (2016) Ecological intensification: local innovation to address global challenges. Sustain Agric Rev **19:** 1–34 (Chap. 1).

Tixier P, Peyrard N, Aubertot J-N, Gaba S, Radoszycki J, Caron-Lormier G, Vinatier F, Mollot G, Sabbadin R (2013). Modelling interaction networks for enhanced ecosystem services in agroecosystems. Adv Ecol Res **49**:437–480.
Tomich TP, Brodt S, Ferris H, Galt R, Horwath WR, Kebreab E, Leveau JHJ, Liptzin D, Lubell M, Merel P, Michelmore R, Rosenstock T, Scow K, Six J, Williams N, Yang L (2013) Agroecology: a review from a global-change perspective. Annu Rev Environ Resour **36**:193–222.
Trenbath B (1993)Intercropping for the management of pests and diseases. Field Crops Res **34**:381–405.
Tscharntke T, Klein AM, Kruess A, Steffan-Dewenter I, Thies C (2005) Landscape perspectives on agricultural intensification and biodiversity–ecosystem service management. Ecol Lett **8**:857–874.
Tscharntke T, Karp DS, Chaplin-Kramer R, Batáry P, DeClerck F, Gratton C, Martin EA (2016) When natural habitat fails to enhance biological pest control– Five hypotheses. Biol Conserv Part B **204**:449–458.
Tylianakis JM, Didham RK, Bascompte J, Wardle DA (2008) Global change and species interactions in terrestrial ecosystems. Ecol Lett **11**:1351–1363.
van der Heijden MG, Bardgett RD, Stralen N (2008) The unseen majority: soil microbes as drivers of plant diversity and productivity in terrestrial ecosystems. Ecol Lett **11**(3):296–310.
van Groenigen JW, Lubbers IM, Vos HMJ, Brown GG, De Deyn GB, van Groenigen KJ (2014) Earthworms increase plant production: a meta-analysis. Sci Rep **4**:63–65.
Vandermeer JH (1992) The ecology of intercropping. Cambridge University Press.
Vanloqueren G, Baret PV (2009) How agricultural research systems shape a technological regime that develops genetic engineering but locks out agroecological innovations. Res Policy **38**:971–983.
Vasseur C, Joannon A, Aviron S, Burel F, Meynard J-M, Baudry J (2013) The cropping systems mosaic: How does the hidden heterogeneity of agricultural landscapes drive arthropod populations? Agric Ecosyst Environ **166**:3–14.
Vos CC, Goedhart PW, Lammertsma DR, Spitzen-Van der Sluijs AM (2007) Matrix permeability of agricultural landscapes: an analysis of movements of the common frog (Rana temporaria). Herpetol J **17**:174–182.
Wardle DA, Bardgett RD, Klironomos JN, Setälä H, van der Putten WH, Wall DH (2004) Ecological linkages between aboveground and belowground biota. Science **304**:1629–1633.
Westphal MI, Field SA, Tyre AJ, Paton D, Possingham HP (2003). Effects of landscape pattern on bird species distribution in the Mt. Lofty Ranges, South Australia. Landsc Ecol **18**:413–426.
Wezel A, Bellon S, Doré T, Francis C, Vallod D, David C (2009) Agroecology as a science, a movement and a practice. A review. Agron Sustain Dev **29**(4): 503–515.
Wimberly MC (2006) Species dynamics in disturbed landscapes: When does a shifting habitat mosaic enhance connectivity? Landsc Ecol **21**:35–46.
Wolfson DJ (2014). Who gets what in environmental policy? Ecol Econ **102**:8–14.

Wretenberg J, Pärt T, Berg Å (2010) Changes in local species richness of farmland birds in relation to land-use changes and landscape structure. Biol Conserv **143**:375–381.
Zak DR, Holmes WE, White DC, Peacock AD, Tilman D (2003) Plant diversity, soil microbial communities, and ecosystem function: are there any links? Ecology **84**:2042–2050.
Zaller JG, Moser D, Drapela T, Schmöger C, Frank T (2009). Parasitism of stem weevils and pollen beetles in winter oilseed rape is differentially affected by crop management and landscape characteristics. Biocontrol **54**:505–514.

Chapter 2
Allelopathy: Principles and Basic Aspects for Agroecosystem Control

Aurelio Scavo, Alessia Restuccia and Giovanni Mauromicale

Abstract Allelopathy is a form of amensalism, an association between organisms in which one is inhibited or destroyed and the other is unaffected through the release into the environment of secondary metabolites called allelochemicals. Allelopathic plant interactions has been known since the 4th century before Christ (BC), but only in recent years they have received an appropriated attention from international scientific community and farmers. Nowadays, in modern agriculture, allelopathy play a key role in maintenance the sustainability of agroecosystems through the adoption of environmentally-friendly strategies such as crop rotation, cover or smother crops, intercropping, crop residue incorporation, mulching and bioherbicides. Crops showing allelopathic properties are numerous: they include arboreal and herbaceous species as well as many weeds. Here we review the general principles and the basic aspects in the field of allelopathy and why they are important in developing innovative sustainable agricultural techniques in agroecosystems for weed control, crop protection, and crop re-establishment with respect of environmental, human and animal health. Particularly, we describe the guidelines for distinguishing allelopathy from competition in field conditions, as well as the chemical nature of allelochemicals. Secondly, we review the volatilization from living parts of the plant, the leaching from aboveground parts of the plant, the decomposition of plant material and the root exudation, which are the major pathways for the release of allelochemicals. Third, we review the influence of abiotic and biotic stress factors on the quantity of allelochemicals released by the donor plant and the effects of allelochemicals on the target plant. Light, temperature, water deficiency, minerals availability, soil characteristics and many biotic factors modify the production of allelochemicals and the sensitivity of target plants. Finally, the interference of alleochemicals with plant physiological processes was also reviewed. Allelopathic compounds rarely act alone, but generally generate "multiple cascating effects". Allelopathic mechanisms influence plant successions,

A. Scavo · A. Restuccia · G. Mauromicale (✉)
Department of Agriculture, Food and Environment (Di3A), University of Catania, Via Valdisavoia N. 5, 95123 Catania, Italy
e-mail: g.mauromicale@unict.it

S. Gaba et al. (eds.), *Sustainable Agriculture Reviews 28*, Ecology for Agriculture 28,
https://doi.org/10.1007/978-3-319-90309-5_2

invasion, spatial vegetation patterns, mutualistic associations, soil nitrogen cycle, crop productivity and crop protection.

Keywords Allelopathy · Competition · Autotoxicity · Allelochemicals
Plant interactions · Secondary metabolites · Phenols · Terpenoids
Mode of action · Mode of release

2.1 Introduction

Allelopathy is an ecological phenomenon of most natural communities and agroecosystems, although it is often unrecognized. It has been observed how orchard replant problems, regeneration of forest species, occurrence of weed-free zones, dominance of exotic plants, spatial vegetation patterns, dynamics of communities, plant productivity and other ecological aspects are strictly linked to allelopathic mechanisms. Root exudates, upon release into the rhizosphere, also play an important role on soil microbial ecology, nutrients biogeochemical cycles and their uptake by plants. One of the most important examples is provided by the improvement of nitrogen use efficiency (NUE) through biological nitrification inhibition (BNI) made by *Nitrosomonas* spp. and *Nitrobacter* spp.

Modern agriculture has to deal with a rapid increase in world population, accompanied by a simultaneous decrease of the available resources. Therefore, in order to feed a growing population, agriculture has pursued the maximization of yields. Cropping intensity resulted in a subsequent higher pest pressure on crops. Particularly, in order to eliminate the presence of weeds, insects and pathogens, synthetic chemicals were used indiscriminately. The wide use of herbicides, for example, resulted in increasing incidence of resistance in weeds to common herbicides, and in environmental pollution and human and animal health. Many of these problems may be effectively resolved through the manipulation of allelopathic mechanisms and their integration to traditional agricultural practices under Integrated Pest and Weed Management System (IPMS, IWMS). The most important modes by which allelopathy can be used in agroecosystems for sustainable crop production refers to (1) the selection of smothering crops, their breeding and inclusion in crop rotations; (2) the use of their residues as living mulches, dead mulches or green manure; (3) the selection of most active allelopathic compounds and their use as bioherbicides.

The term "allelopathy" is derived from the Greek words *allelon*, "of each other", and *pathos*, "to suffer" and literally means "*the injurious effect of one upon another*" (Rizvi et al. 1992). However, the term was coined for the first time by the Austrian plant physiologist Hans Molisch in 1937 in his book *Allelopathie* to include both harmful and beneficial biochemical interactions between all types of plants, including microorganisms.

In his first book, E. L. Rice excluded the beneficial effects, while reconsidered and accepted Molisch's definition in his second monograph: "*any direct or indirect harmful or beneficial effect by one plant (including microorganisms) on another through production of chemical compounds that escape into the environment*". Both positive or stimulatory and negative or inhibitory effects are included in this definition. Rice's definition has been criticized by many authors because it refers to all types of interactions between plants (Watkinson 1998). Instead, several other workers prefer to limit the use of the term to recognize only the negative effects, direct or indirect, produced by a plant, which is identify as the "donor" plant, on another plant called "target" or "afflicted" plant.

According to the definition given by the International Allelopathy Society (IAS) in 1996, allelopathy includes "*any processes involving secondary metabolites produced by plants, microorganisms, viruses and fungi that influence the growth and the development of agricultural and biological systems (excluding animals), including positive and negative effects*" (Torres et al. 1996).

Therefore, allelopathy include plant-plant, plant-microorganisms, microorganisms-plant, microorganisms-microorganisms, plant-insect and plant-higher animal interactions.

2.2 History of Allelopathy

The ability of some plant species to interfere with the germination, growth or development of other plant species has been well documented since antiquity, from more than 2000 years ago. The earliest observations on this phenomenon were made by Theophrastus (370–286 BC), a disciple of Aristotle and the father of botany, who around in 300 BC wrote in his two botanical works, *Historia plantarum* and *De causis plantarum*, about how chickpea (*Cicer arietinum* L.) "exhausted" the soil and destroyed weeds. He also reported the harmful effects of cabbage (the cabbage of Theophrastus refers to something close to the wild form of *Brassica oleracea*, sometimes known as *Brassica cretica*, an edible, but bitter, herb of coastal regions richer in allelochemicals; Willis 2007) on grapevine and proposed that these effects were caused by "odours" of cabbage plants.

The Greek author Bolos Demokritus (460–360 BC) of Mendes in Egypt, in his agricultural work *Georgics*, written around in 200 BC, suggested that trees may be died by sprinkling their roots with a mixture of lupine flowers soaked in hemlock juice.

Cato the Elder (234–140 BC), the famous Roman politician and writer who was a farmer in youth, and later Pliny the Elder (23–79 after Christ) in *Historia naturalis*, noted that walnut trees (*Juglans* spp.) were toxic to other plants and that both chickpea and barley (*Hordeum vulgare* L.) ruined cornlands (Weir et al. 2004). Pliny the Elder has been precede bay Columella, who was a farmer in Cadiz. In his surviving works (*De rerum rusticarum* of 64 AD and *De arboribus*), Columella was the first who spoke about "soil sickness", described as a decrease in soil's fertility due to the repeated cultivation for more years of the crop on the same land.

In the seventeen century, both the English and Japanese literature shown cases of plants do not grow well in the presence of each other due to the production of toxic compounds, for example the Japanese red pine (*Pinus densiflora* Siebold & Zucc.) (Rice 1984).

In 1804, the agronomist Young discovered that clover was apt to fail in some regions of England where it is cultivated constantly due to soil sickness, which accrues over time (Weston 2005).

In 1832, the Swiss botanist De Candolle (1778–1841) proposed that such excretions of roots of some plants could injurious other plants and explain the exhaustion of soil. On the basis of De Candolle's suggestions, in 1881 Stickney and Hoy observed that vegetation under black walnut was very sparse, probably due to the high mineral requirements of the tree.

The interest in the field of allelopathy resumed in the twentieth century, thanks to the development of suitable techniques for the extraction, bioassay and chemical isolation of the involved substances (Willis 2007). For example, in 1907, Schreiner and Reed for the first time isolated soil organic acids released from root exudates of certain plants that strongly inhibited the growth of some adjacent crops. In 1928, Davis was the first to extract and purify from the hulls and roots of walnut the juglone, 5-hydroxy-α-naphtaquinone (Rice 1984).

In the period between 1960 and 1990, much progress has been made in the field of chromatography and spectroscopy, for the isolation and determination of the studied chemical compounds.

Whittaker and Feeny, in 1971, coined the term "allelochemicals". Chou and Waller (1983) describe the biochemical interactions between organisms, both inter- and intra-specific, as "allelochemical interactions".

2.3 Current Status of Allelopathy Research

Despite the allelopathic interactions between plants were known since ancient times, the research on this topic has received great focus only on the end of the twentieth century. Specifically, the number of journal papers using the word "allelopathy", as searched in the database Scopus®, has undergone an exponential growth since the 1970s with a rapid increased in the late 1990s (Fig. 2.1).

Originally considered as a sub-discipline of chemical ecology, currently allelopathy embraces a broad range of disciplines: ecology, biochemistry, chemistry, physiology, agronomy, entomology, microbiology, forestry, soil science, proteomics, genetic, etc.

At least two reasons are involved for the slow progress in this area: one is the difficulty of designing field experiments about that unquestionably prove that a chemical produced by a plant directly affects a neighboring plant (Weir et al. 2004),

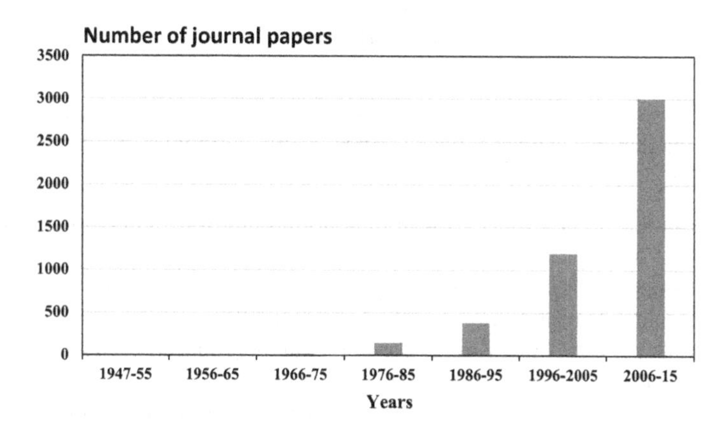

Fig. 2.1 Number of journal papers accessed on Scopus® using the search term 'allelopathy or allelochemical' for each decade of the past 70 years. Note the high increase since the 1970s, particularly from the late 1990s

the so-called "allelopathy paradigm". Another reason refers to the cases in which important papers on allelopathy have been discounted by later work (Duke 2010).

However, great progress have been made in recent years on the study of allelopathic activities and there are numerous examples of high profile research who brought fame to this discipline. Indeed, in addition to the number of journal papers, it should be considered how changed the target of interested journal. In 1994 was found the *Allelopathy Journal*, the first journal exclusively devoted to allelopathy research; nowadays journals such as the *Journal of Chemical Ecology*, *Plant And Soil*, *Phytochemistry*, *Journal of Agricultural and Food Chemistry*, *Weed Science*, *Weed Research*, *Crop Protection*, etc. publish many papers on allelopathy. The most active country in allelopathy research are United States (with over 1000 works) and China (about 825), followed by India (394) and Japan (372) (source: Scopus®).

The increased interest of scientific community on allelopathy, has mean that in 1995 was founded the International Allelopathy Society (IAS) in India, which has a combined membership of over 1000 participants from over 50 countries. The Society hosts one meeting every three years. The first congress was held in Cadiz (Spain, in 1996) and the subsequent in Thunder Bay (Canada, 1999), Tsukuba (Japan, 2002), Wagga (Australia, 2005), Saratoga Springs (NY, USA, 2008), Guangzhou (China, 2011), Vigo (Spain, 2014), Marseille (France, 2017).

Several congresses and symposia were made in Europe and Asia and today exist regional organization focused on allelopathy topic, such as the Asian Allelopathy Society (AAS), or the European Allelopathy Society (EAS). Moreover, other scientific organizations, such as the American Chemical Society and the International Association for Ecology have had allelopathy symposia as part of their programmes (Duke 2010).

2.4 Terminology and Classification

In his first book (1974), Rice reported Grümmer's terminology (1955, Fig. 2.2) for chemical interactions between plants. Specifically, Grümmer recommended the terms:

(i) "antibiotic" to denote a chemical inhibitor produced by a microorganism that is toxic to other microorganisms;
(ii) Waksman's suggested term "phytoncide" for antimicrobial organic substances of plant origin;
(iii) Gaumann's term "marasmins" for compounds produced by microorganisms and harmful to higher plants;
(iv) the term "kolines" for chemical inhibitors produced by higher plants that are toxic to other higher plants species.

However, these terms are rarely adopted by the authors, because compounds released from higher plants may be altered by microorganisms before the altered substance is contacted by another higher plant and it is very difficult to establish the source when a compound of any origin is contacted through the soil medium the same compound. Besides, the same compund is likely to have multiple roles (Einhellig 1995): in fact, there are marasmines that shows harmful effects on other microorganisms, kolines with antimicrobial activity and phytoncides that inhibit the development of plants. For these reasons, nowadays, the generic term "allelochemical" is adopted to denote the chemical compounds produced by plants, microorganisms, viruses and fungi involved in agricultural and biological ecosystems.

With the term "allelopathy" several authors prefers to refers only to detrimental effects, direct or indirect, of higher plants of one species (the donor) on another (the receptor) through the release into the environment of toxic chemical compounds, commonly defined "allelochemicals" (Putnam and Duke 1978). Therefore, the detrimental effects can be direct or indirect and, about that, Aldrich (1984) describes two types of allelopathy: in the first case, the interference is caused by release into the environment of a compound that is active in the form in which is produced and released from the donor plant; this type is called *true* allelopathy. In the second case, the negative action is mediated by soil microorganisms that act on the decaying tissues of the donor plant or by enzymes that, after plant tissues

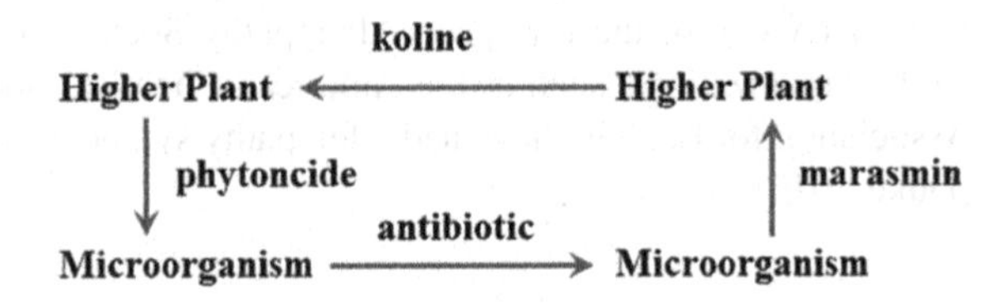

Fig. 2.2 Terms for chemical agents that indicate the type of donor and receiver plants, as shown by the arrows (Einhellig 1995, Modified)

destruction, interact with the pre-allelopathic substances by converting them into allelochemicals; this type is called *functional* allelopathy.

When a given species produces and releases allelochemicals that can causes damage to a different plant species, this phenomenon is called *heterotoxicity*; whereas, when its own germination and development is affected, this allelopathic effect is called *autotoxicity* or *autoallelopathy* (Chon et al. 2006; Kruse et al. 2000; Miller 1996).

2.4.1 Autotoxicity or Autoallelopathy

Autotoxicity is an intraspecific form of allelopathy that was first described by Jensen et al. (1981) on alfalfa (*Medicago sativa* L.). In different cereals and vegetables crops, it was observed a significant reduction in both crop yield and quality due to the "soil sickness". This phenomenon occurs when the same crop (monoculture) or its relative species are cultivated on the same soil successively (Yu 1999).

Autotoxicity has been studied mainly in the Cucurbitaceae family, for example in cucumber, melon and watermelon (de Albuquerque et al. 2010). Other autotoxic effects have been reported also in asparagus (*Asparagus officinalis* L.) (Hartung et al. 1990), rice (*Oryza sativa* L.) (Rice 1979), grape (Vitis sp.) (Brinker and Creasy 1988), wheat (*Triticum aestivum* L.) (Kimber 1973; Lohdi et al. 1987), etc.

Probably autotoxicity derives from natural selection as result of controlling competition between older and younger plants for resources (e.g. light, water, nutrients, etc.) by maintaining a certain distance from them. In alfalfa, an autotoxic zone of about 20 cm radius exists around the old plants. Autoxicity is now identified as the most frequent cause of reseeding failure (Tesar 1993; Miller 1996; Webster et al. 1967).

Biotic and abiotic factors not only influence the production of autotoxins, but modify their effects too (Yu 2001). Autotoxicity involves many chemical compounds that are located in different quantities in the plant's tissues. For example Miller (1996) founded that autotoxic effects of water extracts of plant parts of alfalfa in order are leaf > seed > root > flower > stem. Generally, leaves present the highest concentration of allelochemicals, but often they are also released as root exudates. In the soil, their movement is faster in sandy than in clay soils.

Autotoxicity primarily affects the seed germination and early root elongation, but in severe cases it may causes stand failure. Other typical symptoms are dwarfed, spindly, curling, yellowish-green seedlings with irregular brown-reddish to dark-brown lesions on the tap and lateral roots and only few effective nodules (Webster et al. 1967). However, as reported by Chon et al. (2006), there are ecological advantages linked to autotoxicity. For example, if seeds were not dispersed away from the parent plant leaving the new plant in position to compete for resources, it may encourages geographical distribution of the donor species, serves as an adaptation to induce dormancy and prevents decay of its seeds and propagules (Friedman and Waller 1983).

Crop rotation is the best solution to avoid soil sickness. In particular, rotation intervals higher than 12-months are suggested because allowed natural decomposition of the chemicals. In light-textured soils, irrigation can leach the autotoxic chemical to shorten the rotation interval (Chon et al. 2006).

Under field conditions, finding evidence of autotoxicity is not a problem of simple solution. In fact, despite many cases of auto-heterotoxicity have been reported in literature in laboratory conditions, autotoxicity is rarely present in natural ecosystems (Wilson and Rice 1968; Jackson and Willemsen 1976).

2.4.2 Allelopathy and Competition

Allelopathy is a particular form of amensalism, a negative interaction between two species in which one organism is harmed or inhibited and the other is unaffected.

It is necessary to clarify the difference between the concept of allelopathy and competition to avoid misunderstandings. The first implies the release of inhibiting substances into the environment, whereas in the second a generic resource (e.g. water, light, minerals and space) is removed or reduced by another organism sharing the same habitat. In nature is difficult to separate these mechanisms: in fact, stress caused by competition increase the production of allelochemicals, while growth reduction caused by allelochemicals may reduce the competitive ability of inhibited plant. Besides, mechanisms of interference such as microbial nutrient immobilization, soil characteristics, climatic factors etc. can not be separated under field conditions. Therefore, competition and allelopathy undoubtedly interact in a highly synergism (Willis 1994).

The difficulty in distinguishing and describing separately allelopathic effects from those of competition have induced Muller (1969) to propose the term "interference" to indicate the overall deleterious effects (allelopathy + competition) of one plant on anoher. According to Harper (1977), competition represents a physical interference, while allelopathy a chemical interference. For over 2000 years, allelopathy has been reported in literature with respect to plant interference (Weston 2005). Nowadays, interference is a term widely used in the literature to denote "the total adverse effect that plants exerts on each other when growing in a common ecosystem and it includes competition and allelopathy" (Zimdahl 1999).

Like allelopathy, competition is difficult to demonstrate (Inderjit and Keating 1999). While plant-plant resource competition has been readily accepted by biologists and ecologists, the same has not be the case for allelopathic interactions. According to Blum (2011), the difference in acceptance between competition and allelopathy was due to: (i) the modification in allelopathy definition over time; (ii) the lack in design bioassays of plant-plant allelopathic interactions; (iii) the forceful scepticism of several authors to plant-plant-allelopathic interactions and

(iv) the higher standard of proof required for allelopathic interactions than those for competition.

One of the main challenges in the field of allelopathy is the separation of allelopathic effects from other mechanisms of plant interference, mainly competition (Singh et al. 2001). In recent years several mathematical model were proposed to calculate the contribution of both allelopathy and competition to interference (Liu et al. 2005; An et al. 1993; etc.).

2.4.3 Establishing the Proof of Allelopathy

Although community's scientific attentions on allelopathy are continuously growing, its scientific evidence is not yet accepted by all scientists. This because its mechanisms are largely unknown, from production to release and fate of allelochemicals. Its comprehension is further complicated by vastness and chemical heterogeneity of the substances involved (Nelson 1996). The amount of searches conducted until today on the topic, has induced researchers to assume that it is unusual for a single allelochemical product to be present, in field conditions, in sufficiently high amounts to exteriorise significant effects (Einhellig 1996). Besides, it seems that the effects are often caused by different compounds that act together additively or synergistically; but in some cases these compounds react antagonistically (Seigler 1996). Allelochemicals are introduced into the environment together with a vast number of other compounds, since it is likely that synergistic effects enhance the observed activities (Putnam and Tang 1986).

Moreover, allelochemicals's production and release, their transport and transformation in soil and absorption in the receptor plant, as well as the plant's reaction to the compound, are highly dependent on environmental conditions. In many cases, stress conditions induce the donor plant to produce and release a higher level of allelochemicals and, in poor condition, neighboring plants become more sensitive to these substances.

Establishing the proof of allelopathy and separating allelopathy from other mechanisms of interference such as competition is very difficult. In his book of 1977, page 494, Harper says: "Demostrating this [toxicity in the field] has proved extraordinarily difficult- it is logically impossible to prove that it doesn't happen and perhaps nearly impossible to prove absolutely that it does". Nowadays, the demonstration of allelopathy mechanisms have been achieved by creative experimentation and use of advanced biomolecular analytical techniques (Bais et al. 2003; Vivanco et al. 2004).

To establishing the cause-and-effect relationship in allelopathy, the following events must occur in sequence (Cheng 1992):

1. a phytotoxic chemical is produced;
2. the chemical is transported from the producing organisms to the target plant;

3. the target plant is exposed to the chemical in sufficient quantity and for sufficient time to cause damage.

First of all, it is necessary to determinate if a plant is really allelopathic. Several types of clues can indicate if a species is allelopathic (Duke 2015): (i) if an invasive plant eliminates most native plant species, probably it is allelopathic; (ii) sparse or no vegetation patterning around a particular species can indicate that it is allelopathic, e.g. black walnut; (iii) problems of "soil sickness" are often attributed to buildup of allelochemicals in the soil; (iv) knowledge that a plant species produces one or more potent phytotoxins can be a clue obtained from the phytochemical literature that the species might produce an allelochemical, e.g. the phytotoxic compound sorgoleone produced by all species of *Sorghum* spp.

Secondary, it is necessary to predict if a compound is an allelochemical. In fact, finding a phytotoxic compound in plants does not mean that the compound is necessary an allelochemical. The identification of a substance as allelochimical is actually dependent upon the context rather than on its biosynthetic origin (Berenbaum 1995). According to Inderjit and Duke (2003), a compound may play several roles in nature, including that of an allelochemical, depending on the organisms involved and on the specific environmental parameters affecting the organism. Thus, the exact same compound may sometimes be an allelochemical, and at other times or places play other roles. Modes of release, phytotoxic action, bioactive concentration, persistence and fate in the environment, are all factors influencing the allelopathic nature of a compound, whereby a chemical does not act as an allelochemical in all situations.

Willis (1985) advanced six protocols required to demonstrate allelopathy, based on "Koch's postulates" (Williamson 1990) for demonstrating that a disease is caused by an infectious agent:

(a) a pattern of inhibition of one species or plant by another must be shown;
(b) he putative aggressor plant must produce a toxin;
(c) here must be a mode of toxin release from the plant into the environment;
(d) there must be toxin transport and/or accumulation in the environment;
(e) the afflicted plant must have some means of toxin uptake;
(f) the observed pattern of inhibition cannot be explained solely by physical factors or other biotic factors, especially competition and herbivory.

As mentioned before, it is extremely difficult or impossible to follow these protocols in field conditions since biotic (e.g., soil microflora, root exudates of competitors) and abiotic (e.g., temperature fluctuations, water stress) factors strongly influence allelochemicals's fate.

I consider Macias et al. (2007) guideline useful to refuting the cases of "suspected allelopathy":

(1) plant predominance/distribution/frequency cannot be explained solely on the basis of physical/biotic factors;

(2) the allelopathic plants (donors) should synthesise and release into the environment chemicals that must be or become bioactive;
(3) soil permanence and concentrations should be high enough to produce effects on the germination and/or growth of neughbouring plants, bacteria and/or fungi;
(4) uptake by the target plant and evidence of the detrimental/beneficial effects caused by the chemical/s.

2.5 Chemical Nature of Allelochemicals

According to Reese (1979), allelochemicals are "*non-nutritional chemicals produced by one organism (plants, microorganisms, viruses and fungi) that affects the growth, health, behaviour or population biology of other species*". Most of them are secondary metabolites (Whittaker and Feeny 1971) and are produced as offshoots of primary metabolic pathways of carbohydrates, fats and amino acids. As secondary metabolites, they are of sporadic occurrence and do not play an obvious role in the basic metabolisms of organisms, but serve for defensive adaptation. However, a significant role in allelopathy is also played by certain primary metabolites (Inderjit 1999) as several free amino acids and organic acids.

Allelochemicals, even with a few exceptions, have basically four precursors: acetyl coenzyme A, shikimic acid, mevalonic acid and deoxyxylulose phosphate (Fig. 2.3).

There are many thousands of such compounds, but only a relative limited number of them have been identified as allelochemicals (Rice 1984). Allelochemicals consist of various chemical families. According to Whittaker and

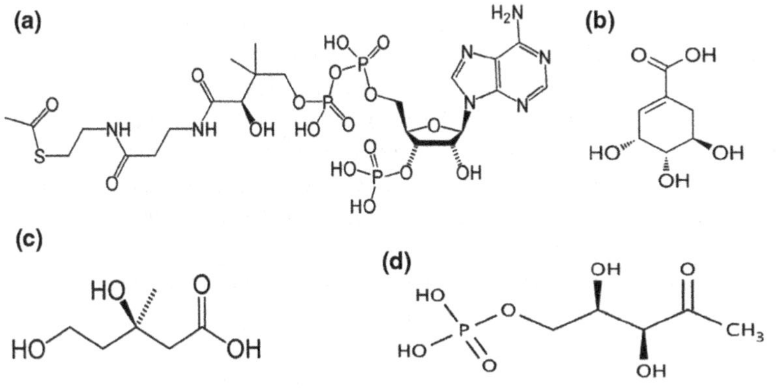

Fig. 2.3 Allelochemical precursors: acetyl coenzyme A (**a**), shikimic acid (**b**), mevalonic acid (**c**) and deoxyxylulose phosphate (**d**) structural formula

Feeny (1971), they could be classified into five major categories: phenylpropanes, acetogenins, terpenoids, steroids and alkaloids. Apart from the phenylpropanes and alkaloids, which originate from amino acids, the rest generally originate from acetate.

Based on the four precursors and on the different structures and properties of these compounds, Rice (1984) classified allelochemicals into 14 chemical classes plus a catchall category (miscellaneous):

a. simple water soluble organic acids, straight chain alcohols, aliphatic aldehydes and ketones;
b. simple unsaturated lactones;
c. long-chain fatty acids and polyacetylenes;
d. naphthoquinones, anthroquinones and complex quinones;
e. simple phenols, benzoic acid and derivatives;
f. cinnamic acid and derivatives;
g. flavonoids;
h. tannins;
i. terpenoids and steroids;
j. amino acids and polypeptides;
k. alkaloids and cyanohydrins;
l. sulphides and glucosides;
m. purines and nucleotides.

Plant growth regulators such as gibberellic acid, ethylene or salicylic acid, are also considered to be allelochemicals. Thanks to the progress of analysis technology in the last decades, it was possible to isolate and identify tens of thousands of allelochemicals and to perform sophisticated structural analysis of these molecules (Cheng and Cheng 2015).

With a few exceptions, allelochemicals produced by higher plants and microorganisms usually arise through either the acetate or the shikimate pathway, or their chemical skeletons come from a combination of these two origins (Fig. 2.4). Generally, plant phenolics originate from the shikimate pathway, while terpenoids from the mevalonate pathway, also known as the isoprenoid pathway. Several types of inhibitors, which originates from amino acids, come through the acetate pathway. Most of compounds that cause allelopathy were derived from amino acids, via the shikimate pathway (Rice 1984). Higher plants presents two pathway for the formation of C5 terpenoid monomers, isopentenyl diphosphate: (i) the glyceraldehyde-3-phosphate/pyruvate pathway in the plastids, and (ii) the cytoplasmic acetate/mevalonate pathway (Lichtenthaler et al. 1997). However, the details of biosynthesis are not always known.

It is possible to group secondary metabolites into three main chemical classes: phenolic compounds, terpenoids and other compounds.

2.5.1 *Phenolic Compounds*

Phenolic compounds falls within the class of most important and common plant allelochemicals in the ecosystem. As shown in Fig. 2.5, they arise from shikimate and phenylalanine pathways (Harborne 1989). Phenolic compounds consisting of a hydroxyl group (-OH) bonded directly to an aromatic hydrocarbon group. In the term “phenolic compound”, structures with different degrees of chemical complexity are included, as shown in Table 2.1, according to the number of carbon atoms of the basic skeleton.

Structural diversity and intraspecific variability are the most significant characteristics of phenolic compounds (Hartmann 1996). Besides, they are water soluble and could easily be leached by rain, whereas leaves are still attached to the plant or, thereafter, from leaf litter (Alsaadawi et al. 1985).

2.5.1.1 Simple phenols

Simple phenols are all monomeric, consisting of only one aromatic ring (Fig. 2.6). Probably, phenol is the precursor of all other phenolic compounds in plants. Schreiner and Reed (1908) reported that vanillin, vanillic acid (a benzoic acid) and hydrokinone are the most general simple phenols with growth-inhibiting allelopathic properties. In particular, hydrokinone is the aglycone of arbutin, and

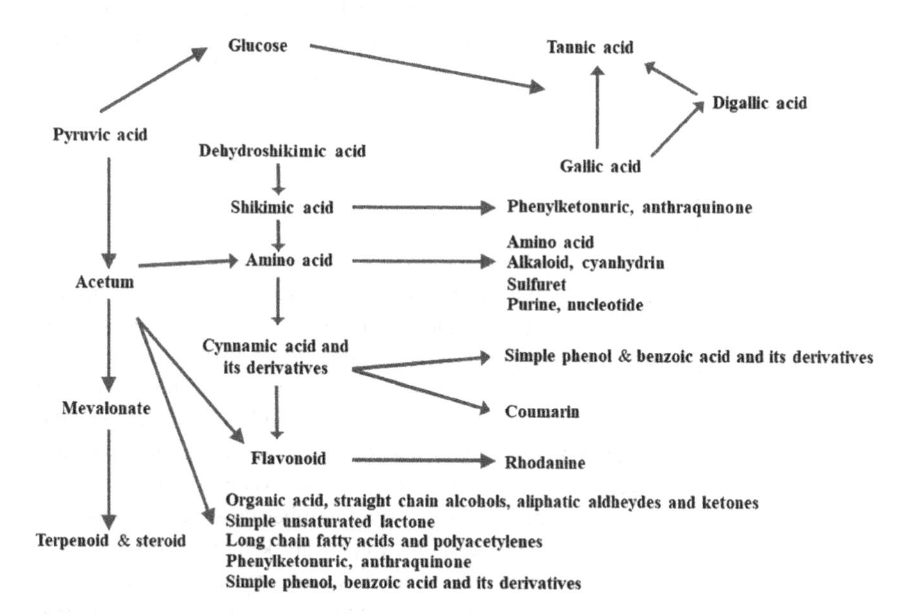

Fig. 2.4 Acetate and shikimate pathway, the biosynthetic pathways of major allelopathic substances (Wang et al. 2006, Modified)

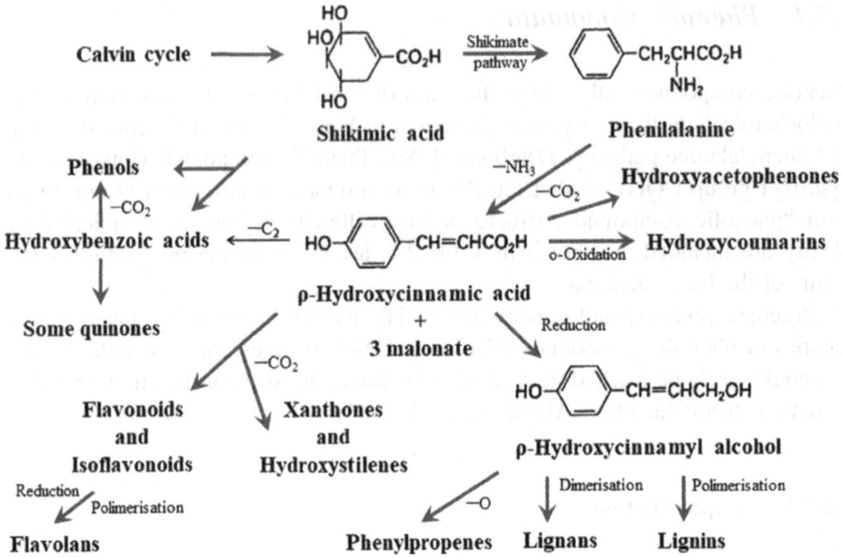

Fig. 2.5 Biosynthetic origin of plant phenolics from shikimate and phenylalanine pathways (Harborne 1989, Modified)

p-hydroxybenzoic acid and vanillic acid are the most commonly identified benzoic acid derivatives involved in allelopathy. Hydrokinone, resorsinol and catechol, which are found in low concentrations in plants, are mostly secreted by insects as a defence mechanism against other insects and animals. Salicylic acid, on the other hand, possesses anaesthetic properties and is the active ingredient of Aspirin® (Pretorius and van der Watt 2011).

2.5.1.2 Flavonoids

Flavonoids have a basic C_6–C_3–C_6 skeleton (Fig. 2.7) in which the A ring is of acetate origin and the B ring of shikimate origin (Neish 1964). Flavonoids are the largest group of natural phenolic compounds in higher plants. In fact, more than 5000 different flavonoids have been described and it is estimated that about 2% of all carbon photosynthesized by plants is converted to flavonoids (Pretorius and van der Watt 2011). They have roles associated with colour and pollination (e.g. flavones, flavonols, chalcones and catechins) and disease resistance, such as phytoalexins. They also have weak oestrogenic activity, for example the isoflavones (Macías et al. 2007). However, only a relative small number have been reported as toxins implicated in allelopathy. The most often cited are kaempferol, which is a yellow colour flavonol presents in apples, onions, citrus fruits, grape fruits etc., quercetin, naringenin, which is a dihydroflavon and ceratiolin, a non-phytotoxic

Table 2.1 The major classes of phenolic compounds in plants (Source: Harborne 1980)

Number of carbon atoms	Basic skeleton	Class	Examples
6	C6	Simple phenols Benzoquinones	Catechol, hydroquinone 2,6-Dimethoxybenzoquinone
7	C6–C1	Phenolic acids	Gallic, salicylic
8	C6–C2	Acetophenones Tyrosine derivatives Phenylacetic acids	3-Acetyl-6-methoxybenzaldehyde Tyrosol p-hydroxyphenylacetic
9	C6–C3	Hydroxycinnamic acids Phenylpropenes Coumarins Isocoumarins Chromones	Caffeic, ferulic Myristicin, eugenol Umbrelliferone, aesculetin Bergenon Eugenin
10	C6–C4	Naphthoquinones	Juglone, plumbagin
13	C6–C1–C6	Xanthones	Mangiferin
14	C6–C2–C6	Stilbenes Anthraquinones	Resveratrol Emodin
15	C6–C3–C6	Flavonoids Isoflavonoids	Quercetin, cyaniding Genistein
18	$(C6–C3)^2$	Lignans Neolignans	Pinoresinol Eusiderin
30	$(C6–C3–C6)^2$	Biflavonoids	Amentoflavone
n	$(C6–C3)_n$ $(C6)_n$ $(C6–C3–C6)_n$	Lignins Catechol melanins Flavolans (Condensed Tannins)	

dihydrochalcone present in the leaves of the dominant shrub of the Florida *Ceratiola ericoides* Michx.

2.5.1.3 Tannins

From the oxidative polymerization of catechins and flavan-3,4-diols derives condensed tannins, also called proanthocyanidins (PAs). When condensed tannins are hydrolysed by concentrated HCl, cyaniding chloride is formed. Another type of tannins are hydrolyzable tannins, which are derivate of gallic and *m*-Digallic acids hydrolysis (Fig. 2.8). Some hydrolyzable tannins derive from a complex mixture of several phenolic acids, whereby many types of hydrolyzable tannin molecules are possible (Rice 1984).

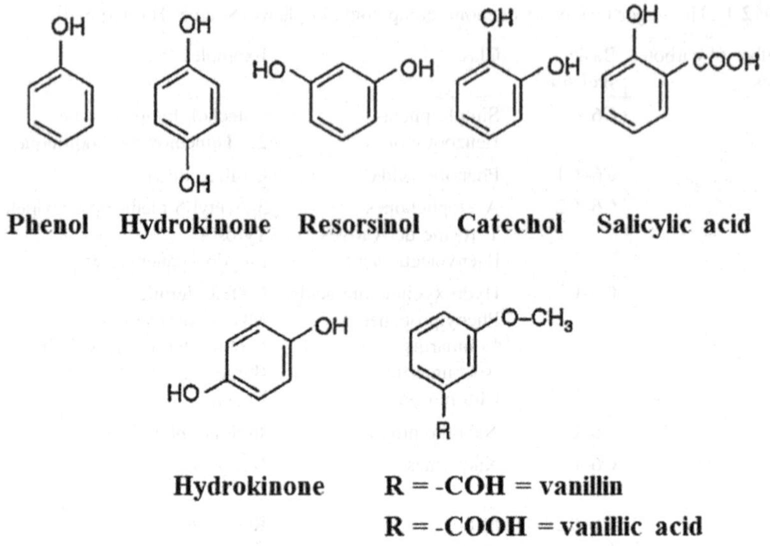

Phenol Hydrokinone Resorsinol Catechol Salicylic acid

Hydrokinone R = -COH = vanillin

R = -COOH = vanillic acid

Fig. 2.6 Some simple phenols with allelopathic potential (Pretorius and van der Watt 2011, Modified)

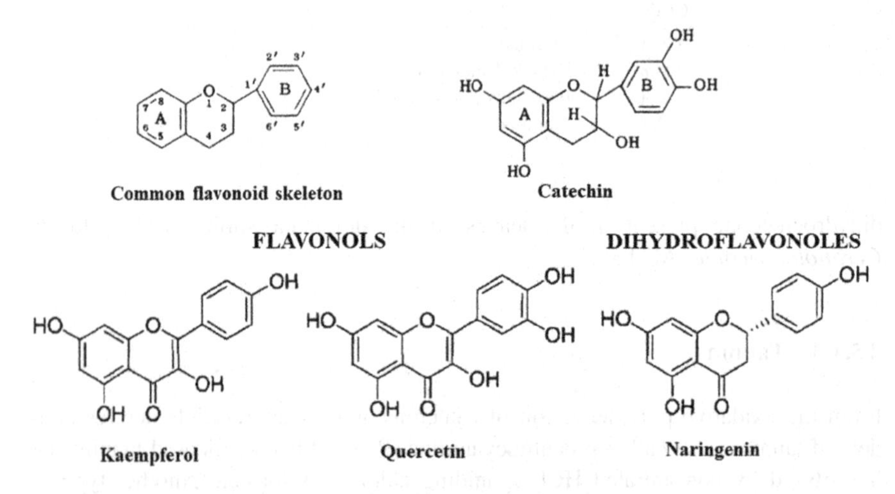

Fig. 2.7 Some allelopathic flavonoids (Macias et al. 2007, Modified)

2.5.1.4 Cinnamic Acid and Derivatives

Cinnamic acid and derivates arise from phenylalanine or tyrosine through the shikimic pathway (Neish 1964; Fig. 2.9). The first step of this pathway is catalyzed

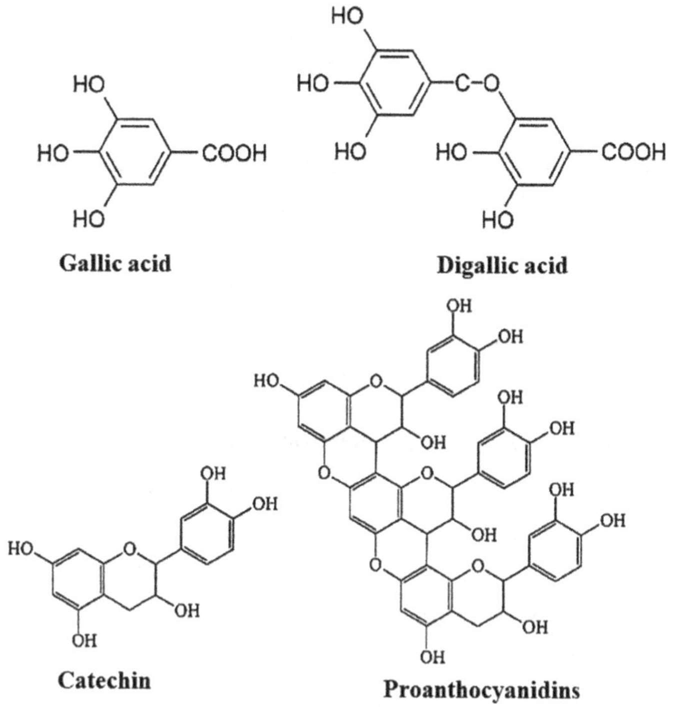

Gallic acid

Digallic acid

Catechin

Proanthocyanidins

Fig. 2.8 Some examples of hydrolyzable and condensed tannins

by the phenylalanine ammonia lyase (PAL), a widely distributed phenylpropanoid enzyme present in green plants, algae, fungi, and even in some prokaryotes (Hyun et al. 2011). They are phenyl propanoids containing 3-carbon side chain coupled to a phenol. They are formed in the biochemical route that yields lignin, the polymeric material that provides mechanical support to the plant cell wall (Xu et al. 2009). Cinnamic acid is a known allelochemical that affects seed germination and plant root growth and therefore influences several metabolic processes. Chlorogenic acids (CGA_s) and isochlorogenic acid are formed as esters between different derivatives of cinnamic acid, caffeic acid specifically, and quinic acid molecules. They are components of the hydroxycinnsmic acids's classes. The main hydroxycinnamic acids with allelopathic properties are ferulic acid, caffeic acid, sinapic acid and ρ-coumaric acid.

2.5.1.5 Coumarins

Coumarins, classified as member of the benzopyrone family, all of which consist of a benzene ring joined to a pyrone ring, are lactones of o-hydroxycinnamic acid (Robinson 1963). They are widely distributed in the Apiaceae, Rutaceae,

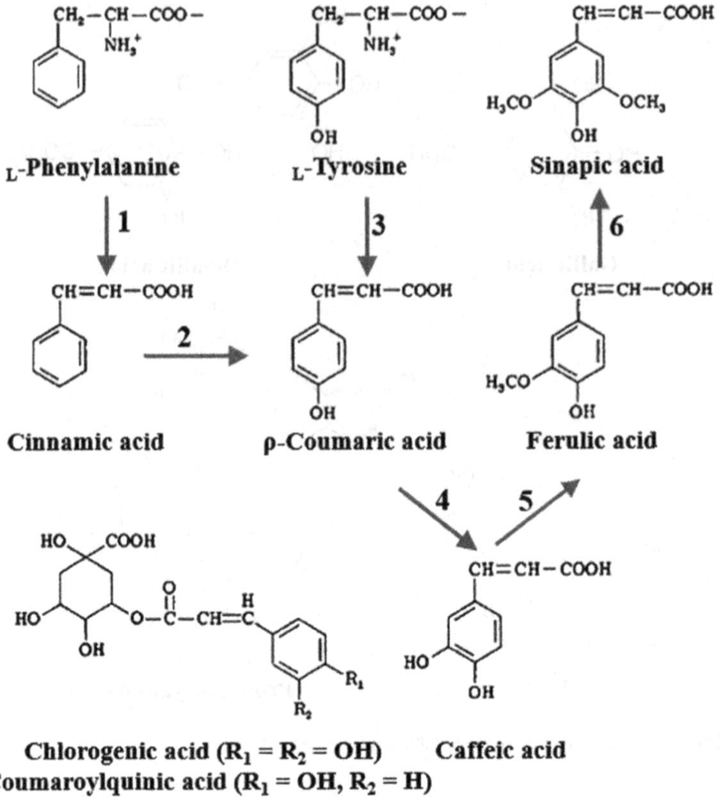

Fig. 2.9 Examples of cinnamic acid derivatives with allelopathic potential (Rice 1984, Modified)

Asteraceae and Fabaceae families. Coumarins, which are almost unknown in the animal kingdom, are present with high frequency in the plant kingdom and occur in all parts of plants, depending on environmental conditions and seasonal changes. It is possible to classified four main coumarin sub-types: the simple coumarins, furanocoumarins, pyranocoumarins and the pyrone-substituted coumarins (Fig. 2.10). The simple coumarins are the hydroxylated, alkoxylated and alkylated derivatives of the parent compound, coumarin, along with their glycosides (Jain and Joshi 2012). Furanocoumarins consist of a five-membered furan ring attached to the coumarin nucleus, divided into linear or angular types with substitution at one or both of the remaining benzoid positions (Ojala 2001). Umbelliferone, esculetin and scopoletin are the most widespread coumarins in nature.

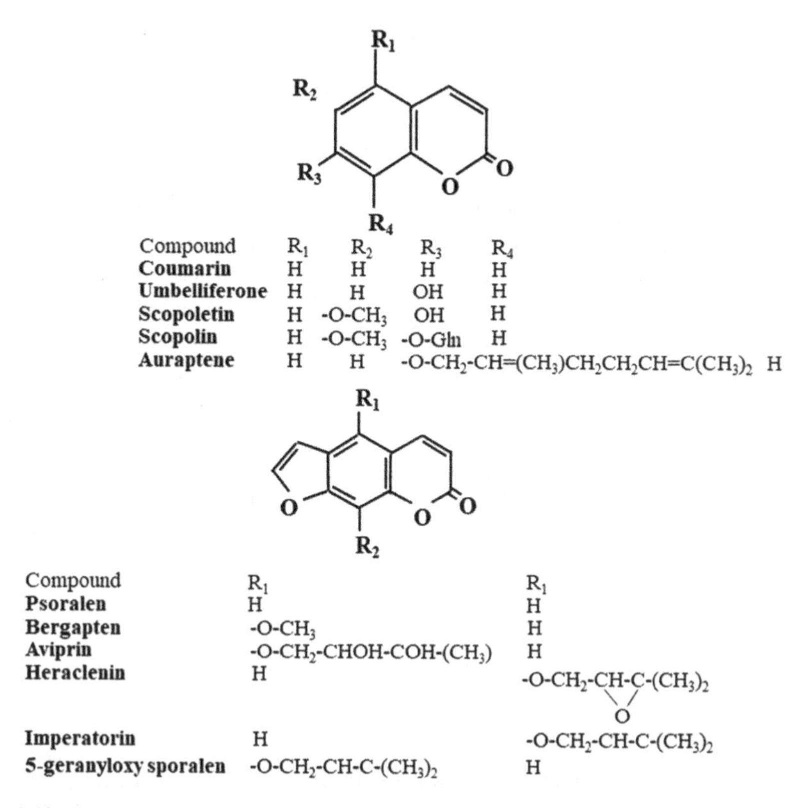

R1
R2
R3
R4
O
O
Compound R1 R2 R3 R4
Coumarin H H H H
Umbelliferone H H OH H
Scopoletin H -O-CH3 OH H
Scopolin H -O-CH3 -O-Gln H
Auraptene H H -O-CH2-CH=(CH3)CH2CH2CH=C(CH3)2 H
R1
O
O
O
R2
Compound R1 R1
Psoralen H H
Bergapten -O-CH3 H
Aviprin -O-CH2-CHOH-COH-(CH3) H
Heraclenin H -O-CH2-CH-C-(CH3)2
O
Imperatorin H -O-CH2-CH-C-(CH3)2
5-geranyloxy sporalen -O-CH2-CH-C-(CH3)2 H

Fig. 2.10 Structures of some allelopathic simple coumarins (up) and furanocoumarins (down) (Razavi 2011, Modified)

2.5.1.6 Lichen Metabolites

Among the second metabolites with allelopathic activities, there are a group of compounds never reported in higher plants: lichen metabolites. In fact, lichens produce secondary metabolites that are unique to the symbiosis (Romagni et al. 2004). Most of these compounds are aromatic and are derived from the polyketide pathway, with a few originating from the shikimic acid and mevalonic acid pathways (Table 2.2). Lichen metabolites can be divided into four main classes: depsides, depsidones, depsones and dibenzofurans (Fig. 2.11). According to Rundel (1978), their ecological roles refer to the protection against damaging light conditions, chemical weathering compounds, allelopathic compounds and antiherbivore defence compounds.

Table 2.2 The major classes of secondary lichen metabolites (Source: Elix 1996)

Biosynthetic Origin	Chemical Class	Examples
Polyketide	Depsides	Lecanoric acid
	Depsones	Picrolichenic acid
	Depsidones	Physodic acid
	Dibenzofurans	Pannaric acid
	Usnic acids	Usnic acid
	Chromones	Sordinone, Eugenitin
	Xanthones	Lichexanthone
	Anthraquinones	Emodin
Mevalonate	Diterpenes	16α-hydroxykaurane
	Triterpenes	Zeorin
	Steroids	Ergosterol
Shikimate	Terphenylquinones	Polyporic acid
	Pulvinic acid	Pulvinic acid

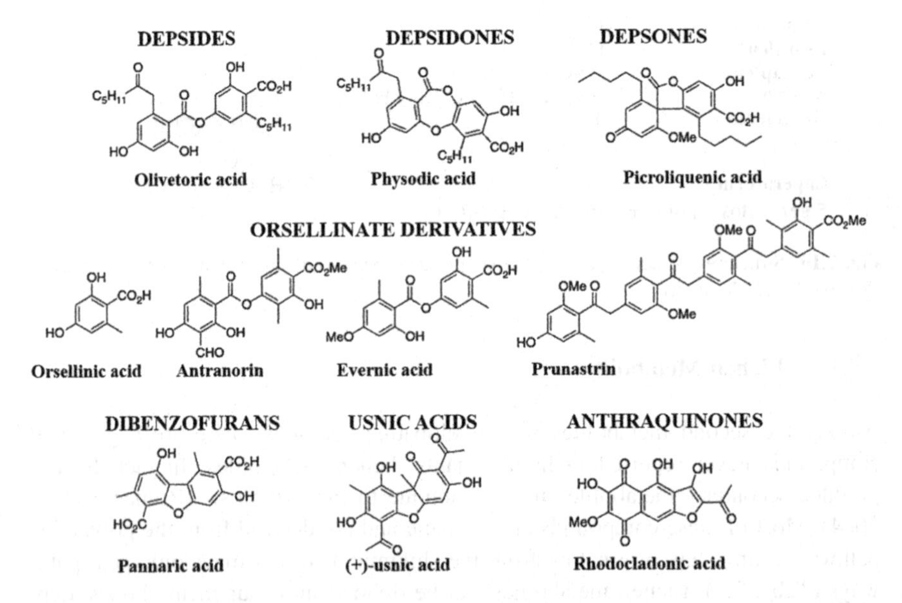

Fig. 2.11 Lichen metabolites reported to have phytotoxic activity (Macías et al. 2007, Modified)

2.5.2 *Terpenoids*

Terpenoids or isoprenoids are secondary metabolites present in many organisms similar to the terpenes, from which they differ because the latter refers only to hydrocarbons. More than 50,000 terpenoids have been isolated from both terrestrial and marine plants, and fungi. After phenolics, they are the second largest group of secondary metabolites implicated in allelopathy. The class of the terpenoids presents a great variety of compounds in which the several structures, most of them multicyclic, differ from one another in their basic carbon skeletons and functional groups. All terpenoids are based on a various but definite number of 5-carbon isoprene units, called also 2-methyl-1,3-butadiene. In plants, there are two independent metabolic pathways that create terpenoids: the classic mevalonic acid (MVA) pathway and the methylerythritol phosphate (MEP) pathway, also known as non-mevalonate pathway or mevalonic acid-independent pathway. The former occurs in the cytosol and produce also cholesterol; the latter takes place entirely in plastids. The MVA pathway provides the precursors for the biosynthesis of sesquiterpenes, phytosterols, brassinosteroids, and triterpenes (Newman and Chappell 1999). Instead, the MEP pathway provides the C_5-building blocks for the biosynthesis of carotenoids, chlorophyll, gibberellins, and monoterpene and diterpene specialized metabolites, which are exclusively or primarily produced in plastids (Lichtenthaler 1999). In either pathways, terpenoids derives through the condensation of the end-products isopentenyl diphosphate (IPP) and its allylic isomer dimethylallyl diphosphate (DMAPP), giving geranyl pyrophosphate (GPP) (Fig. 2.12). Isoprene is formed from DMAPP via the action of the enzyme isoprene synthase which catalyses elimination of diphosphate.

It is possible to classified terpenoids according to the number of isoprene units incorporated in the basic molecular skeleton (Table 2.3) (Fig. 2.13).

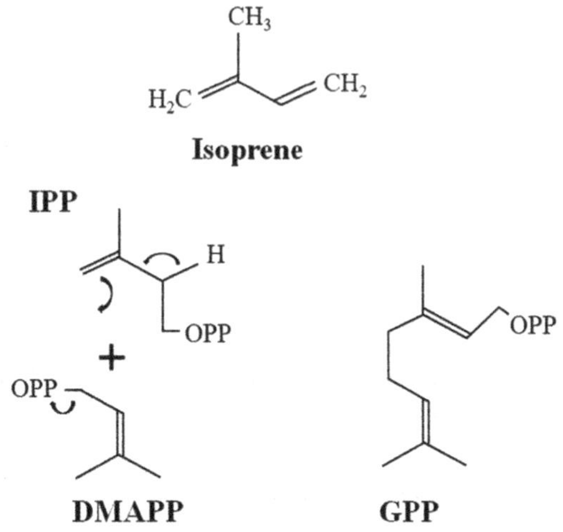

Fig. 2.12 Isoprene (up) and some terpenoids precursors (down) molecular structural formula

Table 2.3 Classification of terpenoids based upon the number of isoprene units

Terpenoids	Isoprene units	Number of carbon atoms
Meroterpenoids	1	C5
Monoterpenoids	2	C10
Sesquiterpenoids	3	C15
Diterpenoids	4	C20
Sesterterpenoids	5	C25
Triterpenoids (es. sterols)	6	C30
Tetraterpenoids (es.carotenoids)	8	C40
Polyterpenoids (es. rubber)	many (>100)	Polymer (>500)

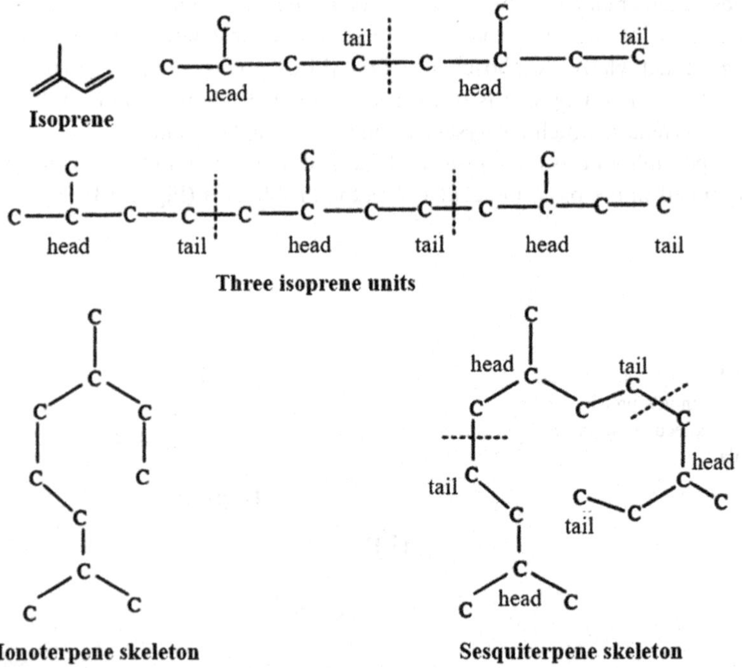

Isoprene

Three isoprene units

Monoterpene skeleton

Sesquiterpene skeleton

Fig. 2.13 Arrangement of isoprene units in mono- and sesquiterpenoids (Bhat et al. 2005, Modified)

2.5.2.1 Monoterpenoids

Monoterpenoids, with sesquiterpenes, are the major components of essential oils. They are volatile compounds that have been described as the predominant terpenoid allelochemicals from higher plants. Most of them are inhibitors of seed germination

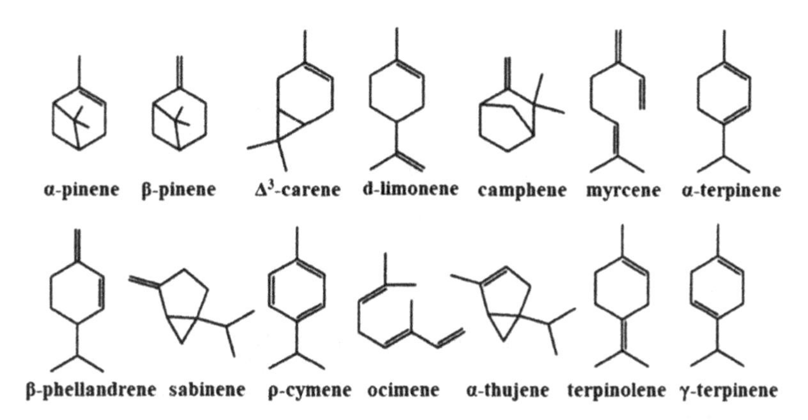

Fig. 2.14 Chemical structures and name of monoterpenes (Shexia 2012, Modified)

and several microorganisms, mainly bacteria and fungi. Besides, they have been proposed as potential starting structures for herbicides (Vaughn and Spencer 1993). For example, it can be observed the high structural similarity between monoterpenes 1,4- and 1,8-cineole and the herbicde cinmethylin (Fig. 2.14). While a few, such as camphor or cineoles, occur in a near pure form, most of terpenoids occur as complex mixtures difficoult to separate. Of the 14 most commonly occurring monoterpenes (α-pinene, β-pinene, Δ(3)-carene, d-limonene, camphene, myrcene, α-terpinene, β-phellandrene, sabinene, ρ-cymene, ocimene, α-thujene, terpinolene, and γ-terpinene), the first six are usually found to be most abundant (Shexia 2012).

2.5.2.2 Sesquiterpene Lactones

Sesquiterpenoids are the terpenoids with 15 carbons containing three isoprene units, plus a lactone ring. Their structures present several acyclic, mono-, bi-, tri-, and tetracyclic systems. They are present in high quantity in several plants, particularly in those of the Compositae family (Fraga 2005). They have a wide range of biological activity, including plant growth-regulating, insect anti-feedant, anti-fungal, and anti-bacterial properties (Picman 1986; Baruah et al. 1994). Some of the most common sesquiterpene lactones are artemisinin, isolated from the plant *Artemisia annua* L., and centaurepensin and cnicin, presents mainly in the members of the families Centaurea (Fig. 2.15).

Sesquiterpene lactones are important allelochemicals involving in the invasive potential of plant species, such as *Centaurea diffusa* Lam. (diffuse knapweed) and *Centaurea maculosa* Lam. (spotted knapweed) in North America.

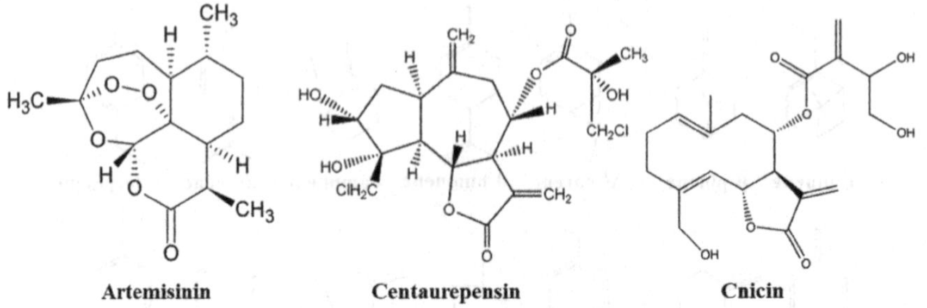

Artemisinin Centaurepensin Cnicin

Fig. 2.15 Chemical structures of some common sesquiterpene lactones

2.5.2.3 Diterpenoids

Diterpenoids have 20 carbon atoms and consist of four isoprene units. Such diterpenoids, such as giberrellins, act as important plant hormones and there are relatively few reported diterpenoid allelochemicals produced by plants.

The most famous are momilactones (Fig. 2.16), rice diterpenes that are exuded from the roots of young rice seedlings due to the infection by blast fungus (*Pyricularia oryzae*) or irradiation with UV light (Cartwright et al. 1981). However, in literature momilactones are reported as phytoalexins, whereas momilactone A and B are the only rice diterpenoids identified as allelopathic agent.

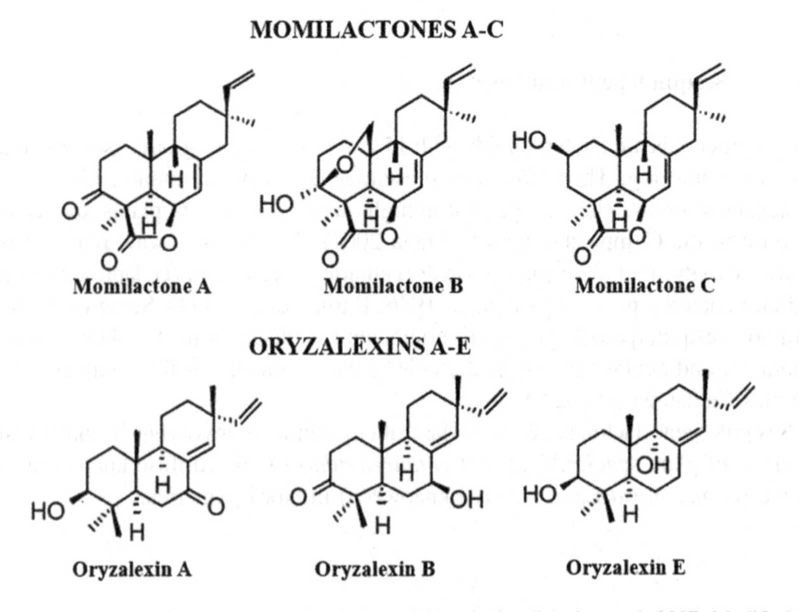

Fig. 2.16 Rice allelochemicals: momilactones and oryzalexins (Macías et al. 2007, Modified)

2.5.2.4 Other Terpenoids

Quassinoids are degradated triterpenes containing six isoprene units reported from the members of the Simaroubaceae family. They present a high structural complexity and, according to their basic skeleton, are categorized into five distinct groups: C-18, C-19, C-20, C-22 and C-25 types shown in Fig. 2.17. Quassinoids present a wide range of biological activities including antitumor, antimalarial, anti-inflammatory, insecticidal, fungicidal and herbicidal. The first quassinoid identified as an allelopathic agent was quassin, a C20 type isolated from the quassia wood in Suriname (*Quassia amara* L.) by Clark's group in 1937. The most important quassinoid is ailanthone, an allelochemical produced by the tree-of-heaven (*Ailanthus altissima* (Mill.) Swingle) which inhibit the growth of other plants. Thanks to ailanthone, the *A. altissima* tree has become a strong invasive species in Europe (Fig. 2.18).

Benzoxazinoids are hydroxamic acids produced by many species, mainly the plants of the Poaceae family (wheat, rye, maize) and in a few species of dicots. These compounds are very instable and, undergo hydrolysis, they contract into the corresponding benzoxazolinones (Macías et al. 2004). The most effective allelopathic compounds are DIBOA, DIMBOA and their breakdown products BOA and MBOA (Barnes and Putnam 1987; Tabaglio et al. 2008).

Other important terpenoids with allelopathic potential belong to the chemical classes of glucosinolates and steroids. Glucosinolates are sulfur- and nitrogen-compounds that occur in most plants of the Brassicales order, e.g. in Brassicaceae and Capparidaceae families, with the role of defence against insects, herbivores and certain microbial pathogens. They are degraded by the endogenous enzyme

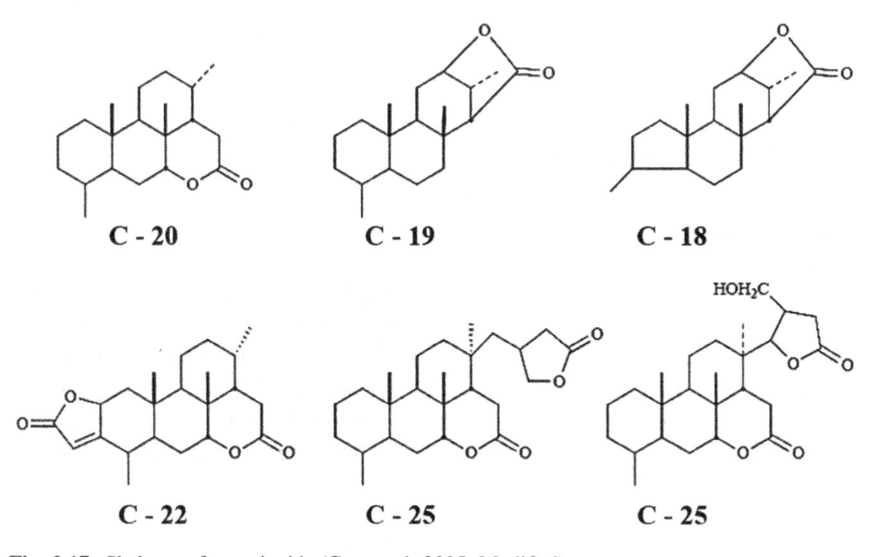

Fig. 2.17 Skeleton of quassinoids (Guo et al. 2005, Modified)

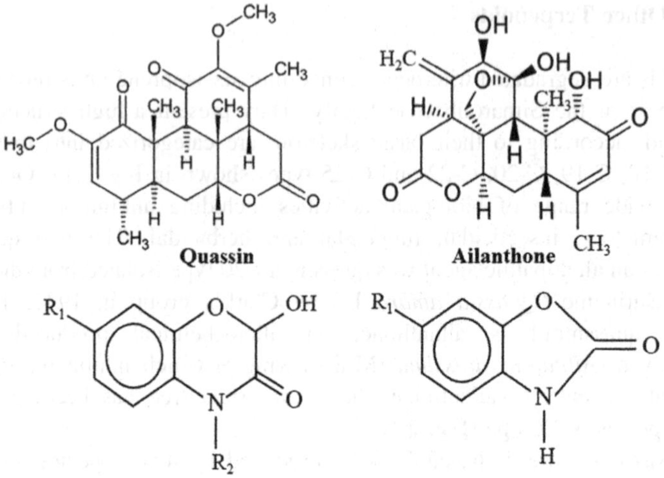

Quassin Ailanthone

DIBOA : $R_1 = H$; $R_2 = OH$
DIMBOA : $R_1 = MeO$; $R_2 = OH$
DINBOA : $R_1 = R_2 = H$
HBOA : $R_1 = H$; $R_2 = OH$
HNBOA : $R_1 = NO_2$; $R_2 = H$

BOA : $R_1 = H$
MBOA : $R_1 = MeO$
NBOA : $R_1 = NO_2$

Fig. 2.18 Chemical structures of some quassinoids and benzoxazinoids mentioned in the text

β-thioglucosidases, called myrosinase, into compounds such as isothiocyanates. Steroids are tetracyclic terpenoids containing 17 carbon atoms with only two methyl groups attached to the ring system. Very few are the steroids linked to plant-plant allelopathy, while there are several examples of antimicrobial activities such as the aglycones digitoxigenin produced by *Digitalis purpurea* L. and strophanthidin, produced by *Convallaria majalis* L. (Robinson 1963).

2.5.3 Other Compounds

2.5.3.1 Alkaloids

Alkaloids are organic compounds containing basic nitrogen atoms in the heterocyclic rings or in side chains. They present an enormous variety of structures and there is not a uniform classification of them. Alkaloids are produced by secondary metabolism of primary metabolites, usually amino acids. These compounds are produced by a large variety of organisms, including bacteria, fungi, plants, and animals. Plant alkaloids often demonstrate defensive activity against a wide variety of predators and competitors among microorganisms, fungi, viruses, invertebrate, vertebrate herbivores and plants (Blum 2004). Today, more than 10,000 alkaloids are known, but only a small number of them present allelopathic activities. There are several works on the role of alkaloids as seed germination inhibitors (Evenari 1949; Wink 1983). Among allelopathic alkaloids there are papaverine, caffeine, emetine, gramine, etc. (Fig. 2.19).

2.5.3.2 Cyanogenic Glycosides

Cyanogenic glycosides are glycosides consisting of a sugar group and a non-sugar group, called aglycone, in this case a cyanide group, which when enzymically hydrolyzed release cyanohydric acid (HCN), a compound extremely toxic. In most cases, hydrolysis is accomplished by the β-glucosidase, producing sugars and a cyanohydrin that spontaneously decomposes to HCN and a ketone or aldehyde (Francisco and Pinotti 2000) (Fig. 2.20).

Cyanogenesis is a plant's protective mechanism against predators such as the herbivores. Cyanohydrins are very common in plant kingdom, but they were also founded in some species of ferns, fungi and bacteria (Harborne 1972). The best known cyanogenic glycosides are dhurrin, presenti in sorghum (*Sorghum vulgare* Pers.) seedlings, amygdalin and prusanin, very common among plants of the Rosaceae, particularly the *Prunus* genus (Fig. 2.21). These compounds are strong germination and growth inhibitors.

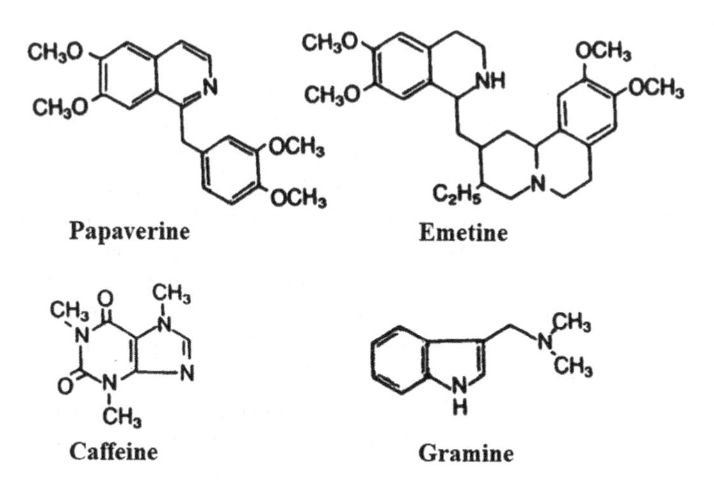

Fig. 2.19 Chemical structures of some alkaloids identified as allelopathic agents

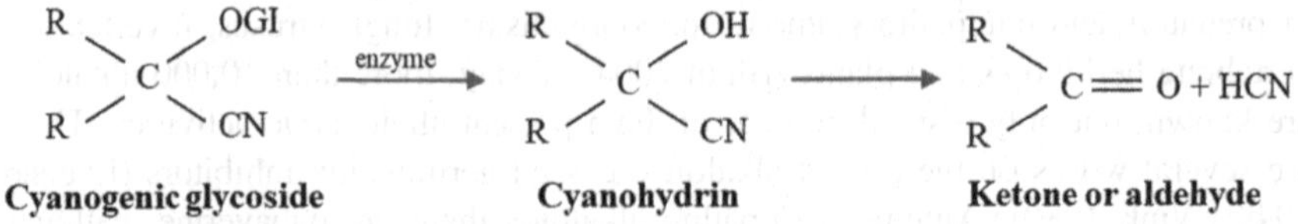

Cyanogenic glycoside —enzyme→ Cyanohydrin → Ketone or aldehyde ($C=O + HCN$)

Fig. 2.20 Pathway of release of HCN by cyanogenic plants (Francisco and Pinotti 2000, Modified)

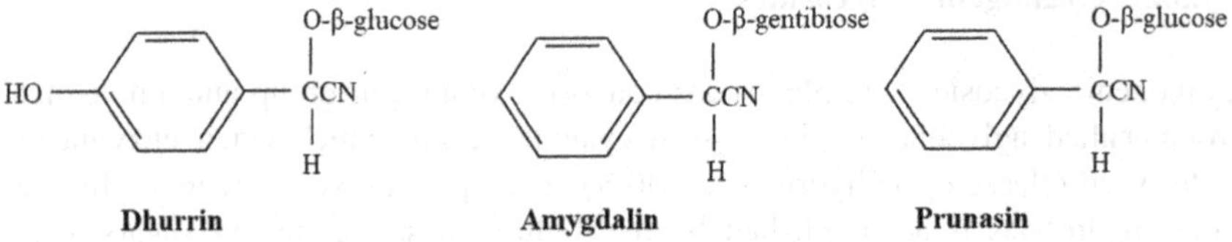

Dhurrin Amygdalin Prunasin

Fig. 2.21 Chemical structures of some cyanogenic glycosides with allelopathic potential

2.6 Modes of Release of Allelochemicals into the Environment

Most of allelochemicals are distributed among many species belonging to several botanical families, but there are particular secondary metabolites that are restricted within a group of taxonomically related species. For example, salicacin, a phenol glucoside, is characteristic of the members of the Salicaceae family while benzoxazinoids are found mainly in the Brassicales order, and so on.

Plant allelochemicals are generally localized and sequestered in glandular or subepidermal layers (Ambika 2013). They can be found, in different concentrations, in several parts of plants: leaves, stems, roots, rhizomes, seeds, flowers or inflorescences, fruits and even pollen (Bertin et al. 2003; Gatti et al. 2004; Kruse et al. 2000). Generally, leaves represent the most consistent source of inhibitors for many plants but, in some cases, this is reversed with roots, which anyway are at least the second source of allelochemicals. Pollen also represent for many species, mainly in Poaceae and Asteraceae, an important source of allelochemicals. Much interesting results the work of Murphy (1999) on pollen allelopathy.

The presence of an allelopathic compound into a plant not necessarily imply a role on the ecosystem due to that compound. To exert an effect, the allelochemical must be released into the environment at a time when it can perform its inhibitory action (Sattin and Tei 2001).

Plants release allelochemicals into the environment through four main pathways (Fig. 2.22): (1) volatilization from living parts of the plant; (2) leaching from aboveground parts of the plant; (3) decomposition of plant material; (4) root exudation.

These pathways varies among species and according to the chemical nature of compounds.

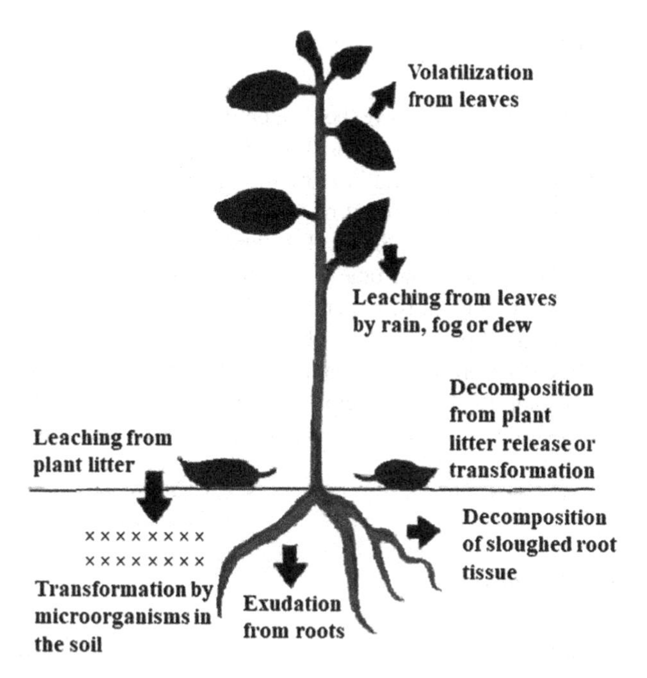

Fig. 2.22 Major pathways of release of allelochemicals into the environment (Rice 1984, Modified)

2.6.1 *Volatilization*

Many plants release volatile inhibitory compounds under vapour form, mainly through leaves, in the atmosphere. Most of these volatile compounds are terpenoids, e.g. monoterpenoids, sesquiterpene lactones, etc. or hormones such as ethylene ($CH_2 = CH_2$), in this case released by fruits. Allelochemicals released by volatilization can be absorbed by plants directly from the atmosphere through gas exchanges or from soil, where they arrive due to rainfall or leaching. Many works has been done on the allelopathic effects of volatile inhibitors (Chu et al. 2014; Nishida et al. 2005; Barney et al. 2005; Kong et al. 2002). Volatile allelochemicals are seed germination and growth inhibitors, present antimicrobial and antifungal activities and are involving in the "old field" succession. This mode of release generally shows its most significant ecological effects under arid and semiarid conditions (Rice 1974).

2.6.2 Leaching

Many kinds of allelochemicals have been identified in the leachates of plants. Leaching is the process that leads to the loss of chemical compounds from the aerial part of the donor plant or the ground litter by means the hydro-solubilization made by rain, fog or irrigation. The quantity of leachates released depends on duration and amount of rainfall and on chemical nature of compounds. Water-soluble compounds are more leached than others. Among allelopathic leachates there are amino acids, phenolic compounds, terpenoids, alkaloids and fatty acids. They can derive from living or dead parts of plants. Buta and Spaulding (1989) founded that allelochemicals leached from excised leaves of tall fescue grass (*Festuca arundinacea* Schreb.) belong to three principal inhibitory compounds, abscisic acid, caffeic acid, and *p*-coumaric acid and founded that abscisic acid was the predominant inhibitor.

2.6.3 Root Exudation

Like for volatilization, many researchers have studied the exudation phenomenon and discovered that living roots of many weed and crop species exude different types of organic compounds such as amino acids, carbohydrates, nucleotides, enzymes, steroids, terpenoids, tannins, fatty acids, alkaloids, vitamins and flavonoids. Root exudates play a fundamental role within the ecological succession of microorganisms in the rhizosphere and influence seed germination, root and shoot growth, nutrient uptake and nodulation (Pandya et al. 1984; Inderjit and Weston 2003; Yu and Matsui 1994). Besides, allelochemicals released by root in the rhizosphere influence resistance to pests and, inevitably, soil characteristic. For example, Hao et al. (2010) founded that rice exudates such as phenolic acids, sugars and free amino acids suppressed the *Fusarium* wilt of watermelon (*Fusarium oxysporum* f. sp. *niveum*), while those of watermelon significantly stimulated *Fusarium* spore germination and sporulation. Root exudation is affected by a variety of factors including the age and species of plant, stress factors such as availability of moisture, temperature and light intensity, mineral nutrition, soil microorganisms (Hale et al. 1971). In cereals, it was observed that an amount between 5 and 21% of plant total photosynthates is released via root exudation (Haller and Stolp 1985; Vivanco et al. 2002).

2.6.4 Decomposition of Plant Material

Considering that it is not easy to distinguish the different modes of release of allelochemicals in the environment, it is believed that decaying and leaching represent the main mode release of allelochemicals. The decomposition of plant residues adds a large quantity of allelochemicals into the rhizosphere (Goel 1987). The process depends on the nature of plant residues, soil characteristics and it is closely associated with microbial activity, which is strongly influenced by temperature and soil water content. These products are converted by soil microflora from nontoxic compounds to toxic ones or into more biologically active products than the parents (Blum and Shafer 1988). Generally, water-soluble inhibitors are easily leached out of plant litter after death when membranes lose their differential permeability (Rice 1974). This pathway of release is often linked to autotoxicity probems, the case of alfalfa (*M. sativa*) for instance, and it is strictly related to the weed management with cover crops, green manure and intercropping.

2.7 Factors Affecting Allelochemical Production

It is important to understand what means when a plant is "stressed". According to Levitt (1980), plant stress can be defined as "*a state in which increasing external demands lead to the destabilization of plant functions, followed by a phase of normalization and improving of the resistance. If the plant is forced out of its tolerance limits and its acclimation capacity is over passed, the result can be a permanent damage or even plant death*". Different types of abiotic and biotic stress factors influence the quantity of allelochemicals released by the donor plant and the effect of an allelochemical on the target plant (Inderjit and Del Moral 1997; Fig. 2.23). Stress factors such as drought, irradiation, light, temperature, nutrient and water availability, diseases and pathogens, competitors, increase allelochemical production in a plant (Einhellig 1996; Reigosa et al. 1999a, b). Allelochemical production is influenced also by morphological, physiological and ecological characteristics such as plant density, life cycle, plant age and habitat (Inderjit and Keating 1999). The *stress hypothesis of Allelopathy* formulated by Reigosa et al. (1999a, b; Reigosa and Pedrol 2002), states that allelopathy can appear and disappear in a place according to environmental changes, so that allelopathy becomes more important when and where plants are under stress.

Even though the production of allelochemicals in a plant can increase in response to stress, it is not clear whether a corresponding release of allelochemicals to the environment also occur (Einhellig 1996; Inderjit and del Moral 1997). However, the sensitivity of target plants to allelochemicals in general is affected by stress and typically it is increased (Einhellig 1996; Reigosa et al. 1999a, b). The combination of several stress factors results in an increase of allelochemical concentrations in donor plants (del Moral 1972). It is important to increase the research on the synergistic effects of stress factors because they generally occur in combinations under field conditions (Rice 1974).

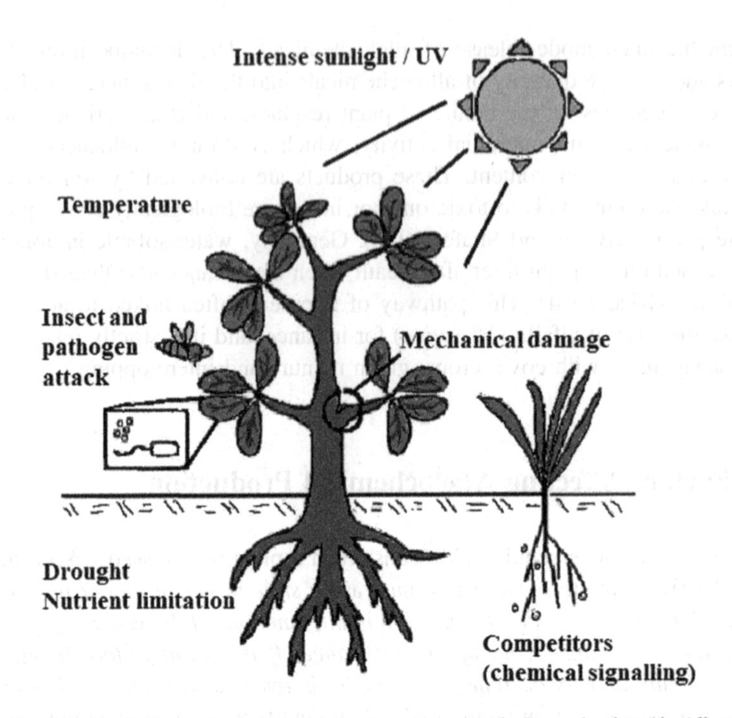

Fig. 2.23 Abiotic and biotic stress factors affecting allelochemicals production (de Albuquerque et al. 2010, Modified)

2.7.1 Light

Amount, intensity and quality of light play an important role on the production of inhibitors. In fact, plants growing in greenhouse produce a less amount of allelochemicals than the same kinds growing out-of-doors (Rice 1984).

Ionizing radiation increases the concentrations of phenolic allelochemicals in sunflower (*Helianthus annuus* L.) and tobacco (*Nicotiana tabacum* var. One Sucker) plants (Fomenko 1968; Koeppe et al. 1970a; Rice 1974).

Also the ultraviolet radiation, generally, increase the amounts of inhibitors produced by donor plants. Del Moral (1972) demonstrated that supplemental UV light increased amounts of total chlorogenic and isochlorogenic acids in sunflower. Furness et al. (2008) reported increased allelopathic influence of houndstongue (*Cynoglossum officinale* L.) on some forage grasses. Li et al. (2009) found that the allelopathic potential of *Zanthoxylum bungeanum* on seed germination rate of alfalfa, lettuce and radish is improved under enhanced UV-B radiation and differed depending on species.

It seems that visible light enhances the synthesis of inhibitory compounds in donor plants. Zucker (1963) was one of the earliest scientists who study the effects of visible light on allelochemical production. Kato-Noguchi (1999) reported that visible light may enhances allelopathic activity of germinating maize due to an increase in the level of DIBOA.

Allelochemical synthesis is influenced also by light quantity, as well as its quality. Photoperiod differently affects short-day plants and long-day ones (Zucker et al. 1965). In most cases, long days increase inhibitory compounds in donor plants regardless of the daylengths required for flowering (Rice 1974).

Therefore, the amount of allelochemicals is generally greater during exposure of ultraviolet light and long daylength.

2.7.2 Temperature

In general, plants under thermal stress tend to produce more allelochemicals but, by contrasts, become more susceptible to them. Quantities produced are higher at lower temperatures, while high temperatures enhance allelochemical effects (Einhellig and Eckrich 1984). Koeppe et al. (1970b) found that chilling (8–9 °C) of tobacco plants increased the concentrations of total chlorogenic acids in old leaves, young leaves and stems, but decreased the concentration in the roots.

2.7.3 Water Deficiency

Water deficiency, as well as all stress factors, result in increased concentrations of allelochemicals. Using NaCl in colture solution to cause water stress on sunflower plants, del Moral (1972) found that, after 31 days of treatment, the concentrations of total chlorogenic and isochlorogenic acids in roots, stems and leaves were increased (Rice 1974). Amount different irrigation levels, Ardi (1986) reported that inhibitory effects of purple nutsedge (*Cyperus rotundus* L.) on sweet corn (*Zea mays*) yield were reduced at the highest waters stress imposed. Tang et al. (1995), studying water deficiency on the allelopathic potential of purple nutsedge, found that both fresh and dry weights of its shoots and roots decrease with increasing waters stress (Inderjit and Keating 1999). Oueslati et al. (2005) reported that barley (*H. vulgare* L.) autotoxicity increase under drought conditions.

2.7.4 Minerals Availability

Many authors (Loche and Chouteau 1963; Lehman and Rice 1972; Mwaja et al. 1995 etc.) demonstrated that the mineral deficiencies of B, Ca, Mg, N, P, K and S, play an important role in the production of inhibitory compounds. This proves that the production of allelochemicals increases under nutritional stress condition.

Loche and Chouteau (1963) found increases in concentrations of scopolin and decreases in those of chlorogenic acid in calcium- and borum-deficient tobacco leaves (Rice 1974). Studying the effect of deficiency of N, K and P on phenolic content in sunflower, Lehman and Rice (1972) reported increased amounts of chlorogenic acid and scopoletin in old leaves, stems and roots of mineral-deficient plants than in controls. Mwaja et al. (1995) studied the effects of three fertility regimes (low, medium and high) on phytotoxicity, biomass production and allelochemical content in rye (*S. cereale*) and concluded that, despite the larger amounts of rye biomass, low fertilisation enhances the phytotoxicity and allelochemical content. Chamacho-Cristóbal et al. (2002) studied the effects of B deficiency on phenols and the activities of the enzymes involved in their biosynthesis in tobacco (*N. tabacum*) plants. They found a positive correlation between phenols concentrations and the activity of phenylalanine ammonia-lyase (PAL) after 5–7 days of B-deficiency. B deficiency, therefore, results in an increase of PAL activity and, in turn, in an enhancement of phenolic levels.

2.7.5 Soil Characteristics

As suggested by Cheng (1989, 1992), the effectiveness of allelochemicals in soil is strictly influenced by physiochemical and biological soil factors.

In general, clay soils, which are characterized by high values of cation exchange capacity (CEC) and anion exchange capacity (AEC), adsorb more allelochemicals on the surface of their colloidal particles than sandy soils. Del Moral (1972), reported that fine-textured soils sorb more amounts of phenolic compounds than sandy-loam soils. Besides, allelochemicals persist for longer duration in clay soils (Einhellig 1987). After soil adsorption, also the transport of allelochemicals (i.e., the movement of allelopathic compounds from roots of donor plants to roots of target plants) depends on soil texture and chemical nature of inhibitors. Transport can be either through the air as vapour or in the soil solution (Cheng 1992). The movement is faster in sandy than in clay soils (Jennings and Nelson 1998).

The chemical characteristics of soil such as pH, organic carbon, nutrients available, ion exchange capacity, oxidation state, also play an important role on the fate of allelochemicals. For example, soil pH can affect the uptake and the immobilization of allelochemicals (Cheng 1992). It must be considered that higher pH can stimulate microbial activity (Aarnio and Marikainen 1994) and, in turn, allelochemical availability. Soil organic carbon can indirectly influence allelochemical stability and persistence (Lehman and Cheng 1988; Huang et al. 1977). Organic matter, in fact, strongly enhance soil microbial activity, exerts a buffer capacity on soil pH and, thanks to its high ion exchange capacity, promote allelochemical sorption and retention. Lehman and Cheng (1988), studyng the reactivity of phenolic acids in several soil, found that they are more stable in forest soils with high organic matter than in cultivated agricultural soils.

2.7.6 Biotic Components

The expression of allelopathy may be influenced by a series of biotic factors such as diseases and pathogen attacks, weed competitors, interactions with herbicides, age of plant organs, plant density and habit (Einhellig 1996).

Despite of detrimental effects, generally pathogens decrease the competitiveness and simultaneously increase the allelopathic activity of their hosts (Mattner 2006). In fact, as consequence of defensive adaptation, the attacks of phytophagous or plant-sucking insects and diseases cause a considerable increase in the release of allelopathic compounds.

Jay et al. (1999) reported that, because of infection with beet western yellows virus (BMYV), oilseed rape (*B. napus* L.) increased glucosinate concentrations in tissues by 14%. Woodhead (1981) found that sorghum plants infected with downy mildew (*Sclerospora sorghi* W. Weston & Uppal) or rust (probably *Puccinia purpurea* Cooke) or sorghum shoot fly (*Atherigona soccata* Rondani), increased phenolic concentrations. Soil in which rusted ryegrass (*Lolium perenne* L.) attacked by *P. coronata* Corda f.sp. *lolii* Brown was grown suppressed clover (*Trifolium repense* L.) biomass by 36% more in comparison with the direct effect of soil in which healthy ryegrass was grown in the greater rainfall areas of south-eastern Australia (Mattner and Parbery 2001: de Albuquerque et al. 2010).

According to Belz (2007), weeds can elicit allelochemical biosynthesis in competing crops as well as insects or pathogens induce plant defences in attacked plants. The author reports examples about the exudation of three major allelochemicals in two allelopathic cultivars of rice (*O. sativa*) due to the presence of *Echinochloa crus-galli* (P.) Beauv. and the release of sorgoleone in a *Sorghum* hybrid after exposure to water-soluble root leachates released from *Abutilon theophrasti* Medik. She suggests that biotic-induced plant defences depends on a direct pest attack and on aerial or rhizosphere signals from healthy or attacked plants.

Lydon and Duke (1993) reported that herbicides, at both lethal and sub-lethal concentrations, influence allelochemical production by both direct and indirect effects. Many works has be done to determine the effect of herbicides on allelochemical biosynthesis. Winkler (1967) reported that levels of scopolin increased after spraying tobacco plants with maleic hydrazide.lydon and Duke (1988) found that redroot piweed (*Amaranthus retroflexus*), ryegrass (*L. perenne*), soybean (*Glycine max*), velvet leaf (*A. theophrasti*) and yellow nutsedge (*C. esculentus*) presented high levels of shikimic acid and certain hydrobenzoic acids due to glyphosate treatments.

Many Authors agree in considering the age of plant organs as an important factor involved in the production of allelopathic compounds. Koeppe et al. (1969, 1970b) found that the age of both tobacco and sunflower leaves influenced the amounts of scopolin, chlorogenic and isochlorogenic acids. To improve weed management, it should be considered the age of donor plant at which release of allelochemicals starts (Inderjit and Keating 1999). For example, Schumacher et al. (1983)

discovered that wild oats (*Avena fatua*) become allelopathic against spring wheat (*T. aestivum*) at the four-leaf stage.

The influence of plant density of target species on the response to alellochemicals it is now widely accepted by the international scientific community. Weidenhamer et al. (1989), studying the density-dependent effects of varying amounts of gallic acid and hydroquinone on *Paspalum notatum* Flüggé and *L. esculentum* grown at different densities, found that the quantity of allelochemicals available to each target species decreases with the increase in target species density.

2.8 Modes of Action of Allelochemicals

Understanding which compounds and which mechanisms of action are involved in allelopathy is important to develop predictive models (Inderjit and Duke 2003). However, this is a question of not easy solution due to the great diversity of chemical families of allelochemicals and to the several molecular target site of phytotoxic compounds. Besides, is important to recognize that, in field situations, allelopathic activity is thought to be often due to joint action of mixtures of allelochemical rather than to one allelochemical (Einhellig 1995). Since the visible effects of allelochemicals on plant processes are only secondary signs of primary changes (Winter 1961), a clear separation of primary from secondary effects is very difficult.

Several Authors divide the mode of action of allelochemicals into indirect and direct action. The influence of secondary metabolites on soil properties and its microbial populations belong, for example, to indirect effects. According to Inderjit and Weiner (2001), indirect allelopathy could be due to (i) degraded or transformed products of released chemicals, (ii) effect of released chemicals on physical, chemical and biological soil factors and (iii) induction of release of biologically active chemicals by a third species.

Allelochemicals can alter:

a. cell division, elongation and structure;
b. membrane stability and permeability;
c. activity of various enzymes;
d. synthesis of plant endogenous hormones;
e. plant respiration;
f. plant photosynthesis and pigment synthesis;
g. protein synthesis and nucleic acid metabolism;
h. mineral uptake;
i. germination (of spores, seeds and pollen) and growth of target plant;
j. water balance of plant and conducting tissue.

These effects are rarely independent of each other. Rather, there is a closely relationship between them, since the same allelopathic compound can generates

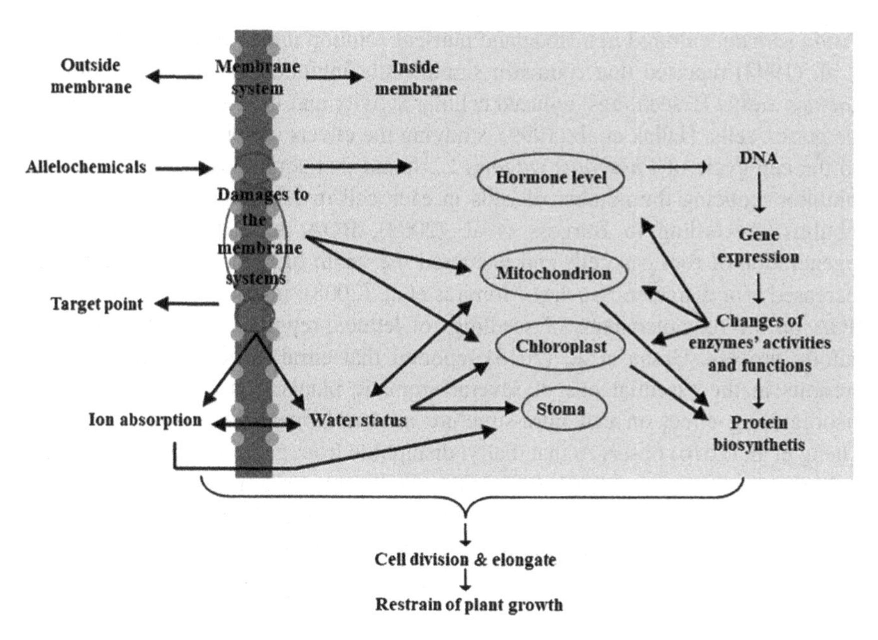

Fig. 2.24 Possible mechanisms of action of allelochemicals in plants and their relationship. Allelochemicals, by altering the permeability of membranes, affect the electrochemical potential gradient across membranes. Once entered inside the cell membrane, allelochemicals cause damage at hormone, mitochondrion, chloroplast and stoma level. These effects are rarely independent of each other, but generally an allelopathic compound generates "multiple cascating effects" (Wang et al. 2006, Modified)

"multiple cascating effects" (Fig. 2.24). Allelopathic effects are not only harmful. In fact, allelochemicals may have beneficial effects in response of their concentrations. A compound may be inhibitory at high concentration, stimulatory at low concentration, or have no effect at other concentrations (Ambika 2013).

2.8.1 *Inhibition of Cell Division, Elongation and Ultra-Structure*

Many works, since middle of twentieth century, have demonstrated that some allelochemicals could inhibit cell elongation, plant root elongation, cell cytology, and therefore affect the development of the whole plant (Li et al. 2010). In last decades, this topic has received great attention by researches (Vaughan and Ord, 1990; Li et al. 1993; Hallak et al. 1999; Burgos et al. 2004; Sanchez-Moreiras et al. 2008; Grana et al. 2013; Cheng et al. 2016). Vaughan and Ord (1990) found that, at high concentrations (1 mM), some phenolic acids inhibited cell division and affected the extension growth of the main root and the number of the lateral roots of

Pisum sativum cultured in a Hoagland nutrient solution under axenic conditions. Li et al. (1993) reported that coumarin significantly inhibited the root elongation of *Lactuca sativa* L. seedlings, reduced cellular activity and increased the thickness of the cortex cells. Hallak et al. (1999), studying the effects of sorghum root exudates on the cell cycle of *Phaseolus vulgaris* L., found that sorgoleone acts as a mitotic inhibitor reducing the number of cells in each cell division period and damaging tubulins. According to Burgoss et al. (2004), BOA and DIBOA reduced the regeneration of root cap cells and increased the width of cortical cells resulting in increased root diameter. Sanchez-Moreiras et al. (2008), investigating the effects of BOA on the root meristems of seedlings of lettuce, reported an inhibition of the mitotic process. Grana et al. (2013) reported that citral, a volatile monoterpene presents in the essential oils of several aromatic plants, has a strong long-term disorganising effect on cell ultra-structure in *Arabidopsis thaliana* L. seedlings. Cheng et al. (2016) observed that diallyl disulphide from garlic (*Allium sativum* L.), at lower concentrations (0.01–0.62 mM) significantly promoted root growth on tomato (*L. esculentum*), whereas higher levels (6.20–20.67 mM) inhibited root growth by affecting both the length and division activity of meristematic cells.

2.8.2 Interference with Cell Membrane Permeability

Several allelopathic agents, especially phenolics, increase cell membrane permeability due to the inhibition of antioxidant enzymes (such as catalases and peroxidases) and to the increase of lipid peroxidation and free radicals level (the so-called reactive oxygen species or ROS) in plasma membranes. These changes in membrane permeability lead to a spillage of cell contents and, therefore, to a slow growth or death of plant tissues (Li et al. 2010).

Baziramakenga et al. (1995) reported the benzoic and cinnamic acids damages cell membrane integrity in intact soybean (*G. max* L. cv. Maple Bell) seedlings due to an increase of lipid peroxidation, which resulted from free radical formation in plasma membranes, inhibition of catalase and peroxidase activities, and sulfhydryl group depletion. Batish et al. (2006) found that 2-Benzoxazolinone (BOA) induces oxidative stress in in both leaves and roots of mung bean (*P. aureus*) as indicated by enhanced lipid peroxidation and accumulation of hydrogen peroxide (H_2O_2). Ladhari et al. (2014), studying the effects of aqueous (15 g/L) and methanol (6 g/L) extracts of *Capparis spinosa* L leaves. and *Cleome arabica* L. siliquae on lettuce, pointed out a disruption in membrane permeability revealed by a strong electrolyte leakage and a trigger in oxidative damage manifested by lipid peroxidation.

2.8.3 Interference with Various Enzyme Activities

Many allelochemicals are known to interfere with the synthesis, functions, contents and activities of various enzymes (Cheng and Cheng 2015). Allelopathic compounds affect the activity of enzymes such as pectolytic enzyme, cellulases, catalases, peroxidases, phosphorylases, ATPases, amylases, invertases, proteinases, decarboxylases, phosphatases, nitrate reductases, etc. (Rice 1984). Several works were published in recent years on this topic (Cheng 2012; Venturelli et al. 2015; Mahdavikia and Saharkhiz 2016). Cheng (2012) found that diethyl phthalate inhibit glutamine synthetase isoenzymes in nitrogen for nitrogen assimilation and antioxidant enzymes in greater duckweed (*Spirodela polyrhiza* L.). Venturelli et al. (2015) reported that cyclic hydroxamic acid (e.g., benzoxazinoids or benzoxazinones such as DIBOA and its methoxylated analog DIMBOA) root exudates inhibit histone deacetylases both in vitro and in vivo and exert their activity through locus-specific alterations of histone acetylation and associated gene expression in *A. thaliana*. Mahdavikia and Saharkhiz (2016) studied the effects of peppermint (*Mentha piperita* L. CV. Mitcham) allelopathic water extracts on the morpho-physiological and biochemical characteristics of tomato (*L. esculentum* Mill. CV. Rio Grande). They concluded that phenolic compounds, at the concentrations of 10% (v/v) extract, showed the maximum inhibitory effect on the amount of proline, soluble sugar and starch, as well as on the activities of tomato's antioxidant enzymes such as ascorbate peroxidase, catalase, peroxidase and superoxide dismutase.

2.8.4 Interference with Synthesis of Plant Endogenous Hormones

Plant endogenous hormones, also known as "phytohormones", are generally present at very low concentrations in plant tissues. Nowadays, they are classified into nine groups (auxins, cytokinins, gibberellins, abscisic acid, ethylene, brassinosteroids, jasmonates, salicylic acid and strigolactones), but more will probably be discovered. Several allelopathic compounds are structurally similar to plant hormones (Olofsdotter 1988) and present similar mechanisms of action. Allelochemicals are able to reduce or inactive the physiological activity of phytohormones and to induce imbalances, thereby altering the normal growth and development of plants. Liu and Hu (2001), studying the effect of ferulic acid (FA) on endogenous hormone level of wheat seedling, found that, at concentrations of 2.50 mmol/L, FA has led to an accumulation of indolacetic acid, gibberellin and cytokinin, but the accumulation of these four hormones induced absisic acid increment. Brunn et al. (1992) reported that some flavonoid aglycones inhibit polar auxin transport, inducing the formation of lateral roots and the suppression of ageotropic growth. However, a few of allelochemicals have phytohormones-protecting activity, mainly indoleacetic acid (IAA), by inhibiting the oxidation of IAA, which leads to the accumulation of auxin and therefore to a greater growth of plant (Andreae 1952; Sondheimer and Griffin 1960; Mato et al. 1994; Cvikrova et al. 1996).

2.8.5 Interference with Respiration

Allelochemicals are able to inhibit most of the processes of respiration, from the O_2 uptake to the three phases of "dark" or mitochondrial respiration: the glycolysis, in which glucose is converted to pyruvate, the Kreb's cycle, which generate CO_2 and NADH and the electron transfer in the mitochondria or oxidatitve phosphorylation, which produce a large amount of ATP (Weir et al. 2004). However, allelochemicals can exert also positive effects on respiration, by stimulating the CO2 production (Lodhi and Nickell 1973). Several allelopathic compounds affect mitochondrial respiration directly. Rasmussen et al. in (1992), pointed out the disruption of mitochondrial functions and the block of electron transport in soybean and corn seedlings caused by sorgoleone. Rye's allelochemicals, BOA and DIBOA, are reported by Burgos et al. (2004) to reduce number of mitochondria, protein synthesis and lipid catabolism in cucumber seedlings. Hejl and Koster (2004) found that juglone affects root oxygen uptake due to the disruption of root plasma membrane functions and, ate the concentrations from 10 to 1000 μM, significantly reduced H^+-ATPase activity in soybean and corn. Unfortunately, many of the allelochemicals effects on mitochondrial respiration are masked by photorespiration that occur in the chloroplasts (Weir et al. 2004).

2.8.6 Inhibition of Photosynthesis and Pigment Synthesis

The adverse effects of allelochemicals on photosynthesis were demonstrated, but the detail mechanisms remains unknown. Allelochemicals can affect the three main process of photosynthesis (Zhou and Yu 2006): (i) the stomatal conductance and, thus, the gas exchanges between plant and atmosphere; (ii) the "light reactions" with refer to the electron transport and (iii) the "dark reactions", also known as the Calvin's cycle, for the carbon reduction.

One of the most important effects of allelochemicals on plant photosynthesis is represented by the acceleration of decomposition of photosynthetic pigments, mainly chlorophyll (Patterson 1981; Sarkar and Chakraborty 2015; Pan et al. 2015). Particularly, allelochemicals can reduce chlorophyll content by enhancing stimulation of Chl degradation, inhibition of Chl synthesis or interfering with the synthesis of porphyrin, which is the precursor of Chl biosynthesis, and porphyrin precursors (Proto, Mg-Proto and pchlide), through the inhibition of Mg-chelatase, the enzyme responsible for the conversion of Proto into Mg-Proto (Zhou and Yu 2006). Besides, Meazza et al. (2002) reported that sorgoleone strongly inhibits the enzyme *p*-hydroxyphenylpyruvate dioxygenase (HPPD), which catalyze the biosynthesis of carotenoids, resulting in foliar bleaching.

There are many works demonstrating a decrease in leaf stomatal conductance due to allelochemicals treatments (Rai et al. 2003; Yu et al. 2003, 2006; Mishra 2015). Generally, the lower stomatal conductance is correlated with a reduction in CO_2 assimilation and intracellular CO_2 concentration (Zhou and Yu 2006). However, is difficult to demonstrate the effective correlation between allelochemicals and stomatal apertures, since opening and closing of stomata are influenced by a series of several factors such as water status of plant, mineral uptake, temperature, wind and relative humidity, photoperiodism, age of leaf, leaf area index, etc.

The most documented mode of action of allelochemicals on photosynthesis is represented by the inhibition of photosystem II (PSII) (Rimando et al. 1998). It is a specialized protein complex, localized in the thylakoid membranes of chloroplasts, that utilizes solar energy to drive the oxidation of water and the reduction of plastoquinone (PQ). Sorgoleone is reported to act in a similar way as triazine herbicides, by disrupting the electron-transfer chain between plastoquinone A (Q_A) and Q_B at the DI protein of PSII (Czarnota et al. 2001). In nature, there are several allelochemicals able to inhibit PSII, such as benzoxazolin-2(3H)-one (BOA), cinnamic acid (CA), capsaicin, the limonoid terpene odoratol and many quinones.

2.8.7 Inhibition of Protein Synthesis and Nucleic Acid Metabolism

Several allelochemicals, mostly phenolics and alkaloids, can influence the protein biosynthesis and the nucleic acid metabolism. According to Wink and Latzbruning (1995), allelochemical alkaloids can inhibit protein biosynthesis and integrate with DNA. The authors concluded that the degree of DNA intercalation is positively correlated with inhibition of DNA polymerase I, reverse transcriptase, and translation at the molecular level. Baziramakenga et al. (1997), studying the allelopathic effects of benzoic, *p*-hydroxy benzoic, vanillic, cinnamic, *p*-coumaric, and ferulic acids on nucleic acid and protein level in soybean seedlings, found that the incorporation of ^{32}P and ^{35}S-methionine into proteins was reduced by all phenolic acids, except for *p*-coumaric acid and vanillic acid at 125 μM.

Allelochemical may generate ROS (reactive oxygen species), such as superoxide anions (O_2^-), hydroxyl (OH^-) or hydroperoxyl (HO_2) radicals, that are able to affect membrane permeability, acid nucleic structure and protein synthesis, leading to cell death (Weir et al. 2004). Allelochemicals can interfere also with the gene expression (He et al. 2012; Ma et al. 2015; Fang et al. 2015), which is often induced in the receiver plants like a form of reaction to the donor plant's attack.

2.8.8 Interference with Mineral Uptake

First of all, it is important to recognize that the mineral salts amount absorbed by the root surface depends on several factors as the ion concentration, the soil pH, the ion availability in the volume soil and the ion requirements for the plant (Lambers et al. 1998). Plants response to biotic and abiotic stress by altering their membrane properties. The modification of membrane activities is strictly correlated to important physiological processes such as cell elongation, seed germination, stomata opening and mineral uptake. Many works show as allelochemicals can affect the uptake of nutrients, which can be exhibited in the form of nutrient deficiency symptoms in growing plants and reduced plant growth (Brooker et al. 1992; Tharayil et al. 2009). It is known that allelopathic inhibition of mineral uptake results from alteration of cellular membrane functions in plant roots (Balke 1985). Allelochemicals can: (i) depolarize the electrochemical potential gradient across membranes, which guide the active absorption of mineral ions; (ii) inhibit the activities of Na^+/K^+ by altering the permeability of membranes and, thus, the absorption and transport of mineral ions; (iii) inhibit electron transport and oxidative phosphorylation, reducing the ATP content of cells; (iv) stimulate the production of superoxide, hydrogen and hydroxyl radicals that cause a direct damage to the membrane, or accumulate free radicals that increase the lipid peroxidation in the plasma membrane inhibiting the correct nutrient uptake from these roots (Fig. 2.25).

The interference of allelopathic compounds with mineral uptake depends on the chemical characteristics of allelochemicals and environmental conditions such as soil moisture, temperature, and especially pH. Particularly, this mode of action is reported to be concentration-dependent and ion- and species-specific (Inderjit and Keating 1999).

Besides, some allelochemicals may act as natural chelators, enhancing the availability of minerals for the plant. Several phenolic compounds are reported to bind with Fe, Mg, Al and Ca, and thus increase the availability of phosphate which otherwise forms complex with these metal ions (Appel 1993). In soil with high concentrations of Al, the chelation of metal cations may increase the plant's resistance (Jabran et al. 2013).

2.8.9 Interference with Plant-Water Relationships

Many allelochemicals are able to affect water balance of the target plant by corking and clogging of xylem elements, reducing stomatal conductance of water, lowering water potential of plant and, thus, decreasing water uptake by roots. Barkosky and Einhellig (2003), investigating the effect of *p*hydroxybenzoic acid (*p*HBA) on growth and plant-water relationships of soybean seedlings, found that, at concentration of 0.75 mM, *p*HBA had significantly lowered stomatal conductance,

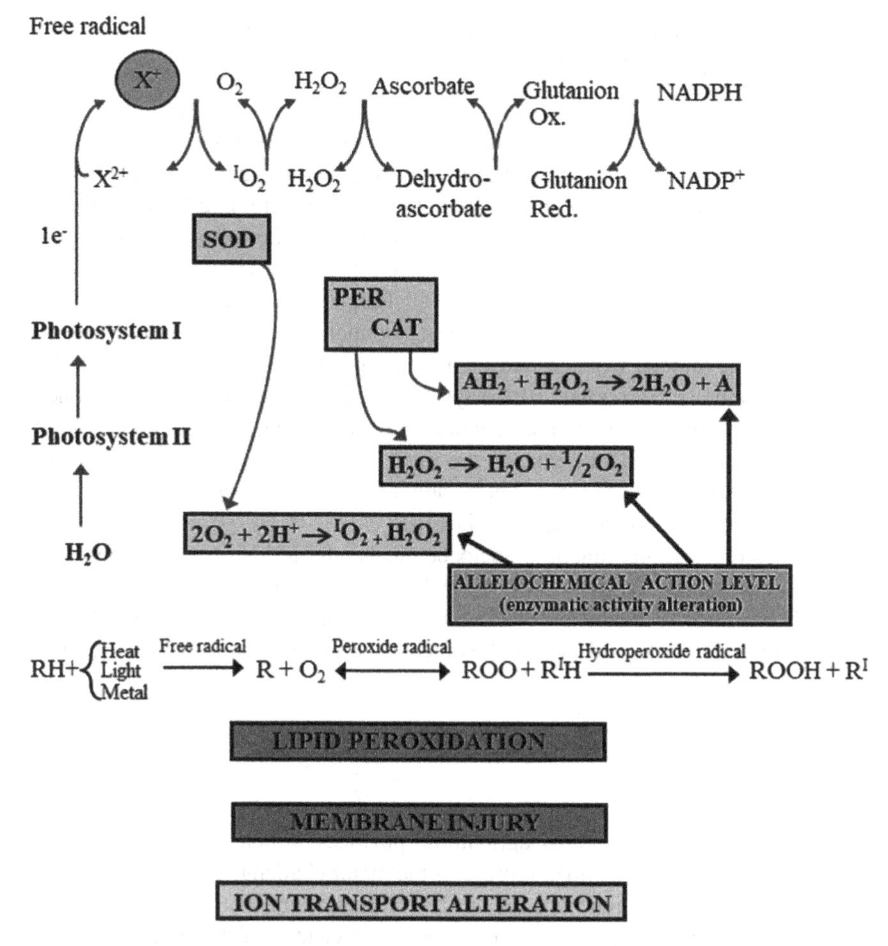

Fig. 2.25 Possible allelochemical mode of action at enzymatic level with the subsequent ion uptake alteration. If the allelopathic compound causes an alteration in the activity of the enzymes implied in the oxygen metabolism, an accumulation of hydroperoxide radicals can occur. These radicals can be toxic to the membrane by a direct damage or by the formation of free radicals that induce an increase in lipid peroxidation (Sánchez-Moreiras and Weiss 2001, Modified)

water potential and water use efficiency. The Authors concluded that the impact of *p*HBA on water relationships is an important mechanism of action causing a reduction in plant growth. Other phenolics compounds such as ferulic, *p*-coumaric, caffeic, hydrocinnamic, salicylic, *p*-hyroxybenzoic, gallic, and chlorogenic acids, as well as hydroquinone, vanillin, and umbelliferone altered normal water balance of target plant by reducing leaf water potential, turgor pressure, conductance, or changing tissue carbon-isotope ratio (Einhellig 2004; Barkosky et al. 2000; Einhellig et al. 1985).

2.8.10 Inhibition of Germination and Interference with Growth of Plants

The inhibition of germination and the effects on growth of both crop and weed species are secondary expressions of primary effects such as the interferences with cell division and elongation, with cell membrane permeability and the alteration of plant respiration and photosynthesis. There is a large number of publications upon the effects of allellochemicals on growth and germination (Reigosa et al. 1999a, b; Turk and Tawaha 2003; Vokou et al. 2003; Reigosa and Pazos-Malvido 2007; Scavo et al. 2018; etc.). In recent years, new potential allelochemicals from several donor plants have been isolated and identified, and the number of works on this topic is continuously growing. Bioassays using Petri dishes are the commonest technique for proving allelopathic mechanism of action on seed germination and growth of target plants. Plant extracts are made from any part of the donor plant, commonly leaves and roots. They can be aqueous, hydroalcholic or fractions from different solvents (de Albuquerque et al. 2010).

In addition to seed germination, allelochemicals inhibit that of pollen grains and spores (Murphy and Aarssen 1995; Roshchina and Melnikova 1998; Roshchina 2009). These three mechanisms of inhibition represent a means available for the species with high ecological potential to increase their environmental distribution.

2.9 Conclusion

The development of eco-friendly agricultural practices, which are able to increase crop production at the same time, represents the major challenge of new millennium agriculture. Allelopathic mechanisms are an important tool that may contribute to the improvement of the genetic diversity and maintenance of ecosystem stability. Allelopathy may be employed also in cropping systems for enhancing soil fertility and yields, as well as for weed and pest control through a chemical-free management. However, many aspects of the allelopathic phenomenon are still unknown. Therefore, the scientific community is called for further efforts to better understand the pathways for release of allelochemicals into the environment by the donor plant, the effects of these inhibitory compounds on target plants physiological processes and on soil microbial population, and factors affecting their production. Only after acquiring a better knowledge of basic principles, it is possible to develop new strategies for weed and pest control in sustainable and organic farming agriculture.

References

Aarnio T, Martikainen PJ (1994) Mineralization of carbon and nitrogen in acid forest soil treated with forest and slow-release nutrients. Plant Soil 164:187–193. https://doi.org/10.1007/BF00010070

Aldrich JD (1984) Weed-crop ecology: principles and practices. Breton Publishers, pp 215–241

AlSaadawi IS, Al-Uqaili JK, Al-Hadithy SM, AlRubeaa AJ (1985) Effects of gamma radiation on allelopathic potential of *Sorghum bicolor* against weeds and nitrification. J Chem Ecol 11:1737–1745. https://doi.org/10.1007/BF01012123

Ambika SR (2013) Multifaceted attributes of allelochemicals and mechanism of allelopathy. In: Cheema ZA, Farooq M, Wahid A (eds) Allelopathy: current trends and future applications, vol 16. Springer, Berlin, pp 389–405. https://doi.org/10.1007/978-3-642-30595-5_16

An M, Johnson IR, Lovett JV (1993) Mathematical modelling of allelopathy: biological response to allelochemicals and its interpretation. J Chem Ecol 19(10):2379–2388. https://doi.org/10.1007/BF00979671

Andreae WA (1952) Effects of scopoletin on indoleacetic acid metabolism. Nature 170:83–84. https://doi.org/10.1038/170083a0

Appel HM (1993) Phenolics in Ecological interactions: the importance of oxidation. J Chem Ecol 19:1521–1552. https://doi.org/10.1007/BF00984895

Ardi (1986) Interference between sweet corn (*Zea mays* L.) and purple nutsedge (*Cyperus rotundus* L.) at different irrigation levels. Master thesis, University of Hawaii, Honolulu

Bais HP, Vepachedu R, Gilroy S, Callaway RM, Vivanco JM (2003) Allelopathy and exotic plant invasion: from molecules and genes to species interactions. Science 301(5638):1377–1380. https://doi.org/10.1126/science.1083245

Balke NE (1985) Effects of allelochemicals on mineral uptake and associated physiological processes. In: Thompson AC (ed) The chemistry of allelopathy. ACS symposium series 268. American Chemical Society, Washington, D.C., pp 161–178. https://doi.org/10.1021/bk-1985-0268.ch011

Barkosky RR, Einhellig FA (2003) Allelopathic interference of plant-water relationships by para-hydroxybenzoic acid. Bot Bull Acad Sinica 44:53–58

Barkosky RR, Einhellig FA, Butler JL (2000) Caffeic acid induced changes in plant-water relationships and photosynthesis in leafy spurge (*Euphorbia esula*). J Chem Ecol 26:2095–2109. https://doi.org/10.1023/A:1005564315131

Barnes JP, Putnam AR (1987) Role of benzoxazinones in allelopathy by rye (*Secale cereale* L). J Chem Ecol 13(4):889–906. https://doi.org/10.1007/BF01020168

Barney JN, Hay AG, Weston LA (2005) Isolation and characterization of allelopathic volatiles from mugwort (*Artemisia vulgaris*). J Chem Ecol 31(2):247–265. https://doi.org/10.1007/s10886-005-1339-8

Baruah NC, Sarma JC, Barua NC, Sarma S, Sharma RP (1994) Germination and growth inhibitory sesquiterpene lactones and a flavone from *Tithonia diversifolia*. Phytochemistry 36(1):29–36. https://doi.org/10.1016/S0031-9422(00)97006-7

Batish DR, Singh HP, Setia N, Kaur S, Kohli RK (2006) 2-Benzoxazolinone (BOA) induced oxidative stress, lipid peroxidation and changes in some antioxidant enzyme activities in mung bean (*Phaseolus aureus*). Plant Physiol Biochem 44(11–12):819–827. https://doi.org/10.1016/j.plaphy.2006.10.014

Baziramakenga R, Leroux GD, Simard RR, Nadeau P (1997) Allelopathic effects of phenolic acids on nucleic acid and protein levels in soybean seedlings. Can J Botany 75(3):445–450. https://doi.org/10.1139/b97-047

Baziramakenga R, Leroux GD, Simard RR (1995) Effects of benzoic and cinnamic acids on membrane permeability of soybean roots. J Chem Ecol 21(9):1271–1285. https://doi.org/10.1007/BF02027561

Belz RG (2007) Allelopathy in crop/weed interactions-an update. Pest Manag Sci 63(4):308–326. https://doi.org/10.1002/ps.1320

Berenbaum MR (1995) Turnabout is a fair play: secondary roles for primary compounds. J Chem Ecol 21(7):925–940. https://doi.org/10.1007/BF02033799

Bertin C, Yang X, Weston LA (2003) The role of root exudates and allelochemicals in the rhizophere. Plant Soil 256(1):67–83. https://doi.org/10.1023/A:1026290508166

Bhat SV, Nagasampagi BA, Sivakumar M (2005) Chemistry of natural products. Springer, Berlin, 2:116

Blum MS (2004) The importance of alkaloidal functions. In: Macías FA, Galindo JCG, Molinillo JMG, Cutler HG (eds) Allelopathy: chemistry and mode of action of allelochemicals, vol 8. CRC Press, Boca Raton, pp 163–181. https://doi.org/10.1201/9780203492789.ch8

Blum U (2011) Plant-plant allelopathic interactions: Phenolic acids, cover crops and weed emergence. Springer, New York. https://doi.org/10.1007/978-94-007-0683-5

Blum U, Shafer SR (1988) Microbial populations and phenolic acids in soil. Soil Biol Biochem 20 (6):793–800. https://doi.org/10.1016/0038-0717(88)90084-3

Brinker AM, Creasy LL (1988) Inhibitors as a possible basis for grape replant problem. J Am Soc Hortic Sci 113:304–309

Brooker FL, Blum U, Fiscus EL (1992) Short-term effect of ferulic acid on ion uptake and water relations in cucumber seedlings. J Exp Bot 43(5):649–655. https://doi.org/10.1093/jxb/43.5.649

Brunn SA, Muday GK, Haworth P (1992) Auxin transport and the interaction of phytotropins: probing the properties of a phytotropin binding protein. Plant Physiol 98(1):101–107. https://doi.org/10.1104/pp.98.1.101

Burgos NR, Talbert RE, Kim KS, Kuk YI (2004) Growth inhibition and root ultrastructure of cucumber seedlings exposed to allelochemicals from rye (*Secale cereale*). J Chem Ecol 30 (3):671–689. https://doi.org/10.1023/B:JOEC.0000018637.94002.ba

Buta JG, Spaulding DW (1989) Allelochemicals in tall fescue-abscisic and phenolic acids. J Chem Ecol 15(5):1629–1636. https://doi.org/10.1007/BF01012389

Cartwright DW, Langcake P, Pryce RJ, Leworthy DP, Ride JP (1981) Isolation and characterization of two phytoalexins from rice as momilactones A and B. Phytochemistry 20 (3):535–537. https://doi.org/10.1016/S0031-9422(00)84189-8

Chamacho-Cristóbal JJ, Anzellotti D, Gonzàlez-Fontez A (2002) Changes in phenolic metabolism of tobacco plants during short-term boron deficiency. Plant Physiol Biochem 40(12):997–1002. https://doi.org/10.1016/S0981-9428(02)01463-8

Cheng F, Cheng Z (2015) Research progress on the use of plant allelopathy in agriculture and the physiological and ecological mechanisms of allelopathy. Front Plant Sci 6:1020. https://doi.org/10.3389/fpls.2015.01020

Cheng F, Cheng Z, Meng H, Tang X (2016) The garlic allelochemical diallyl disulfide affects tomato root growth by influencing cell division, phytohormone balance and expansin gene expression. Front Plant Sci 7:1199. https://doi.org/10.3389/fpls.2016.01199

Cheng HH (1989) Assessment of the fate and transport of allelochemicals in the soil. In: Chou CS, Waller GR (eds) Phytochemical ecology: allelochemicals, mycotoxins and insect pheromones and allomones. Academia Sinica Monogr Series No. 9. Institute of Botany, Taipei, ROC, pp 209–216

Cheng HH (1992) A conceptual framework for assessing allelochemicals in the soil environment. In: Rizvi SJK, Rizvi V (eds) Allelopathy: basic and applied aspects. Chapman & Hall, London, pp 21–29. https://doi.org/10.1007/978-94-011-2376-1_3

Cheng TS (2012) The toxic effects of diethyl phthalate on the activity of glutamine synthetase in greater duckweed (*Spirodela polyrhiza* L.). Aquat Toxicol 124–125:171–178. https://doi.org/10.1016/j.aquatox.2012.08.014

Chon SU, Jennings JA, Nelson CJ (2006) Alfalfa (*Medicago sativa* L.) autotoxicity: Current status. Allelopathy J 18:57–80

Chou CH, Waller GR (1983) Allelochemicals and pheromones. Institute of Botany, Academia Sinica Monograph Series No. 5, Taipei, Taiwan

Chu C, Mortimer PE, Wang H, Wang Y, Liu X, Yu S (2014) Allelopathic effects of *Eucalyptus* on native and introduced tree species. Forest Ecol Manag 323:79–84. https://doi.org/10.1016/j.foreco.2014.03.004
Clark EP (1937) Quassin I: the preparation and purification of quassin and neoquassin, with information concerning their molecular formulas. J Am Chem Soc 59(5):927–931. https://doi.org/10.1021/ja01284a046
Cvikrova M, Hrubcova M, Eder J, Binarova P (1996) Changes in the levels of endogenous phenolics, aromatic monoamines, phenylalanine ammonia-lyase, peroxidase, and auxin oxidase activities during initiation of alfalfa embryogenic and nonembryogenic calli. Plant Physiol Biochem 34(6):853–861
Czarnota MA, Paul RN, Dayan FE, Nimbal CI, Weston LA (2001) Mode of action, localization of production, chemical nature, and activity of sorgoleone: a potent PS II inhibitor in *Sorghum* spp. root exudates. Weed Technol 15(4):813–825 10.1614/0890-037X(2001) 015[0813:MOALOP]2.0.CO;2
Davis RF (1928) The toxic principles of *Juglans nigra* as identified with synthetic juglone and its toxic effects on tomato and alfalfa plants. Am J Bot 15:620
de Albuquerque MB, dos Santos RC, Lima LM, Filho PAM, Nogueira JMC, da Câmara CAG, Ramos AR (2010) Allelopathy, an alternative tool to improve cropping systems. A Rev Agron Sustain Dev 31(2):379–395. https://doi.org/10.1051/agro/2010031
del Moral R (1972) On the variability of chlorogenic acid concentration. Oecologia 9(3):289–300. https://doi.org/10.1007/BF00345238
Duke SO (2010) Allelopathy: current status of research and future of the discipline: a commentary. Allelopathy J 25(1):17–30
Duke SO (2015) Proving allelopathy in crop–weed interactions. Weed Sci 63(sp1):121–132. https://doi.org/10.1614/WS-D-13-00130.1
Einhellig FA (1987) Interactions among allelochemicals and other stress factors of the plant environment. In: Waller GR (ed) Allelochemicals: role in agriculture and forestry. ACS Symposium Series, vol 330 (Chap. 32), Washington DC, pp 343–357. https://doi.org/10.1021/bk-1987-0330.ch032
Einhellig FA (1995) Allelopathy—current status and future goals. In: Inderjit, Dakshini KMM, Einhellig FA (eds) Allelopathy: organisms, processes and applications. ACS Symposium Series, vol 582 (Chap. 1), Washington DC, pp 1–24. https://doi.org/10.1021/bk-1995-0582.ch001
Einhellig FA (1996) Interactions involving allelopathy in cropping systems. Agron J 88(6):886–893. https://doi.org/10.2134/agronj1996.00021962003600060007x
Einhellig FA (2004) Mode of allelochemical action of phenolic compounds. In: Macías FA, Galindo JCG, Molinillo JMG, Cutler HG (eds) Allelopathy: chemistry and mode of action of allelochemicals, vol 11. CRC Press, Boca Raton, pp 217–238. https://doi.org/10.1201/9780203492789.ch11
Einhellig FA, Eckrich PC (1984) Interaction of temperature and ferulic acid stress on grain sorghum and soybeans. J Chem Ecol 10(1):161–170. https://doi.org/10.1007/BF00987653
Einhellig FA, Stille Muth M, Schon MK (1985) Effects of allelochemicals on plant-water relationships, vol 268 (Chap. 12). ACS Symposium Series, pp 179–195. https://doi.org/10.1021/bk-1985-0268.ch012
Elix JA (1996) Biochemistry and secondary metabolites. In: Nash III TH (ed) Lichen biology, University Press, Cambridge, UK, pp 154–180
Evenari M (1949) Germination inhibitors. Bot Rev 15:153–194. https://doi.org/10.1007/BF02861721
Fang C, Li Y, Li C, Li B, Ren Y, Zheng H, Zeng X, Shen L, Lin W (2015) Identification and comparative analysis of microRNAs in barnyardgrass (*Echinochloa crus-galli*) in response to rice allelopathy. Plant Cell Environ 38(7):1368–1381. https://doi.org/10.1111/pce.12492
Fomenko BS (1968) Effect of ionizing radiation on the metabolism of some phenols in the shoots of plants differing in their radiosensitivity. Biol Nauk 11:45–50

Fraga BM (2005) Natural sesquiterpenoids. Nat Prod Rep 22:465–486. https://doi.org/10.1039/b501837b

Francisco IA, Pinotti MHP (2000) Cyanogenic glycosides in plants. Braz Arch Biol Technol 43 (5):487–492. https://doi.org/10.1590/S1516-89132000000500007

Friedman J, Waller GR (1983) Caffeine hazards and their prevention in germinating seeds of coffee *Coffea arabica* L. J Chem Ecol 9(8):1099–1106. https://doi.org/10.1007/BF00982214

Furness NH, Adomas B, Dai Q, Li S, Upadhyaya MK (2008) Allelopathic influence of houndstongue (*Cynoglossum officinale*) and its modification by UV-B radiation. Weed Technol 22(1):101–107. https://doi.org/10.1614/WT-06-015.1

Gatti AB, Perez SCJG, Lima MIS (2004) Atividade alelopatica de extratos aquosos de *Aristolochia esperanzae* O. Kuntze na germinacao e no crescimento de *Lactusa sativa* L. e *Raphanus sativus* L. Acta Bot Bras 18(3):459–472. https://doi.org/10.1590/s0102-33062004000300006

Goel U (1987) Allelopathic effect of weeds associated with gram (*Cicer arietinum*) on its seed germination and seedling growth. Acta Bot Indica 15:129–130

Grana E, Sotelo T, Diaz-Tielas C, Araniti F, Krasuska U, Bogatek R et al (2013) Citral induces auxin and ethylene-mediated malformations and arrests cell division in *Arabidopsis thaliana* roots. J Chem Ecol 39(2):271–282. https://doi.org/10.1007/s10886-013-0250-y

Grümmer G (1955) Die gegenseitige Beeinflussung höherer Pflanzen-Allelopathie. Fischer, Jena

Guo Z, Vangapandu S, Sindelar RW, Walker LA, Sindelar RD (2005) Biologically active quassinoids and their chemistry: potential leads for drug design. Curr Med Chem 12(2):173–190. https://doi.org/10.2174/0929867053363351

Hale MG, Foy CL, Shay FG (1971) Factors affecting root exudation. Adv Agron 23:89–109. https://doi.org/10.1016/S0065-2113(08)60151-0

Hallak AMG, Davide LC, Souza IF (1999) Effects of sorghum (*Sorghum bicolor* L.) root exudates on the cell cycle of the bean plant (*Phaseolus vulgaris* L.) root. Genet Mol Biol 22(1):95–99. https://doi.org/10.1590/S1415-47571999000100018

Haller T, Stolp H (1985) Quantitative estimation of root exudation of maize plants. Plant Soil 86 (2):207–216. https://doi.org/10.1007/BF02182895

Hao W, Ren L, Ran W, Shen Q (2010) Allelopathic effects of root exudates from watermelon and rice plants on *Fusarium oxysporum* f.sp. *niveum*. Plant Soil 336(1):485–497. https://doi.org/10.1007/s11104-010-0505-0

Harborne JB (1972) Cyanogenic glycosides and their function. In: Phytochemical ecology. Academic Press, London, pp 104–123

Harborne JB (1980) Plant phenolics. In: Bell EA, Charlwood B (eds) Encyclopedia of plant physiology, vol 8. New Series. Springer, Berlin, pp 329–395

Harborne JB (1989) General procedures and measurement of total phenolics. In: Harborne JB (ed) Methods in plant biochemistry: plant phenolics, vol I. Academic Press Limited, London, pp 1–28

Harper JL (1977) Population biology of plants. Academic Press, London, p 892

Hartmann T (1996) Diversity and variability of plant secondary metabolism: a mechanistic view. Entomol Exp Appl 80(1):177–188. https://doi.org/10.1111/j.1570-7458.1996.tb00914.x

Hartung AC, Nair MG, Putnam AR (1990) Isolation and characterization of phytotoxic compounds from asparagus (*Asparagus officinalis*) roots. J Chem Ecol 16:1707–1718. https://doi.org/10.1007/BF01014102

He H, Wang H, Fang C, Wu H, Guo X, Liu C, Lin Z, Lin W (2012) Barnyard grass stress up regulates the biosynthesis of phenolic compounds in allelopathic rice. J Plant Physiol 169 (17):1747–1753. https://doi.org/10.1016/j.jplph.2012.06.018

Hejl AM, Koster KL (2004) Juglone disrupts root plasma membrane H^+-ATPase activity and impairs water uptake, root respiration, and growth in soybean (*Glycine max*) and corn (*Zea mays*). J Chem Ecol 30:453–471. https://doi.org/10.1023/B:JOEC.0000017988.20530.d5

Huang PM, Wang TSC, Wang MK, Wu MH, Hsu NW (1977) Retention of phenolic acids by non-crystalline hydroxy-aluminium and -iron compounds and clay mineral of soil. Soil Sci 123 (4):213–219. https://doi.org/10.1097/00010694-197704000-00001

Hyun MW, Yun YH, Kim JY, Kim SH (2011) Fungal and plant phenylalanine ammonia-lyase. Mycobiology 39(4):257–265. https://doi.org/10.5941/MYCO.2011.39.4.257

Inderjit, del Moral L (1997) Is separating resource competition from allelopathy realistic? Bot Rev 63(2):221–230. https://doi.org/10.1007/bf02857949

Inderjit Duke SO (2003) Ecophysiological aspects of allelopathy. Planta 217(4):529–539. https://doi.org/10.1007/s00425-003-1054-z

Inderjit, Keating IK (1999) Allelopathy: principles, procedures, processes, and promises for biological control. Adv Agron 67:141–231. https://doi.org/10.1016/s0065-2113(08)60515-5

Inderjit, Weiner J (2001) Plant allelochemical interference or soil chemical ecology? Perspect Plant Ecol Evol Syst 4(1):3–12. https://doi.org/10.1078/1433-8319-00011

Inderjit, Weston LA (2003) Root exudates: an overview. In: de Kroon H, Visser EJW (eds) Root ecology, vol 168. Springer, Berlin, pp 235–255. https://doi.org/10.1007/978-3-662-09784-7_10

Jabran K, Farooq M, Aziz T, Siddique KHM (2013) Allelopathy and crop nutrition. In: Cheema ZA, Farooq M, Wahid A (eds.) Allelopathy: current trends and future applications. Springer, Berlin, pp 337–348. Chap. 14. https://doi.org/10.1007/978-3-642-30595-5

Jackson JR, Willemsen RW (1976) Allelopathy in the first stages of secondary succession on the Piedmont of New Jersey. Amer J Bot 63(7):1015–1023

Jain PK, Joshi H (2012) Coumarin: chemical and pharmacological profile. J Appl Pharm Sci 2 (6):236–240. https://doi.org/10.7324/JAPS.2012.2643

Jay CN, Rossall S, Smith HG (1999) Effects of beet western yellows virus on growth and yield of oilseed rape (*Brassica napus*). J Agr Sci 133(02):131–139. https://doi.org/10.1017/S0021859699006711

Jennings JA, Nelson CJ (1998) Influence of soil texture on alfalfa autotoxicity. Agron J 90:54–58. https://doi.org/10.2134/agronj1998.00021962009000010010x

Jensen EH, Hartman BJ, Lundin F, Knapp S, Brookerd B (1981) Autotoxicity of Alfalfa. Univ Nevada Agric Exp Stn Bull Rep 144. Max C. Fleischmann College of Agriculture, Univ. Nevada, Reno, NY

Kato-Noguchi H (1999) Effect of light irradiation on allelopathic potential of germinating maize. Phytochemistry 52(6):1023–1027. https://doi.org/10.1016/S0031-9422(99)00365-9

Kimber R (1973) Phytotoxicity from plant residues. The relative effect of toxins and nitrogen immobilization on the germination and growth of wheat. Plant Soil 38(3):543–555. https://doi.org/10.1007/BF00010694

Koeppe DE, Rohrbaugh LM, Rice EL, Wender H (1970a) The effect of x-radiation on the concentration of scopolin and caffeoylquinic acids in tobacco. Radiat Bot 10(3):261–265. https://doi.org/10.1016/S0033-7560(70)80019-9

Koeppe DE, Rohrbaugh LM, Rice EL, Wender SH (1970b) The effect of age and chilling temperatures on the concentration of scopolin and caffeoylquinic acids in tobacco. Physiol Plant 23(2):258–266. https://doi.org/10.1111/j.1399-3054.1970.tb06415.x

Koeppe DE, Rohrbaugh LM, Wender SH (1969) The effect of varying UV intensities on the concentration of scopolin and caffeoylquinic acids in tobacco and sunflower. Phytochemistry 8 (5):889–896. https://doi.org/10.1016/S0031-9422(00)85879-3

Kong C, Hu F, Xu X (2002) Allelopathic potential and chemical constituents of volatile from *Ageratum conyzoides* under stress. J Chem Ecol 28(6):1173–1182. https://doi.org/10.1023/A:1016229616845

Kruse M, Strandberg M, Strandberg B (2000) Ecological effects of allelopathic plants— a review. National Environmental Research Institute, Silkeborg, Denmark, pp 66. NERI Technical Report No. 315

Ladhari A, Omezzine F, Haouala R (2014) The impact of Tunisian *Capparidaceae* species on cytological, physiological and biochemical mechanisms in lettuce. S Afr J Bot 93:222–230. https://doi.org/10.1016/j.sajb.2014.04.014

Lambers H, Chapin III FS, Pons TL (1998) Mineral nutrition. In: Plant physiological ecology. Springer, New York, pp 255–320. https://doi.org/10.1007/978-0-387-78341-3

Lehman RG, Cheng HH (1988) Reactivity of phenolics acids in soil and formation of oxidation products. Soil Sci Soc Am J 52(5):1304–1309. https://doi.org/10.2136/sssaj1988.03615995005200050017x

Lehman RH, Rice EL (1972) Effect of deficiencies of nitrogen, potassium and sulfur on chlorogenic acids and scopolin in sunflower. Am Midl Nat 87(1):71–80. https://doi.org/10.2307/2423882

Levitt J (1980) Responses of plants to environmental stresses, 2nd edn, vol 1. Chilling, Freezing, and High Temperature Stresses. Academic Press, NewYork

Li H-H, Inoue M, Nishimura H, Mizutani J, Tsuzuki E (1993) Interactions of trans-cinnamic acid, its related phenolic allelochemicals, and seed germination of lettuce. J Chem Ecol 19(8):1775–1787

Li H, Pan K, Liu Q, Wang J (2009) Effect of enhanced ultraviolet-B on allelopathic potential of *Zanthoxylum bungeanum*. Sci Horticult 119(3):310–314. https://doi.org/10.1016/j.scienta.2008.08.010

Li ZH, Wang Q, Ruan X, Pan CD, Jiang DA (2010) Phenolics and plant allelopathy. Molecules 15 (12):8933–8952. https://doi.org/10.3390/molecules15128933

Lichtenthaler HK (1999) The 1-deoxy-D-xylulose-5-phosphate pathway of isoprenoid biosynthesis in plants. Annu Rev Plant Physiol Plant Mol Biol 50:47–65. https://doi.org/10.1146/annurev.arplant.50.1.47

Lichtenthaler HK, Rohmer M, Schwender J (1997) Two independent biochemical pathways for isopentenyl diphosphate and isoprenoid biosynthesis in higher plants. Plant Physiol 101 (3):643–652. https://doi.org/10.1111/j.1399-3054.1997.tb01049.x

Liu DL, An M, Johnson IR, Lovett JW (2005) Mathematical modelling of allelopathy: IV: Assessment of contributions of competition and allelopathy to interference by barley. Nonlinear Biol Toxicol Med 3(2):213–224. https://doi.org/10.2201/nonlin.003.02.003

Liu XF, Hu XJ (2001) Effects of allelochemical ferulic acid on endogenous hormone level of wheat seedling. Chin J Ecol Agric 9(1):96–98

Loche J, Chouteau J (1963) Incidences des carences en Ca, Mg, or P sur l'accumulation des polyphenol dans la feuille de tabac. C R Hebd Seances Acad Agr Fr 49:1017–1026

Lodhi MAK, Nickell GL (1973) Effects of *Celtis laevigata* on growth, water content and carbon exchange rate of three grass species. Bull Torrey Bot Club 100(3):159–165. https://doi.org/10.2307/2484627

Lohdi MAK, Bilal R, Malik KA (1987) Allelopathy in agroecosystems: wheat phytotoxicity and its possible roles in crop rotation. J Chem Ecol 13(8):1881–1891. https://doi.org/10.1007/BF01013237

Lydon J, Duke SO (1988) Glyphosate induction of elevated levels of hydroxybenzoic acids in higher plants. J Agric Food Chem 36(4):813–818. https://doi.org/10.1021/jf00082a036

Lydon J, Duke SO (1993) The role of pesticides on host allelopathy and their effects on allelopathic compounds. In: Altman J (ed) Pesticide interactions in crop production: beneficial and deleterious effects. CRC, Boca Raton, pp 37–56

Ma DW, Wang YN, Wang Y, Zhang H, Liao Y, He B (2015) Advance in allelochemical stress induced damage to plant cells. Acta Ecol Sin 35(5):1640–1645. https://doi.org/10.5846/stxb201311122718

Macías FA, Molinillo JM, Varela RM, Galingo JC (2007) Review: Allelopathy—a natural alternative for weed control. Pest Manag Sci 63(4):327–348. https://doi.org/10.1002/ps.1342

Macías FA, Oliveros-Bastida A, Marín A, Castellano D, Simonet A, Molinillo JM (2004) Degradation studies on benzoxazinoids. Soil degradation dynamics of 2,4-dihydroxy-7-methoxy-2(H)-1,4-benzoxazin-3(4H)- one (DIMBOA) and its degradation products, phytotoxic allelochemicals from *Gramineae*. J Agric Food Chem 52(21):6402–6413. https://doi.org/10.1021/jf0488514

Mahdavikia F, Saharkhiz MJ (2016) Secondary metabolites of peppermint change the morphophysiological and biochemical characteristics of tomato. Biocatal Agric Biotechnol 7:127–133. https://doi.org/10.1016/j.bcab.2016.05.013

Mato MC, Mendez J, Vazquez A (1994) Polyphenolic auxin protectors in buds of juvenile and adult chestnut. Physiol Plant 91(1):23–26. https://doi.org/10.1111/j.1399-3054.1994.tb00654.x

Mattner SW (2006) The impact of pathogens on plant interference and allelopathy. In: Inderjit, Mukerji KG (eds.) Allelochemicals: Biological control of plant pathogens and diseases, vol 2. Springer, Netherlands, pp 79–101. https://doi.org/10.1007/1-4020-4447-x_4

Mattner SW, Parbery DG (2001) Rust-enhanced allelopathy of perennial Ryegrass against White clover. Agron J 93(1):54–59. https://doi.org/10.2134/agronj2001.93154x

Meazza G, Scheffler BE, Tellez MR, Rimando AM, Romagni JG, Duke SO, Nanayakkara D, Khan IA, Abourashed EA, Dayan FE (2002) The inhibitory activity of natural products on plant p-hydroxyphenylpyruvate dioxygenase. Phytochemistry 60(3):281–288. https://doi.org/10.1016/S0031-9422(02)00121-8

Miller DA (1996) Allelopathy in forage crop systems. Agron J 88(6):854–859. https://doi.org/10.2134/agronj1996.00021962003600060003x

Mishra A (2015) Allelopathic effect of *Cassia tora* extract on transpiration rate in *Mangifera indica* and *Syzgium cumini*. Indian J Appl Res 5(10):516–517. https://doi.org/10.13140/RG.2.1.3421.5525

Muller CH (1969) Allelopathy as a factor in ecological process. Vegetatio 18:348–357

Murphy SD (1999) Pollen allelopathy. In: Inderjit, Dakshini KMM, Foley CL (eds) Principles and practices in plant ecology: allelochemical interactions, vol. 9. CRC Press, Boca Raton, pp 129–148

Murphy SD, Aarssen LW (1995) Allelopathic pollen extract from *Phleum pratense* L. (*Poaceae*) reduces germination, in vitro, of pollen of sympatric species. Int J Plant Sci 156(4):425–434. https://doi.org/10.1086/297264

Mwaja VN, Masiunas JB, Weston LA (1995) Effect of fertility on biomass, phytotoxicity, and allelochemical content of cereal rye. J Chem Ecol 21:81–96. https://doi.org/10.1007/BF02033664

Neish AC (1964) Major pathways of biosynthesis of phenols. In: Harborne JB (ed) Biochemistry of phenolic compounds. Academic Press, New York, pp 295–359

Nelson CJ (1996) Allelopathy in cropping systems foreword. Agron J 88:991–996. https://doi.org/10.2134/agronj1996.00021962003600060002x

Newman JD, Chappell J (1999) Isoprenoid biosynthesis in plants: carbon partitioning within the cytoplasmic pathway. Crit Rev Biochem Mol Biol 34(2):95–106. https://doi.org/10.1080/10409239991209228

Nishida N, Tamotsu S, Nagata N, Saito C, Sakai A (2005) Allelopathic effects of volatile monoterpenoids produced by *Salvia leucophylla*: inhibition of cell proliferation and DNA synthesis in the root apical meristem of *Brassica campestris* seedlings. J Chem Ecol 31(5):1187–1203. https://doi.org/10.1007/s10886-005-4256-y

Ojala T (2001) PhD thesis, University of Helsinki, Helsinki, Finland, pp 95–106

Olofsdotter M (1998) Allelopathy for weed control in organic farming. In: El-Bassam N, Behl RK, Prochnow B (eds) Sustainable agriculture for food, energy and industry: strategies towards achievement. James and James (Science Publishers) Ltd., London, pp 453–457

Oueslati O, Ben-Hammouda M, Ghorbal MH, Guezzah M, Kremer RJ (2005) Barley autotoxicity as influenced by varietal and seasonal variation. J Agron Crop Sci 191(4):249–254. https://doi.org/10.1111/j.1439-037X.2005.00156.x

Pan L, Li XZ, Yan ZQ, Guo HR, Qin B (2015) Phytotoxicity of umbelliferone and its analogs: Structure-activity relationships and action mechanisms. Plant Physiol Biochem 97:272–277. https://doi.org/10.1016/j.plaphy.2015.10.020

Pandya SM, Dave VR, Vyas KG (1984) Effect of *Celosia argentea* Linn. on root nodules and nitrogen contents of three legume crops [*Cajanus cajan*, *Vigna aconitifolia* and *Phaseolus aureus*]. Sci Cult 50:161–162

Patterson DT (1981) Effects of allelochemicals on growth and physiological responses of soybean (*Glycine max*). Weed Sci 29(1):53–59

Picman AK (1986) Biological activities of sesquiterpene lactones. Biochem Syst Ecol 14(3):255–281. https://doi.org/10.1016/0305-1978(86)90101-8

Pretorius JC, van der Watt E (2011) Natural products from plants: commercial prospects in terms of antimicrobial, herbicidal and bio-stimulatory activities in an integrated pest management system. In: Dubey NK (ed) Natural products in plant pest management. CABI, Preston, pp 42–90. Chap. 3 https://doi.org/10.1079/9781845936716.0042

Putnam AR, Duke WB (1978) Allelopathy in agroecosystems. Annu Rev Phytopathol 16:431–451. https://doi.org/10.1146/annurev.py.16.090178.002243

Putnam AR, Tang CS (1986) Allelopathy: State of the science. In: Putnam AR, Tang CS (eds) The science of allelopathy. Wiley, New York, pp 1–22

Rai VK, Gupta SC, Singh B (2003) Volatile monoterpenes from *Prinsepia utilis* L. leaves inhibit stomatal opening in *Vicia faba* L. Biol Plantarum 46(1):121–124. https://doi.org/10.1023/a:1022397730599

Rasmussen JA, Hejl AM, Einhellig FA, Thomas JA (1992) Sorgoleone from root exudate inhibits mitochondrial functions. J Chem Ecol 18(2):197–207. https://doi.org/10.1007/BF00993753

Razavi SM (2011) Plant coumarins as allelopathic agents. Int J Biol Chem 5(1):86–90. https://doi.org/10.3923/ijbc.2011.86.90

Reese JC (1979) Interaction of allelochemicals with nutrients in herbivore food. In: Rosenthal GP, Janzen DH (eds) Herbivores: their interaction with secondary plant metabolites. Academic Press, New York, pp 309–330

Reigosa MJ, Pazos-Malvido E (2007) Phytotoxic effects of 21 plant secondary metabolites on *Arabidopsis thaliana* germination and root growth. J Chem Ecol 33(7):1456–1466. https://doi.org/10.1007/s10886-007-9318-x

Reigosa MJ, Pedrol N (2002) Allelopathy from molecules to ecosystems. Scientific Publishers Inc., Enfield, NH

Reigosa MJ, Sanchez-MoreiraS A, Gonzales L (1999a) Ecophysiological approach in allelopathy. Crit Rev Plant Sci 18(5):577–608. https://doi.org/10.1080/07352689991309405

Reigosa MJ, Souto XC, Gonz'lez L. (1999) Effect of phenolic compounds on the germination of six weeds species. Plant Growth Regul 28(2):83–88. https://doi.org/10.1023/a:1006269716762

Rice EL (1974) Allelopathy. Academic Press, New York

Rice EL (1979) Allelopathy. An update. Bot Rev 45(1):15–109

Rice EL (1984) Allelopathy, 2nd edn. Academic Press, New York

Rimando AM, Dayan FE, Czarnota MA, Weston LA, Duke SO (1998) A new photosystem II electron transport inhibitor from *Sorghum bicolor*. J Nat Prod 61(7):927–930. https://doi.org/10.1021/np9800708

Rizvi SJH, Haque H, Singh VK, Rizvi V (1992) A discipline called allelopathy. In: Rizvi SJH, Rizvi V (eds) Allelopathy, basic and applied aspects. Chapman & Hall, London, 1–10. https://doi.org/10.1007/978-94-011-2376-1_1

Robinson T (1963) The organic constituents of higher plants. Burgess Pub. Co., Minneapolis

Romagni JG, Rosell RC, Nanayakkara NPD, Dayan FE (2004) Ecophysiology and potential modes of action for selected lichen secondary metabolites. In: Macías FA, Galindo JCG, Molinilo JMG, Cutler HC (eds) Allelopathy: chemistry and mode of action of allelochemicals, vol. 1. CRC Press, Boca Raton, pp 13–33. https://doi.org/10.1201/9780203492789.ch1

Roshchina VV (2009) Effects of proteins, oxidants and antioxidants on germination of plant microspores. Allelopathy J 23:37–50

Roshchina VV, Melnikova EV (1998) Allelopathy and plant reproductive cells: participation of acetylcholine and histamine in signaling in the interactions of pollen and pistil. Allelopathy J 5 (2):171–182

Rundel PW (1978) The ecological role of secondary lichen substances. Biochem Syst Ecol 6 (3):157–170. https://doi.org/10.1016/0305-1978(78)90002-9

Sanchez-Moreiras AM, de la Peña TC, Reigosa MJ (2008) The natural compound benzoxazolin-2 (3H)-one selectively retards cell cycle in lettuce root meristems. Phytochemistry 69(11):2172–2179. https://doi.org/10.1016/j.phytochem.2008.05.014

Sanchez-Moreiras AM, Weiss O (2001) Root uptake and release of ions. In: Manuel J. Reigosa Roger (ed.) Handbook of Plant Ecophysiology Techniques. Kluwer Academic Publishers, 25:413–427. https://doi.org/10.1007/0-306-48057-3_25

Sarkar E, Chakraborty P (2015) Allelopathic effect of *Amaranthus spinosus* Linn. on growth of rice and mustard. Journal of Tropical Agriculture 53(2):139–148

Sattin M, Tei F (2001) Malerbe componente dannosa degli agroecosistemi. In: Catizone P, Zanin G (eds) Malerbologia. Pàtron Editore, Bologna, pp 171–245

Scavo A, Restuccia A, Pandino G, Onofri A, Mauromicale G (2018) Allelopathic effects of *Cynara cardunculus* L. leaf aqueous extracts on seed germination of some Mediterranean weed species. Ital J Agron 11. https://doi.org/10.4081/ija.2018.1021

Schreiner O, Reed HS (1908) The toxic action of certain organic plant constituents. Bot Gaz 45 (2):73–102. https://doi.org/10.1086/329469

Schumacher WJ, Thill DC, Lee GA (1983) Allelopathic potential of wild oat (*Avena fatua*) on spring wheat (*Triticum aestivum*) growth. J Chem Ecol 9(8):1235–1246. https://doi.org/10.1007/BF00982225

Seigler DS (1996) Chemistry and mechanisms of allelopathic interactions. Agron J 88(6):876–885. https://doi.org/10.2134/agronj1996.00021962003600060006x

Shexia Ma (2012) Production of secondary organic aerosol from multiphase monoterpenes. In: Hayder Abdul-Razzak (ed) Atmospheric Aerosols-Regional Characteristics-Chemistry and Physics. InTech. https://doi.org/10.5772/48135

Singh HP, Batish DR, Kohli RK (2001) Allelopathy in agroecosystems: An overview. J Crop Prod 4(2):1–41. https://doi.org/10.1300/J144v04n02_01

Sondheimer E, Griffin DH (1960) Activation and inhibition of indolacetic acid oxidase activity from peas. Science 131(3401):672. https://doi.org/10.1126/science.131.3401.672

Tabaglio V, Gavazzi C, Schulz M, Marocco A (2008) Alternative weed control using the allelopathic effect of natural benzoxazinoids from rye mulch. Agron Sustain Dev 28(3):397–401. https://doi.org/10.1051/agro:2008004

Tang CS, Cai WF, Kohl K, Nishimoto RK (1995) Plant stress and allelopathy. In: Inderjit, Dakshini KMM, Einhellig FA (eds) Allelopathy: Organisms, Processes, and Applications. ACS Symposium Series, Washington DC 582:142–157. https://doi.org/10.1021/bk-1995-0582.ch011

Tesar MB (1993) Delayed seeding of alfalfa avoids autotoxiciy after plowing or glyphosate treatment of established stands. Agron J 85(2):256–263. https://doi.org/10.2134/agronj1993.00021962008500020018x

Tharayil N, Bhowmik P, Alpert P, Walker E, Amarasiriwardena D, Xing B (2009) Dual purpose secondary compounds: phytotoxin of *Centaurea diffusa* also facilities nutrient uptake. New Phytol 181(2):424–434. https://doi.org/10.1111/j.1469-8137.2008.02647.x

Torres A, Oliva RM, Castellano D, Cross P (1996) First world congress on allelopathy: A science of the future. SAI (University of Cadiz), Cadiz, p 278

Turk MA, Tawaha AM (2003) Allelopathic effect of black mustard (*Brassica nigra* L.) on germination and growth of wild oat (*Avena fatua* L.). Crop Prot 22(4):673–677. https://doi.org/10.1016/S0261-2194(02)00241-7

Vaughan D, Ord BG (1990) Influence of phenolic acids on morphological changes in roots of *Pisum sativum*. J Sci Food Agr 52(3):289–299. https://doi.org/10.1002/jsfa.2740520302

Vaughn SF, Spencer GF (1993) Volatile monoterpenes as potential parent structures for new herbicides. Weed Sci 41:114–119

Venturelli S, Belz RG, Kämper A, Berger A, Von Horn K, Wegner A et al (2015) Plants release precursors of histone deacetylase inhibitors to suppress growth of competitors. Plant Cell 27 (11):3175–3189. https://doi.org/10.1105/tpc.15.00585

Vivanco JM, Bais HP, Stermitz TR, Thelen GC, Callaway RM (2004) Biogeochemical variation in community response to root allelochemistry: novel weapons and exotic invasion. Ecol Lett 7 (4):285–292. https://doi.org/10.1111/j.1461-0248.2004.00576.x

Vivanco JM, Guimarães RL, Flores HE (2002) Underground plant metabolism: the biosynthetic potential of roots. In: Kafkafi U, Waisel Y, Eshel A, (eds) Plant roots: the hidden half, 2nd edn. Marcel Dekker, New York, pp 1045–1070. http://dx.doi.org/10.1201/9780203909423.ch58

Vokou D, Douvli P, Blionis GJ, Halley JM (2003) Effects of monoterpenoids, acting alone or in pairs, on seed germination and subsequent seedling growth. J Chem Ecol 29(10):2281–2301. https://doi.org/10.1023/A:1026274430898

Wang Q, Ruan X, Li Z, Pan C (2006) Autotoxicity of plants and research of coniferous forest autotoxicity. Sci Sil Sin 43(6):134–142. https://doi.org/10.11707/j.1001-7488.20070624

Watkinson AR (1998) Reply from A. R Watkinson. Trends Ecol Evol 13(10):407. http://dx.doi.org/10.1016/S0169-5347(98)01454-2

Webster GR, Kahn SU, Moore AW (1967) Poor growth of alfalfa (*Medicago sativa*) on some Alberta soils. Agron J 59(1):37–41. https://doi.org/10.2134/agronj1967.00021962005900010011x

Weidenhamer JD, Hartnett DC, Romeo JT (1989) Density-dependent phytotoxicity: distinguishing resource competition and allelopathic interference in plants. J Appl Ecol 26(2):613–624. https://doi.org/10.2307/2404086

Weir TL, Park SW, Vivanco JM (2004) Biochemical and physiological mechanisms mediated by allelochemicals. Curr Opin Plant Biol 7(4):472–479. https://doi.org/10.1016/j.pbi.2004.05.007

Weston LA (2005) History and current trends in the use of allelopathy for weed management. HortTechnol 15(3):529–534

Whittaker RH, Feeny PP (1971) Allelochemics: chemical interactions among species. Science 171 (3973):757–770. https://doi.org/10.1126/science.171.3973.757

Williamson GB (1990) Allelopathy, Koch's postulates, and the neck riddle. In: Grace JB, Tilman D (eds) Perspectives on Plant Competition. Academic Press Inc, San Diego, CA, pp 143–162

Willis RJ (1985) The historical bases of the concept of allelopathy. Hist Biol 18:71–102. https://doi.org/10.1007/BF00127958

Willis RJ (1994) Terminology and trends in allelopathy. Allelopathy J 1:6–28

Willis RJ (2007) The history of allelopathy. Springer, Netherlands. https://doi.org/10.1007/978-1-4020-4093-1

Wilson RE, Rice EL (1968) Allelopathy as expressed by *Helianthus annuus* and its role in old-field succession. Bull Torrey Bot Club 95(5):432–448. https://doi.org/10.2307/2483475

Wink M (1983) Inhibition of seed germination by quinolizidine alkaloid. Aspects of allelopathy in *Lupinus albus* L. Planta 158(4):365–368. https://doi.org/10.1007/BF00397339

Wink M, Latz-Brüning B (1995) Allelopathic properties of alkaloids and other natural products. In: Inderjit. Dashiki KMM, Einhellig FA (eds) Allelopathy: organisms, processes, and applications. ACS Symposium Series, Washington DC, 582:117–126. https://doi.org/10.1021/bk-1995-0582.ch008

Winkler BC (1967) Quantitative analysis of coumarins by thin layer chromatography, related chromatographic studies, and the partial identification of a scopoletin glycoside present in tobacco tissue culture. Ph.D. Dissertation, University of Oklahoma, Norman

Winter AG (1961) New physiological and biological aspects in the inter-relationships between higher plants. Symp Soc for Exp Biol 15:229–244

Woodhead S (1981) Environmental and biotic factors affecting the phenolic content of *Sorghum bicolor*. J Chem Ecol 7:1035–1047. https://doi.org/10.1007/BF00987625

Xu Z, Zhang D, Hu J, Zhou X, Ye X, Reichel K, Stewart N, Syrenne R, Yang X, Gao P et al (2009) Comparative genome analysis of lignin biosynthesis gene families across the plant kingdom. BMC Bioinform 10(Suppl. 11):S3. https://doi.org/10.1186/1471-2105-10-S11-S3

Yu JG (2001) Autotoxic potential of cucurbic crops: phenomenon, chemicals, mechanisms and means to overcome. J Crop Prod 4(2):335–348. https://doi.org/10.1300/J144v04n02_15

Yu JQ (1999) Autotoxic potential of vegetable crops. In: Narwal SS (ed) Allelopathy update: basic and applied aspects. Scientific Publisher Inc, New Hampshire, USA 2:149–162

Yu JQ, Matsui Y (1994) Phytotoxic substances in the root exudates of *Cucumis sativus* L. J Chem Ecol 20:21–31. https://doi.org/10.1007/BF02065988

Yu JQ, Ye SF, Zhang MF, Hu WH (2003) Effects of root exudates, aqueous root extracts of cucumber (*Cucumis sativus* L.) and allelochemicals on photosynthesis and antioxidant enzymes in cucumber. Biochem Syst Ecol 31:129–139. https://doi.org/10.1016/S0305-1978(02)00150-3

Yu J, Zhang Y, Niu C, Li J (2006) Effects of two kinds of allelochemicals on photosynthesis and chlorophyll fluorescence parameters of *Solanum melongena* L. seedlings. Chin. J Appl Ecol 17 (9):1629–1632

Zhou YH, Yu JQ (2006) Allelochemicals and Photosynthesis. In: Reigosa MJ, Pedrol N, Gonzàlez L (eds) Allelopathy: a physiological process with ecological implications. Springer, 6:127–139. https://doi.org/10.1007/1-4020-4280-9_6

Zimdahl RL (1999) Fundamentals of weed science, 2nd edn. Academic Press, San Diego, California, USA

Zucker M (1963) The influence of light on synthesis of protein and chlorogenic acid in potato tuber tissue. Plant Physiol 38(5):575–580. https://doi.org/10.1104/pp.38.5.575

Zucker M, Nitsch C, Nitsch JP (1965) The induction of flowering in Nicotiana. II. Photoperiodic alteration of the chlorogenic acid concentration. Am J Bot 52(3):271–277

Chapter 3
Conservation Biological Control of Insect Pests

Ryan J. Rayl, Morgan W. Shields, Sundar Tiwari and Steve D. Wratten

Abstract The human population is predicted to reach 9 billion by 2050. To achieve food security for this growing population, agricultural intensification is occurring, with increasing use of pesticides to reach the necessary yields. However, there is strong evidence that suggests pesticides cannot provide the agro-ecosystem growth and stability needed for the increasing demand. It is well established that pesticides are harmful to human, animal and environmental health. This information has reached the public, causing some governments to create policies that require the reduction of pesticide inputs in agro-ecosystems. Consequently, there is a need to manage pests using alternative techniques. One such approach is to enhance an ecosystem service (ES) which is conservation biological control (CBC). This is defined as manipulating the agro-ecosystem to enhance natural enemy fitness, populations and efficacy to suppress pest numbers. The problem is, not every study that has added multiple resources to agro-ecosystems is successful. Such resources may act synergistically and provide multiple ES delivery or include elements of redundancy, competition, or generate ecosystem dis-services. Here, we synthesize current reviews, provide a critical analysis and indicate future strategies. The key area that needs more focused research is understanding why floral resources are not always successful in the enhancement of natural enemy populations that lead to top-down pest suppression. Associated with this challenge, there are large knowledge gaps in natural enemy non-consumptive effects on prey and how these can be manipulated and used in CBC. For example, adding flowering plants to an agro-ecosystem is likely to impact several invertebrate and vertebrate communities,

R. J. Rayl (✉) · M. W. Shields · S. Tiwari · S. D. Wratten (✉)
Bio-Protection Research Centre, Lincoln University, Cnr Springs and Ellesmere Junction Roads, PO Box 85084, Lincoln 7647, New Zealand
e-mail: ryanrayl88@gmail.com

S. D. Wratten
e-mail: Steve.Wratten@lincoln.ac.nz

M. W. Shields
e-mail: Morgan.Shields@lincolnuni.ac.nz

S. Tiwari
e-mail: Sundar.Tiwari@lincolnuni.ac.nz

S. Gaba et al. (eds.), *Sustainable Agriculture Reviews 28*, Ecology for Agriculture 28,
https://doi.org/10.1007/978-3-319-90309-5_3

not always positively. These effects may, however, only be short term and localized. Existing landscape complexity has potential to supplement local effects but this is a highly multivariate approach. Major impediments to CBC being widely deployed certainly do not include cost. A typical 100 m × 2.3 m vineyard inter-row with buckwheat seeds costs only US $2.00. Key limitations to uptake of CBC by farmers and growers include: individual government policies which are inimical to agro-ecological approaches; the marketing power of agro-chemical companies; farmer innate conservatism; and most importantly, a weak emphasis of delivery systems and pathways to implementation. The most effective way to addressing the latter is 'farmer field schools', led by 'farmer teachers'. Outputs do not lead to outcomes unless multiple delivery systems and pathways of implementation are involved and developed at the beginning of the research.

Keywords Agro-ecosystem · Conservation biological control · Delivery system Ecosystem service · Ecosystem dis-service · Food security · Implementation pathway · Natural enemies · Pest control · Pesticides

3.1 Introduction

Projections for the human population indicate that it will reach 9 billion globally by 2050 (Godfray et al. 2012; Tscharntke et al. 2012; Ingram et al. 2013). Pesticides, fertilizers and irrigation have been key components of this growth to date (Cooper and Dobson 2007; Tilman et al. 2011), as in the 'Green Revolution' (Tilman 1998). Prophylaxis has been a common approach (Unruh et al. 2012; Schmitz and Barton 2014) and reports to the United Nations (de Schutter 2010; United Nations 2017) confirm that this use of pesticide has caused external costs such as reduced human health and environmental issues (Sandhu et al. 2012; Tscharntke et al. 2012). This has led to a decrease in support for the use of pesticides, especially in the European Union (United Nations 2017).

Demand for low-input alternatives to pesticides has arisen from issues with pesticide resistance, increasing costs and regulation, and emerging consumer pressure (Jetter and Paine 2004; Grygorczyk et al. 2014; Wollaeger et al. 2015; Gurr et al. 2017; United Nations 2017). This has triggered a growing interest for policymakers to create programs that support sustainable growth (Fabinyi et al. 2016; Gartaula et al. 2016), sometimes called 'sustainable intensification' (Godfray et al. 2010). One of these alternatives is manipulating the agro-ecosystem to make it more favorable for natural enemies; this is known as conservation biological control (CBC) (Begg et al. 2017; Gurr et al. 2017; Gurr and You 2016). Many techniques can be used under the umbrella of CBC to manipulate the agro-ecosystem in this way; some of these include floral resource augmentation (Fig. 3.1) (Pywell et al. 2015; Tschumi et al. 2015; Gurr et al. 2016), 'beetle banks' (a type of banker plant) (Thomas et al. 1991; MacLeod et al. 2004) or artificial food sprays (Seagraves et al. 2011; Tena et al. 2015). A key and well-used acronym which encompasses most aspects of how the environment can

Fig. 3.1 *Lobularia maritima* '*Benthamii* White' (alyssum) amongst spinach and lettuce crops in Auckland, New Zealand. The alyssum was planted to provide SNAP (Shelter, Nectar, Alternative food and Pollen) for parasitoid wasps that parasitize larvae of a leaf mining fly pest. Photo: Ryan Rayl

be manipulated under CBC is SNAP: Shelter, Nectar, Alternative food and Pollen (Fig. 3.1) (Gurr et al. 2017). In recent work, the nectar and pollen components can be summarized by another acronym: BAP. This captures buckwheat, (sweet) alyssum and phacelia, which are often the most effective floral additions to crops, founded by prior laboratory bioassays. Although CBC is popular with the scientific community and many studies have explored different aspects of it, many unanswered questions remain. For example, to what extent do added resources act synergistically to provide multiple ES delivery or do they include elements of redundancy and competition, generating ecosystem dis-services?

3.2 The Practicalities of Conservation Biological Control

When applying CBC in an agro-ecosystem, the natural enemy-pest community needs to be assessed and manipulated in a range of ways. The acronym, ARMED, can be helpful in this process: Assess, Rank, Manipulate, Evaluate, Deploy (Shields et al. in press). A range of sampling methods exist to *Assess* which species comprise the invertebrate community in a particular crop and, to some extent at least, to which guilds the different taxa belong. The taxa can then be *Ranked* in a number of ways. The simplest being abundance. Populations and efficacy of some individuals or guilds can then be *Manipulated*, the effects of which then need to be *Evaluated*. After evaluation, the protocols developed to manipulate the arthropods need to be *Deployed* appropriately in agro-environment schemes, with effective delivery systems and pathways to implementation. In that context, Wratten et al. (in press) showed that although the number of agro-ecological publications has increased exponentially in recent decades, this has not usually led to many on-farm outcomes, with the exception of work by Khan et al. (2011) and Gurr et al. (2016). It is important to note that attempts at the ARMED process may reveal ecosystem dis-services (EDS) which at least partially negate the benefits (Tscharntke et al. 2005; Bianchi et al. 2006; Tschumi et al. 2015; Rusch et al. 2016; Gurr et al. 2017). EDS in the context of CBC are specific unintended negative impacts from the added biodiversity and its processes (Zhang et al. 2007; Gurr et al. 2017), such as added plants becoming weeds (Shields et al. 2016).

Although this focus on floral resources has generated many successful manipulations that have reduced pest numbers (Géneau et al. 2012; Tschumi et al. 2015), much less work has been done in terms of alternative hosts or prey living on the provided flowering plants (Messelink et al. 2014; Gurr et al. 2015; Gillespie et al. 2016). This aspect is represented by the A in SNAP. Furthermore, if some insect species feed on both the deployed plants and the adjacent crop, then targeting that potential EDS needs to be quantified.

3.2.1 Non-consumptive Effects in Conservation Biological Control

Non-consumptive effects (NCE) are changes in prey behavior and physiology that improve their predator avoidance (Buchanan et al. 2017; Hermann and Landis 2017), which can impact pest management and influence entire agro-ecosystems through trophic cascades (Hermann and Landis 2017). This emerging field of study has already found strong evidence that NCE may have a substantial role in reducing pest damage and needs to be considered when developing protocols for CBC. Many studies have shown that prey respond to predators by changing either their behaviour or physiology to reduce the risk of predation. These include increased predator avoidance (Wratten 1976; Hoefler et al. 2012; Lee et al. 2014), reduced feeding (Rypstra and Buddle 2013; Kaplan et al. 2014; Thaler et al. 2014), reduced oviposition (Wasserberg et al. 2013; Sendoya et al. 2009) and changes in host plant preference (Wilson and Leather 2012; Sidhu and Wilson Rankin 2016).

To manipulate NCE in CBC, the specific mechanisms of predator detection must be understood. These are predominantly chemical cues (Gonthier 2012; Hoefler et al. 2012; Gonzalvez and Rodriguez-Girones 2013; Hermann and Thaler 2014) such as aphids (*Rhopalosiphum padi* L.) having reduced colonisation on plants where ladybird beetle (*Coccinella septempunctata* L.) larvae had previously foraged (Ninkovic et al. 2013). Another example is reduced feeding of Colorado beetles (*Leptinotarsa decemlineata* Say) when exposed to the predatory stink bug, *Podisus maculiventris* Say (Hermann and Thaler 2014). However, visual detection of predators, while often undervalued, may contribute substantially to predator avoidance (Sendoya et al. 1996). For instance, pollinator visitation decreased when models that look like crab spiders were on the flowers (Antiqueira and Romero 2016). Also, dried ants pinned to plants reduced butterfly oviposition (Sendoya et al. 2009). One of the most important NCE that should be investigated in CBC manipulations is pests modifying their choice of host plant from the preferred hosts to one of lower nutrition, based on their perceived predation risk (Hermann and Landis 2017). For instance, in a mesocosm experiment, the grasshopper, *Melanoplus femurrubrum* De Geer, switched from preferred grasses to less nutritional forbs when the spider predator, *Pisaurina mira* Walckenaer, was present (Schmitz 1998). The potential for manipulating NCE in CBC habitat manipulation and IPM protocols is enormous, such as in push-pull systems (Hermann and Landis 2017). However, very few studies have been conducted in the field at time scales longer than 1 week or investigated the impacts of NCE on agro-ecosystem functions and across different spatial scales (Hermann and Landis 2017).

3.2.2 The Importance of Long-Term Studies

Communities and landscapes change over time and with those changes comes the shifting of the natural enemy-pest community. Most studies to date have examined short-term effects (<3 years) in these systems. Gillespie et al. (2016) and Gurr et al. (2017) have both stressed the need for more long-term studies. Both agree that short-term studies cannot accurately capture population trends of the organisms in and around the agro-ecosystem. Providing SNAP in the form of annual flowering plants for one season only is the most common practice (Fig. 3.1). Naturally-occurring SNAP, for example through perennial shelter, can operate more long term. Other examples are manipulated hedgerows (Holland et al. 2016), beetle banks (Thomas et al. 1991; MacLeod et al. 2004) and long-term plantings of some biofuel crops (Fig. 3.2) (Porter et al. 2009; Littlejohn et al. 2015). Short-term rotational coppice (Langer 2001; Rusch et al. 2014) and plantings of the giant hybrid grass *Miscanthus* x *giganteus* (Greef et Deu) can provide refugia for natural enemies (Fig. 3.2) (Semere and Slater 2007; Shields et al. in press) as well as providing other non-biological control refugia (Littlejohn et al. 2015). However, there are many ecological mechanisms which impede the delivery of biological control from refugia and these are reviewed in Tscharntke et al. (2016). A good example is that in Europe, carabid beetles and other predatory fauna disperse into

Fig. 3.2 *Miscanthus* x *giganteus* (Greef et Deu) shelterbelt on a dairy farm providing refugia for natural enemies, Canterbury, New Zealand. Photo: Chris Littlejohn

the adjacent crop from hedges and beetle banks in the spring (Thomas et al. 1991; Holland et al. 2016), while in New Zealand, those habitats provide refugia all year round with little emigration from them (McLachlan and Wratten 2003).

3.3 Spatial Scales from Plots to Agro-Ecosystems

Research strongly suggests that monocultural agro-ecosystems are detrimental as they can lead to higher pest populations (Kremen and Miles 2012; Nilsson et al. 2016; Rusch et al. 2016). Although much research supports agro-ecosystem diversification, recent evidence suggests that this can be a complex problem to address and is not as straightforward as increasing diversity in these agro-ecosystems. There is now considerable evidence suggesting that agro-ecosystem management should take place at the landscape scale because it can provide better arthropod management (Tscharntke et al. 2007; Karp et al. 2012; Chaplin-Kramer et al. 2013; Roubos et al. 2014). This is because it was found that at a landscape scale of at least 2.5 km, there was a strong relationship between natural habitat and the abundance of predatory flies (Chaplin-Kramer et al. 2013).

While it is true that diverse cropping systems generally do lead to more natural enemies, fewer pests and less crop damage, in some cases yield can significantly decline (Jonsson et al. 2010; Tscharntke et al. 2016; Begg et al. 2017; Gurr et al. 2017). This was demonstrated by Letourneau et al. (2011) in a meta-analysis using hundreds of case studies. The message from that and many other studies is that when, for example, flowering plants are added to an agricultural or horticultural system, part of the selection criteria is that their effect on yield should be minimal (e.g. Balmer et al. 2014; Iverson et al. 2014; Pywell et al. 2015; Shields et al. 2016). This has certainly been the case when buckwheat (*Fagopyrum esculentum* Moench.) and phacelia (*Phacelia tanacetifolia* Benth.) are sown between vine rows (Fig. 3.3) (Berndt et al. 2006; Barnes et al. 2009) and when flowering sesame and other species are used in rice (Gurr et al. 2016).

Plant diversification often positively influences arthropod diversity (Parker et al. 2010; Haddad et al. 2011; Iverson et al. 2014; Hatt et al. 2016; Begg et al. 2017), but also the type of diversification can affect different aspects of the agro-ecosystem, including negative influences. The proportion of natural habitat surrounding the farm directly influences the abundance of natural enemies (Chisholm et al. 2014) but can also increase pest abundance in some cases

Fig. 3.3 *Phacelia tanacetifolia* in the inter-row of vines, Waipara Valley, New Zealand. Photo: Jean-Luc Dufour

(Tylianakis and Romo 2010). This can be attributed to such factors as intra-guild predation, which can include the killing and consumption of potential competitors, i.e. other natural enemy species. Other factors associated with natural habitat that can contribute to increased pest abundance are barriers to movement and resources, such as hedgerows (Mauremooto et al. 1995), as well as abiotic factors (Thomas et al. 1991). These factors can benefit the second trophic level, i.e. the herbivore pest, more than the third trophic level (the natural enemies).

3.4 Interactions with the Landscape

It is important to note that there are many indirect effects that the habitat has on the natural enemy-pest community. Plants can indirectly influence the diversity of parasitoids (parasitic wasps) without affecting pest populations (Tylianakis and Binzer 2014). Changing land use has also been connected to differing densities of natural enemies (Zhou et al. 2014; Begg et al. 2017) and the presence of secondary pests can influence the abundance of other natural enemies (Bompard et al. 2013). Secondary pests are organisms that do not cause substantial damage unless their natural enemies are removed (Maxwell and Jennings 1980). Another challenge is that the target pests, or others, can also feed on floral resources and increase their fitness (Gurr et al. 2017). Ecosystem dis-services (EDS) of this type can diminish CBC effectiveness. It has also been suggested that non-crop species can compete for pollinators when the crop requires pollination but actual studies of this are rare (Free 1993; Holzschuh et al. 2012). These indirect effects on the natural enemy-pest community are likely to be specific for each agro-ecosystem and need to be considered as such when developing habitat management protocols for CBC.

Chemical fertilizers, herbicides, conventional tillage and pesticide application are all primary techniques in agricultural systems. These can stifle agro-ecosystem stability and decrease the abundance and diversity of natural enemies (Altieri 1999; Begg et al. 2017). Furthermore, moderate shade, adequate labor and appropriate but not prophylactic use of artificial inputs can be combined with ecological engineering (Gurr et al. 2004) to provide high biodiversity and sustainable crop yields (Clough et al. 2011). Many studies confirm that a 'whole system' approach may be necessary to create a sustainable agro-ecosystem (Tscharntke et al. 2007, 2012; Gurr et al. 2017) and that synergies among ecosystem functions may make the system more stable and easier to manage (Iverson et al. 2014; Turner et al. 2014). Specifically, up to 8% of the field production area can be used for wildlife habitat with no negative effect on yields and in most cases where such habitat was present, higher yields were reported, up to 30% in some crops (Pywell et al. 2015). To manage agro-ecosystems effectively by taking advantage of synergies among ecosystem functions, multiple models may be necessary as they can potentially predict variables accurately enough to be useful in agro-ecosystem management (Turner et al. 2014).

3.5 Conservation Biological Control in Changing Climates

CBC manipulates the biotic and abiotic environment to enhance natural enemy efficacy, populations and pest suppression, but a rising concern in the scientific community that can upset CBC functions is changing climates. Evidence is building that suggests that climate change is becoming an imminent global threat to food security (de Schutter 2010; Tilman et al. 2011; Woodward and Porter 2016; Myers et al. 2017; United Nations 2017). One factor is that pests' host range, fecundity and damage potential can increase with the changing climates (Tylianakis et al. 2008; War et al. 2016). Invertebrates are ectothermic organisms and their physiology and resultant fitness are strongly influenced by microclimate. This makes them very sensitive to climate change at the population level and changes can occur over very short periods of time (Boggs 2016). In general, a changing climate favors generalist species, which suggests that specialists' populations are likely to decline or become extinct (Hof and Svahlin 2015; Van Dyck et al. 2015).

Through the above processes, higher temperatures can cause negative impacts on agro-ecosystems. This is because insect outbreaks can increase in frequency due to these changes and associated increased frequency of droughts (Fig. 3.4). Changes in precipitation patterns and increasing of CO_2 in the environment (Murdock et al. 2013; Hof and Svahlin, 2015; War et al. 2016) also impact on invertebrates. For example, an increase in temperature and CO_2 can lengthen the larval period in some

Fig. 3.4 A wilting tomato crop during drought in Chitwan, Nepal. Photo: Sundar Tiwari

species and causes greater mortality (Sharma et al. 2015). High CO_2 levels can reduce populations of the aphid parasitoid, *Diaeretiella rapae (*M'Intosh), by approximately 50% and result in a shorter lifespan for adults (Roth and Lindroth 1995). Not only does elevated CO_2 decrease natural enemy effectiveness, there can be a difference in the relative effect on predator and parasitoid strategies. Under these conditions, generalists can maintain effectiveness while specialists experience a reduction in fitness (Chen et al. 2005). These CO_2 effects also occur with increased temperature (Hemerik et al. 2015).

Changing climates will also have major impacts on crop production, interactions between plants and invertebrates, and plant physiology. Factors such as temperature, solar radiation, relative humidity, precipitation and wind speed directly influence plant physiological processes (Olesen and Bindi 2002). Climate affects the phenology of plants, invertebrates and their local abundance and distributions (Hegland et al. 2009). Higher CO_2 concentrations change photosynthesis rates and respiration, and can subsequently impact crop production (Hegland et al. 2009). Similarly, CO_2 and nitrogen enrichment may in some instances increase nectar quality and the abundance of flowers (Hegland et al. 2009). On the other hand, climate changes can reduce plant defenses against pests, thereby making them vulnerable to attack (Dhaliwal et al. 2004). Examples of these changes to crops include: rice yields which are lower with increasing night temperatures (Peng et al. 2004), wheat yields that show reductions with temperatures above 32 °C (Gregory et al. 1999) and populations of the moth pest, *Helicoverpa armigera* (Hubner), have shifted from tropical to temperate regions where they impact legumes and related crops (Sharma 2005). These plant and invertebrate physiological responses to changing climates lead to shifts in geographical distributions, changes in food availability and increased competition among organisms. With these changes and the warming of the environment, fragile niches could be destroyed and the resulting agro-ecosystems are likely to have reduced natural enemy diversity with increased pest outbreaks and successful invasions as a result.

3.6 Multiple Ecosystem Services

CBC is one of many ES, including pollination, nutrient cycling and soil aeration among others. ES are processes and functions derived from nature that benefit humans directly or indirectly (Bennett and Chaplin-Kramer 2016; Costanza et al. 2017). Multiple ES can arise from the deployment of SNAP (see Sect. 3.1) and other agri-environment interventions but have been greatly undervalued (Sandhu et al. 2013). Although they can lead to many benefits beyond those originally intended, these are rarely quantified. For example, phacelia can enhance biological control but is also greatly favored by bees (Sprague et al. 2016). Furthermore, endemic New Zealand flowering plants in vineyards (Fig. 3.5) can enhance multiple ES such as weed suppression, mineralization of plant material and soil water retention while leading to increased natural enemy abundance and efficacy (Shields et al. 2016).

Fig. 3.5 An endemic New Zealand plant (*Hebe chathamica* Cockayne et Allan) under vines, providing multiple ecosystem services to the vineyard. Photo: Jean Jack

However, this is not the norm as the CBC literature has been largely limited to single ES studies which ignore the other ES, heavily restricting end-user attractiveness and adoptability. These single ES studies range from field research on natural enemy-pest communities and floral resources (Scarratt et al. 2008; Gurr et al. 2012) to laboratory studies looking at natural enemy flower preferences (Sivinski et al. 2011), oviposition rates and longevity when they are fed nectar from different sources (Lee and Heimpel 2008). While work on these specific aspects of CBC are helpful in understanding parts of the full system, recent research suggests that designing protocols that potentially enhance multiple ES would have a higher value and possibly a greater likelihood of adoption (Crowder and Jabbour 2014; Iverson et al. 2014; Geertsema et al. 2016; Gurr et al. 2016). For instance, landscape-scale manipulations can be successful in this respect (Tscharntke et al. 2007; Rusch et al.

2016) but can sometimes lead to results which are not related to the original aim (Jonsson et al. 2012).

An example of highly successful enhancement of multiple ES is the 'push-pull' work in East Africa (Khan et al. 2000). This economically viable approach has been adopted by over 30,000 farmers and provides multiple ES such as pest management, soil fertility, nitrogen fixation and animal fodder. These ES provide farmers with multiple avenues of income with low inputs such as increased maize yields (by 2.5 t/ha) and sorghum yields (by 1 t/ha), as well as higher milk production (Khan et al. 2011). The integration of multiple ES in agro-ecosystems with approaches such as 'push-pull' is considered by many to be a key component of future food production (de Schutter 2010; Khan et al. 2011; Tscharntke et al. 2012; Iverson et al. 2014; Gurr et al. 2016; Costanza et al. 2017; Gurr et al. 2017).

3.7 Ecosystem Dis-Services

Habitat manipulation intended to enhance CBC can have specific unintended negative impacts; these are known as ecosystem dis-services (EDS) (Zang et al. 2007; Gurr et al. 2017). For instance, added vegetation such as floral resources (i.e., SNAP) may benefit pests more than the natural enemies (Lynch et al. 2001; Brimner and Boland 2003; Winding et al. 2004; Tscharntke et al. 2005; Carvalho 2006; Zehnder et al. 2007; Cullen et al. 2008; Wolfenbarger et al. 2008; Roubos and Liburd 2009). An example is that the fecundity of the moth pest, *Epiphyas postvittana* (Walker), is enhanced with an increase in the availability of some flowering plants (Begum et al. 2006). This indicates the importance selecting plant species that minimize EDS which could be in the form of competition for resources (Gurr et al. 2017) and pollination (Holzschuh et al. 2012) between the added vegetation and crop species. Also the added vegetation could be allelopathic towards the crop (Zhang et al. 2007). Furthermore, the wider complexities of food webs need to be considered, for instance honeydew-producing pests such as mealybugs are 'farmed' by ants that are predators of many natural enemies of the mealybug and may reduce biological control of this pest (Daane et al. 2007). Many non-crop plants are also hosts of these (Gutierrez et al. 2008) and other pests which complicates the selective process for vegetation to be used in CBC habitat manipulation (Winkler et al. 2010).

Potential EDS can be reduced by considering them when designing CBC experiments such as measuring *E. postvittana* development on floral and non-floral plant tissues while investigating multiple ES (Shields et al. 2016). Furthermore, modeling the key parameters involved in natural enemy traits can be employed to inform biological control (Kean et al. 2003). Complex modeling could be used to develop sophisticated CBC service providing protocols (SPPs) (Wratten et al. in press) that prevent EDS. However, to achieve this, substantial knowledge of how habitat management practices reduce pest damage and what are the associated EDS need to be determined (Gurr et al. 2017).

3.8 Implementation of Conservation Biological Control: From Outputs to Outcomes

One of the largest barriers to implementation of CBC research results to farming practices is relaying information to end-users (primarily farmers) and their willingness to uptake and deploy that knowledge. A specific barrier that reduces uptake of this is the perceived negative impacts on the livelihood of the local community when CBC advice is provided 'remotely', for example at government level (Bennett and Dearden 2014). The solution to this is improved communication between policy-makers, scientists and farmers, and improved farmer education about CBC (Bennett and Dearden 2014; Murage et al. 2015; Wyckhuys et al. 2017). This may be achieved using information and communication technologies (ICT) which are increasingly being used with hand-held devices such as tablets and mobile phones that can provide multi-media access to CBC information with video, SMS and voice-based information delivery (Aker 2011; Wyckhuys et al. 2017). These can be two-way information delivery systems where farmers and growers can communicate issues to local government (Aker 2011; Vong et al. 2013). For example, in Vietnam, rural telecentres have been implemented to provide technologically isolated agricultural communities with access to a two-way information channel that

Fig. 3.6 A farmer field school in action in a rice crop in Birendranagar, Nepal. Photo: Sundar Tiwari

provides information and solutions to problems that local farmers have conveyed (Vong et al. 2013). These ICT communication pathways are still largely under-valued, but have enormous 'penetration' potential, particularly where other infrastructure is underdeveloped. However, actual adoption of CBC through these delivery systems is still limited (Aker 2011; Wyckhuys et al. 2017).

Although modern 'delivery systems' such as information on social media can be used, and probably have more 'penetration' than leaflets and farmer meetings etc., these need to complement direct 'pathways to implementation' involving on-farm, face-to-face communication. Trends in social science research in agriculture suggest that working with the famers directly, including having demonstration or trial plots, can be successful (Estrada-Carmona et al. 2014). This needs complementary approaches as well, including farmer field schools (Fig. 3.6), led not by scientists but by 'farmer teachers' which are direct and proven pathways to implementation (Warner 2007; Khan et al. 2011; Ferguson and Lovell 2014; Kiptot and Franzel 2014; Waddington et al. 2014; Wyckhuys et al. 2017).

3.9 The Future of Conservation Biological Control

Many international agenices are indicating that modern, high-impact farming, with associated dependance on fossil fuels and a high level of external costs (negative impacts on human health and the environment) cannot continue (de Schutter 2010; Sandhu et al. 2012; Tscharntke et al. 2012; Pywell et al. 2015; Gurr et al. 2016; United Nations 2017). For example, the 'Green Revolution' was a 'solution' to higher yields (Gaud 1968). However, it became out of favor because of its negative socio-economic effects and the external costs associated with its adoption and practices (Pearse 1980; Tilman 1998; Iverson et al. 2014). Now, potentially oxymoronic ideas such as 'sustainable intensification' are frequently advocated (e.g. Godfray et al. 2010; Pywell et al. 2015; Gurr et al. 2017). Whatever the proposed improvements, CBC needs to be part of a paradigm change in terms of how agriculture is practised, with a much greater emphasis on enhancing the contribution that ES can make to sustainability, incuding reduced dependence on fossil-fuel based inputs. The key challenges will be: (1) understanding the food-web dynamics of which CBC is a substantial part; (2) establishing clear pathways from a research outputs to achieving real outcomes; and (3) moving from high-level policy goals to practical on-farm changes. The latter is made more difficult by the relentless pursuit of GDP as part of economic growth (Costanza et al. 2014) coupled with key governments pursuing neo-liberal policies in which interventions in society by the state are transferred to private enterprise. Compounding the above is the intense and well-funded marketing by agro-chemical companies, supporting prophylactic use of their products, coupled with a low educational standard in developing countries. Converting science outputs to outcomes, including the deployment of proven pathways to implementation remains a key challenge in CBC, and in all agro-ecological approaches.

References

Aker JC (2011) Dial "A" for agriculture: a review of information and communication technologies for agricultural extension in developing countries. Agric Econ 42(6):631–647

Altieri MA (1999) The ecological role of biodiversity in agroecosystems. Agric Ecosyst Environ 74:19–31. https://doi.org/10.1111/j.1574-0862.2011.00545.x

Antiqueira PAP, Romero GQ (2016) Floral asymmetry and predation risk modify pollinator behavior, but only predation risk decreases plant fitness. Oecologia 181:475–485. https://doi.org/10.1007/s00442-016-3564-y

Balmer O, Géneau CE, Belz E, Weishaupt B et al (2014) Wildflower companion plants increase pest parasitation and yield in cabbage fields: experimental demonstration and call for caution. Biol Control 76:19–27. https://doi.org/10.1016/j.biocontrol.2014.04.008

Barnes AM, Wratten SD, Sandhu HS (2009) Harnessing biodiversity to improve vineyard sustainability. Outlooks Pest Manag 20(6):250–255. https://doi.org/10.1564/20dec04

Begg GS, Cook SM, Dye R, Ferrante M et al (2017) A functional overview of conservation biological control. Crop Prot 97:145–158. https://doi.org/10.1016/j.cropro.2016.11.008

Begum M, Gurr GM, Wratten SD, Hedberg PR, Nicol HI (2006) Using selective food plants to maximize biological control of vineyard pests. J Appl Ecol 43:547–554. https://doi.org/10.1111/j.1365-2664.2006.01168.x

Bennett EM, Chaplin-Kramer R (2016) Science for the sustainable use of ecosystem services. F1000 Research, 5, 2622. https://doi.org/10.12688/f1000research.9470.1

Bennett NJ, Dearden P (2014) Why local people do not support conservation: Community perceptions of marine protected area livelihood impacts, governance and management in Thailand. Marine Policy 44:107–116. https://doi.org/10.1016/j.marpol.2013.08.017

Berndt LA, Wratten SD, Scarratt SL (2006) The influence of floral resource subsidies on parasitism rates of leafrollers (Lepidoptera: Tortricidae) in New Zealand vineyards. Biol Control 37:50–55. https://doi.org/10.1016/j.biocontrol.2005.12.005

Bianchi FJJ, Booij CJH, Tscharntke T (2006) Sustainable pest regulation in agricultural landscapes: a review on landscape composition, biodiversity and natural pest control. Proc R Soc Lond B Biol Sci 273(1595):1715–1727. https://doi.org/10.1098/rspb.2006.3530

Boggs CL (2016) The fingerprints of global climate change on insect populations. Curr Opin Insect Sci 17:69–73. https://doi.org/10.1016/j.cois.2016.07.004

Bompard A, Jaworski CC, Bearez P, Desneux N (2013) Sharing a predator: can an invasive alien pest affect the predation on a local pest? Popul Ecol 55:433–440. https://doi.org/10.1007/s10144-013-0371-8

Brimner TA, Boland GJ (2003) A review of the non-target effects of fungi used to biologically control plant diseases. Agr Ecosyst Environ 100:3–16. https://doi.org/10.1016/S0167-8809(03)00200-7

Buchanan AL, Hermann SL, Lund M, Szendrei Z (2017) A meta-analysis of non-consumptive predator effects in arthropods: the influence of organismal and environmental characteristics. Oikos 126(9):1233–1240. https://doi.org/10.1111/oik.04384

Carvalho FP (2006) Agriculture, pesticides, food security and food safety. Environ Sci Policy 9:685–692. https://doi.org/10.1016/j.envsci.2006.08.002

Chaplin-Kramer R, de Valpine P, Mills NJ, Kremen C (2013) Detecting pest control services across spatial and temporal scales. Agr Ecosyst Environ 181:206–212. https://doi.org/10.3410/f.723898378.793501007

Chen F, Ge F, Parajulee MN (2005) Impact of elevated CO_2 on tri- trophic interactions of *Gossypium hirsutum*, *Aphis gossypii*, and *Leis axyridis*. Environ Entomol 34:37–46. https://doi.org/10.1603/0046-225X-34.1.37

Chisholm PJ, Gardiner MM, Moon EG, Crowder DW (2014) Tools and techniques for investigating impacts of habitat complexity on biological control. Biol Control 75:48–57. https://doi.org/10.1016/j.biocontrol.2014.02.003

Clough Y, Barkmann J, Juhrbandt J, Kessler M, Wanger TC, Anshary A et al (2011) Combining high biodiversity with high yields in tropical agroforests. Proc Natl Acad Sci 108:8311–8316. https://doi.org/10.1073/pnas.1016799108

Costanza R, Kubiszewski I, Giovannini E, Lovins H et al (2014) Development: time to leave GDP behind. Nature 505:283–285. https://doi.org/10.1038/505283a

Costanza R, de Groot R, Braat L, Kubiszewski I et al (2017) Twenty years of ecosystem services: how far have we come and how far do we still need to go? Ecosyst Serv 28:1–16. https://doi.org/10.1016/j.ecoser.2017.09.008

Cooper J, Dobson H (2007) The benefits of pesticides to mankind and the environment. Crop Prot 26:1337–1348. https://doi.org/10.1016/j.cropro.2007.03.022

Crowder DW, Jabbour R (2014) Relationships between biodiversity and biological control in agroecosystems: current status and future challenges. Biol Control 75:8–17. https://doi.org/10.1016/j.biocontrol.2013.10.010

Cullen R, Warner KD, Jonsson M, Wratten SD (2008) Economics and adoption of conservation biological control. Biol Control 45:272–280. https://doi.org/10.1016/j.biocontrol.2008.01.016

Daane KM, Sime KR, Fallon J, Cooper ML (2007) Impacts of Argentine ants on mealybugs and their natural enemies in California's coastal vineyards. Ecol Entomol 32:583–596. https://doi.org/10.1111/j.1365-2311.2007.00910.x

de Schutter OD (2010) Report submitted by the special rapporteur on the right to food. United Nations General Assembly. http://www2.ohchr.org/english/issues/food/docs/A-HRC-16-49.pdf. Accessed 8 Nov 2017

Dhaliwal G, Aroroa R, Dhawan A (2004) Crop losses due to insect pests in Indian agriculture: an update. Indian J Ecol 31(1):1–7

Estrada-Carmona N, Hart AK, DeClerck FAJ, Harvey CA, Milder JC (2014) Integrated landscape management for agriculture, rural livelihoods, and ecosystem conservation: An assessment of experience from Latin America and the Caribbean. Landsc Urban Plan 129:1–11. https://doi.org/10.1016/j.landurbplan.2014.05.001

Fabinyi M, Dressler WH, Pido MD (2016) Fish, trade and food security: moving beyond "availability" discourse in marine conservation. Hum Ecol 45(2):177–188. https://doi.org/10.1007/s10745-016-9874-1

Ferguson RS, Lovell ST (2014) Permaculture for agroecology: Design, movement, practice, and worldview. A review. Agron Sustain Dev 34:251–274. https://doi.org/10.1007/s13593-013-0181-6

Free JB (1993) Insect pollination of crops, 2nd edn. Academic Press, London, United Kingdom, p 684

Gartaula H, Patel K, Johnson D, Devkota R, Khadka K, Chaudhary P (2016) From food security to food wellbeing: examining food security through the lens of food wellbeing in Nepal's rapidly changing agrarian landscape. Agric Hum Values 34(3):573–589. https://doi.org/10.1007/s10460-016-9740-1

Gaud WS (1968) The green revolution: Accomplishments and apprehensions. Speech given before the Society of International Development, Washington, D.C. USA. http://www.agbioworld.org/biotech-info/topics/borlaug/borlaug-green.html Accessed 1 Nov 2017

Geertsema W, Rossing WA, Landis DA, Bianchi FJ et al (2016) Actionable knowledge for ecological intensification of agriculture. Front Ecol Environ 14:209–216. https://doi.org/10.1002/fee.1258

Géneau CE, Wäckers FL, Luka H, Daniel C, Balmer O (2012) Selective flowers to enhance biological control of cabbage pests by parasitoids. Basic Appl Ecol 13:85–93. https://doi.org/10.1016/j.baae.2011.10.005

Gillespie MAK, Gurr GM, Wratten SD (2016) Beyond nectar provision: The other resource requirements of parasitoid biological control agents. Entomol Exp Appl 159:207–221. https://doi.org/10.1111/eea.12424

Godfray HCJ, Beddington JR, Crute IR, Haddad L et al (2010) Food security: the challenge of feeding 9 billion people. Science 327(5967):812–818. https://doi.org/10.1126/science.1185383

Godfray HCJ, Beddington JR, Crute IR, Haddad L et al (2012) The challenge of food security. Science 327:812. https://doi.org/10.1126/science.1185383

Gonzalvez FG, Rodriguez-Girones MA (2013) Seeing is believing: information content and behavioural response to visual and chemical cues. Proc R Soc Lond Biol Sci 280:20130886. https://doi.org/10.1098/rspb.2013.0886

Gonthier DJ (2012) Do herbivores eavesdrop on ant chemical communication to avoid predation? PLoS ONE 7:e28703. https://doi.org/10.1371/journal.pone.0028703

Gregory PJ, Ingram JS, Campbell B, Goudriaan J et al (1999) Managed production systems. In: Walker B, Steffen W, Canadell J, Ingram J (eds) The terrestrial biosphere and global change: implications for natural and managed ecosystems. Cambridge University Press, New York, pp 229–270

Gurr GM, Lu Z, Zheng X, Xu H et al (2016) Multi-country evidence that crop diversification promotes ecological intensification of agriculture. Nat Plants 2:16014. https://doi.org/10.1038/nplants.2016.14

Gurr G, Wratten SD, Altieri MA (eds) (2004) Ecological engineering for pest management: advances in habitat manipulation for arthropods. CSIRO Publishing, Collingwood, Australia, p 186

Gurr G, Wratten SD, Landis DA, You M (2017) Habitat management to suppress pest populations: progress and prospects. Annu Rev Entomol 62:91–109. https://doi.org/10.1146/annurev-ento-031616-035050

Gurr GM, Wratten SD, Snyder WE (eds) (2012) Biodiversity and insect pests: key issues for sustainable management. Wiley, Sussex, UK, p 360

Gurr GM, You M (2016) Conservation biological control of pests in the molecular era: New opportunities to address old constraints. Front Plant Sci 6:1255. https://doi.org/10.3389/fpls.2015.01255

Gurr GM, Zhu ZR, You M (2015) The big picture: prospects for ecological engineering to guide the delivery of ecosystem services in global agriculture. Rice Planthoppers: ecology, management, socio economics and policy. Zhejiang University Press, Hangzhou, China, p 231

Gutierrez AP, Daane KM, Ponti L, Walton VM, Ellis CK (2008) Prospective evaluation of the biological control of vine mealybug: refuge effects and climate. J Appl Ecol 45:524–536. https://doi.org/10.1111/j.1365-2664.2007.01356.x

Grygorczyk A, Turecek J, Lesschaeve I (2014) Consumer preferences for alternative pest management practices used during production of an edible and a non-edible greenhouse crop. J Pest Sci 87:249–258. https://doi.org/10.1007/s10340-013-0544-4

Haddad NM, Crutsinger GM, Gross K, Haarstad J, Tilman D (2011) Plant diversity and the stability of foodwebs. Ecol Lett 14:42–46. https://doi.org/10.1111/j.1461-0248.2010.01548.x

Hatt S, Lopes T, Boeraeve F, Chen J, Francis F (2016) Pest regulation and support of natural enemies in agriculture: experimental evidence of within field wildflower strips. Ecol Eng 98:240–245. https://doi.org/10.1016/j.ecoleng.2016.10.080

Hegland SJ, Nielsen A, Lázaro A, Bjerknes AL, Totland O (2009) How does climate warming affect plant-pollinator interactions? Ecol Lett 12:184–195. https://doi.org/10.1111/j.1461-0248.2008.01269.x

Hemerik L, Gebauer K, Meyhöfer R (2015) Comparison of the effect of predicted climate change on two agricultural pest-parasitoid systems. Proc Environ Sci 29:75–76. https://doi.org/10.1016/j.proenv.2015.07.166

Hermann SL, Landis DA (2017) Scaling up our understanding of non-consumptive effects in insect systems. Curr Opin Insect Sci 20:54–60. https://doi.org/10.1016/j.cois.2017.03.010

Hermann S, Thaler J (2014) Prey perception of predation risk: volatile chemical cues mediate non-consumptive effects of a predator on a herbivorous insect. Oecologia 176:669–676. https://doi.org/10.1007/s00442-014-3069-5

Hoefler CD, Durso LC, McIntyre KD (2012) Chemical-mediated predator avoidance in the European house cricket (*Acheta domesticus*) is modulated by predator diet. Ethology 118:431–437. https://doi.org/10.1111/j.1439-0310.2012.02028.x

Hof AR, Svahlin A (2015) The potential effect of climate change on the geographical distribution of insect pest species in the Swedish boreal forest. Scand J For Res 31:29–39. https://doi.org/10.1080/02827581.2015.1052751

Holland JM, Bianchi FJ, Entling MH, Moonen AC, Smith BM, Jeanneret P (2016) Structure, function and management of semi-natural habitats for conservation biological control: a review of European studies. Pest Manag Sci 72(9):1638–1651. https://doi.org/10.1002/ps.4318

Holzschuh A, Dudenhöffer J-H, Tscharntke T (2012) Landscapes with wild bee habitats enhance pollination, fruit set and yield of sweet cherry. Biol Cons 153:101–107. https://doi.org/10.1016/j.biocon.2012.04.032

Ingram JSI, Wright HL, Foster L, Aldred T et al (2013) Priority research questions for the UK food system. Food Secur 5:617–636. https://doi.org/10.1007/s12571-013-0294-4

Iverson AL, Marín LE, Ennis KK, Gonthier DJ et al (2014) Do polycultures promote win-wins or trade-offs in agricultural ecosystem services? A meta-analysis. J Appl Ecol 51(6):1593–1602. https://doi.org/10.1111/1365-2664.12334

Jonsson M, Wratten SD, Landis DA, Tompkins JML, Cullen R (2010) Habitat manipulation to mitigate the impacts of invasive arthropod pests. Biol Invasions 12(9):2933–2945. https://doi.org/10.1007/s10530-010-9737-4

Jonsson M, Buckley HL, Case BS, Wratten SD, Hale RJ, Didham RK (2012) Agricultural intensification drives landscape-context effects on host–parasitoid interactions in agroecosystems. J Appl Ecol 49(3):706–714. https://doi.org/10.1111/j.1365-2664.2012.02130.x

Jetter K, Paine TD (2004) Consumer preferences and willingness to pay for biological control in the urban landscape. Biol Control 30:312–322. https://doi.org/10.1016/j.biocontrol.2003.08.004

Kaplan I, McArt SH, Thaler JS (2014) Plant defenses and predation risk differentially shape patterns of consumption, growth, and digestive efficiency in a guild of leaf-chewing insects. PLoS ONE 9:20120948. https://doi.org/10.1371/journal.pone.0093714

Karp DS, Rominger AJ, Zook J, Ranganathan J, Ehrlich PR, Daily GC (2012) Intensive agriculture erodes β-diversity at large scales. Ecol Lett 15:963–970. https://doi.org/10.1111/j.1461-0248.2012.01815.x

Kean J, Wratten S, Tylianakis J, Barlow N (2003) The population consequences of natural enemy enhancement, and implications for conservation biological control. Ecol Lett 6:604–612. https://doi.org/10.1046/j.1461-0248.2003.00468.x

Khan ZR, Pickett JA, Berg JVD, Wadhams LJ, Woodcock CM (2000) Exploiting chemical ecology and species diversity: stem borer and striga control for maize and sorghum in Africa. Pest Manag Sci 56(11), 957–962. https://doi.org/10.1002/1526-4998(200011)56:11<957::AID-PS236>3.0.CO;2-T

Khan Z, Midega C, Pittchar J, Pickett J, Bruce T (2011) Push—pull technology: a conservation agriculture approach for integrated management of insect pests, weeds and soil health in Africa. Int J Agric Sustain 9:162–170. https://doi.org/10.3763/ijas.2010.0558

Kiptot E, Franzel S (2014) Voluntarism as an investment in human, social and financial capital: Evidence from a farmer-to-farmer extension program in Kenya. Agric Hum Values 31:231–243. https://doi.org/10.1007/s10460-013-9463-5

Kremen C, Miles A (2012) Ecosystem services in biologically diversified versus conventional farming systems: benefits, externalitites, and trade-offs. Ecol Soc 17:1–23. https://doi.org/10.5751/ES-05035-170440

Langer V (2001) The potential of leys and short rotation coppice hedges as reservoirs for parasitoids of cereal aphids in organic agriculture. Agr Ecosyst Environ 87:81–92. https://doi.org/10.1016/S0167-8809(00)00298-X

Lee DH, Nyrop JP, Sanderson JP (2014) Non-consumptive effects of the predatory beetle *Delphastus catalinae* (Coleoptera: Coccinellidae) on habitat use patterns of adult whitefly *Bemisia argentifolii* (Hemiptera: Aleyrodidae). Appl Entomol Zool 49:599–606. https://doi.org/10.1007/s13355-014-0294-7

Lee JC, Heimpel GE (2008) Floral resources impact longevity and oviposition rate of a parasitoid in the field. J Anim Ecol 77:565–572. https://doi.org/10.1111/j.1365-2656.2008.01355.x

Letourneau DK, Armbrecht I, Rivera BS, Lerma JM et al (2011) Does plant diversity benefit agroecosystems? A synthetic review. Ecol Appl 21:9–21. https://doi.org/10.1890/09-2026.1
Littlejohn CP, Curran TJ, Hofmann RW, Wratten SD (2015) Farmland, food, and bioenergy crops need not compete for land. Solutions 6(3):36–50
Lynch L, Hokkanen H, Babendreier D, Bigler F et al (2001) Insect biological control and non-target effects: a European perspective. In: Wajnberg E, Scott JK, Quimby PC (eds) Evaluating indirect ecological effects of biological control. CABI, New York, USA, pp 99–125
MacLeod A, Wratten SD, Sotherton NW, Thomas MB (2004) 'Beetle banks' as refuges for beneficial arthropods in farmland: long-term changes in predator communities and habitat. Agric For Entomol 6(2):147–154. https://doi.org/10.1111/j.1461-9563.2004.00215.x
Mauremooto JR, Wratten SD, Worner SP, Fry GLA (1995) Permeability of hedgerows to predatory carabid beetles. Agr Ecosyst Environ 52(2):141–148. https://doi.org/10.1016/0167-8809(94)00548-S
Maxwell FG, Jennings PR (1980) Breeding plants resistant to insects. Wiley, New York, USA, p 683
McLachlan ARG, Wratten SD (2003) Abundance and species richness of field-margin and pasture spiders (Araneae) in Canterbury, New Zealand. N Z J Zool 30:57–67. https://doi.org/10.1080/03014223.2003.9518324
Messelink GJ, Bennison J, Alomar O, Ingegno BL et al (2014) Approaches to conserving natural enemy populations in greenhouse crops: current methods and future prospects. Biocontrol 59:377–393. https://doi.org/10.1007/s10526-014-9579-6
Murage W, Pittchar JO, Midega CO, Onyango CO, Khan ZR (2015) Gender specific perceptions and adoption of the climate-smart push–pull technology in eastern Africa. Crop Prot 76:83–91. https://doi.org/10.1016/j.cropro.2015.06.014
Murdock TQ, Taylor SW, Flower A, Mehlenbacher A et al (2013) Pest outbreak distribution and forest management impacts in a changing climate in British Columbia. Environ Sci Policy 26:75–89. https://doi.org/10.1016/j.envsci.2012.07.026
Myers SS, Smith MR, Guth S, Golden CD et al (2017) Climate change and global food systems: potential impacts on food security and undernutrition. Annu Rev Public Health 38:259–277. https://doi.org/10.1146/annurev-publhealth-031816-044356
Nilsson U, Porcel M, Swiergiel W, Wivstad M (2016) Habitat manipulation – as a pest management tool in vegetable and fruit cropping systems, with the focus on insects and mites. Centre for Organic Food & farming, Uppsala, Sweden, p 52
Ninkovic V, Feng Y, Olsson U, Pettersson J (2013) Ladybird footprints induce aphid avoidance behavior. Biol Control 65:63–71. https://doi.org/10.1016/j.biocontrol.2012.07.003
Olesen JE, Bindi M (2002) Consequences of climate change for European agricultural productivity, land use and policy. Eur J Agron 16:239–262. https://doi.org/10.1016/S1161-0301(02)00004-7
Parker JD, Salminen JP, Agrawal AA (2010) Herbivory enhances positive effects of plant genotypic diversity. Ecol Lett 13:553–563. https://doi.org/10.1111/j.1461-0248.2010.01452.x
Pearse A (1980) Seeds of plenty, seeds of want: social and economic implications of the green revolution. In. Utting P (ed) Revisiting sustainable development. United Nations Institute for Social Development, Geneva, Switzerland, pp 139–157
Peng S, Huang J, Sheehy JE, Laza RC et al (2004) Rice yields decline with higher night temperature from global warming. Proc Natl Acad Sci USA 101(27):9971–9975. https://doi.org/10.1073/pnas.0403720101
Porter J, Costanza R, Sandhu H, Sigsgaard L, Wratten S (2009) The Value of Producing Food, Energy, and Ecosystem Services within an Agro-Ecosystem. AMBIO J Hum Environ 38 (4):186–193. https://doi.org/10.1579/0044-7447-38.4.186
Pywell RF, Heard MS, Woodcock BA, Hinsley S et al (2015) Wildlife-friendly farming increases crop yield: evidence for ecological intensification. Proc R Soc B Biol Sci 282:20151740. https://doi.org/10.1098/rspb.2015.1740

Roth SK, Lindroth RL (1995) Elevated atmospheric CO2: effects on phytochemistry, insect performance and insect-parasitoid interactions. Glob Change Biol 1:173–182. https://doi.org/10.1111/j.1365-2486.1995.tb00019.x

Roubos CR, Liburd OE (2009) Effect of trap color on captures of grape root borer (Lepidoptera: Sesiidae) males and non-target insects. J Agric Urban Entomol 25:99–109. https://doi.org/10.3954/1523-5475-25.2.99

Roubos CR, Rodriguez-Saona C, Isaacs R (2014) Mitigating the effects of insecticides on arthropod biological control at field and landscape scales. Biol Control 75:28–38. https://doi.org/10.1016/j.biocontrol.2014.01.006

Rusch A, Birkhofer K, Bommarco R, Smith HG, Ekbom B (2014) Management intensity at field and landscape levels affects the structure of generalist predator communities. Oecologia 175 (3):971–983. https://doi.org/10.1007/s00442-014-2949-z

Rusch A, Chaplin-Kramer R, Gardiner MM, Hawro V et al (2016) Agricultural landscape simplification reduces natural pest control: a quantitative synthesis. Agr Ecosyst Environ 221:198–204. https://doi.org/10.1016/j.agee.2016.01.039

Rypstra AL, Buddle CM (2013) Spider silk reduces insect herbivory. Biol Let 9:20120948. https://doi.org/10.1098/rsbl.2012.0948

Sandhu H, Porter J, Wratten S (2013) Experimental assessment of ecosystem services in agriculture. In: Wratten S, Sandhu H, Cullen C, Costanza R (eds) Ecosystem services in agricultural and urban landscapes. Wiley, United Kingdom, Oxford, pp 122–135

Sandhu HS, Crossman ND, Smith FP (2012) Ecosystem services and Australian agricultural enterprises. Ecol Econ 74:19–26. https://doi.org/10.1016/j.ecolecon.2011.12.001

Scarratt SL, Wratten SD, Shishehbor P (2008) Measuring parasitoid movement from floral resources in a vineyard. Biol Control 46:107–113. https://doi.org/10.1016/j.biocontrol.2008.03.016

Schmitz OJ (1998) Direct and indirect effects of predation and predation risk in old-field interaction webs. Am Nat 151:327–342. https://doi.org/10.1086/286122

Schmitz OJ, Barton BT (2014) Climate change effects on behavioral and physiological ecology of predator-prey interactions: Implications for conservation biological control. Biol Control 75:87–96. https://doi.org/10.1016/j.biocontrol.2013.10.001

Seagraves MP, Kajita Y, Weber DC, Obrycki JJ, Lundgren JG (2011) Sugar feeding by coccinellids under field conditions: the effects of sugar sprays in soybean. Biocontrol 56 (3):305–314. https://doi.org/10.1007/s10526-010-9337-3

Semere T, Slater FM (2007) Invertebrate populations in *Miscanthus* (*Miscanthus* × *giganteus*) and reed canary-grass (*Phalaris arundinacea*) fields. Biomass Bioenerg 31:30–39. https://doi.org/10.1016/j.biombioe.2006.07.002

Sendoya SF, Freitas AVL, Oliveira PS (1996) Ants as selective agents on herbivore biology: effects on the behaviour of a non-myrmecophilous butterfly. J Anim Ecol 65:205–210. https://doi.org/10.2307/5723

Sendoya F, Freitas VL, Oliveira PS (2009) Egg-laying butterflies distinguish predaceous ants by sight. Am Naturlist 174:134–140. https://doi.org/10.1086/599302

Sharma HC (2005) Heliothis/Helicoverpa management: The emerging trends and strategies for future research. Oxford & IBH Publishing Co., Pvt. Ltd, New Delhi, India, p 482

Sharma M, Ghosh R, Pande S (2015) Dry root rot (*Rhizoctonia bataticola* (Taub.) Butler): An emerging disease of chickpea–where do we stand? Arch Phytopathol Plant Prot 48:797–812. https://doi.org/10.1080/03235408.2016.1140564

Shields MW, Tompkins JM, Saville DJ, Meurk CD, Wratten S (2016) Potential ecosystem service delivery by endemic plants in New Zealand vineyards: successes and prospects. PeerJ 4:e2042. https://doi.org/10.7717/peerj.2042

Shields MW, Buckley HL, Wratten SD (In press) Complementary approaches to evaluating the biological control refuge potential of a biofuel feedstock grass. Agric Ecosyst Environ

Sidhu SC, Wilson Rankin EE (2016) Honey bees avoiding ant harassment at flowers using scent cues. Environ Entomol 45:420–426. https://doi.org/10.1093/ee/nvv230

Sivinski J, Wahl D, Holler T, Dobai SA, Sivinski R (2011) Conserving natural enemies with flowering plants: Estimating floral attractiveness to parasitic Hymenoptera and attraction's relationship to flower and plant morphology. Biol Control 58:208–214. https://doi.org/10.1016/j.biocontrol.2011.05.002

Sprague R, Boyer S, Stevenson GM, Wratten SD (2016) Assessing pollinators' use of floral resource subsidies in agri-environment schemes: An illustration using *Phacelia tanacetifolia* and honeybees. PeerJ 4:e2677. https://doi.org/10.7717/peerj.2677

Tena A, Pekas A, Cano D, Wäckers FL, Urbaneja A (2015) Sugar provisioning maximizes the biocontrol service of parasitoids. J Appl Ecol 52:795–804. https://doi.org/10.1111/1365-2664.12426

Thaler JS, Contreras H, Davidowitz G (2014) Effects of predation risk and plant resistance on Manduca sexta caterpillar feeding behaviour and physiology. Ecol Entomol 39:210–216. https://doi.org/10.1111/een.12086

Thomas MB, Wratten SD, Sotherton NW (1991) Creation of 'island' habitats in farmland to manipulate populations of beneficial arthropods: predator densities and emigration. J Appl Ecol 28(3):906–917. https://doi.org/10.2307/2404216

Tilman D (1998) The greening of the green revolution. Nature 396(6708):211–212

Tilman D, Balzer C, Hill J, Befort BL (2011) Global food demand and the sustainable intensification of agriculture. Proc Natl Acad Sci 108(50):20260–20264. https://doi.org/10.1073/pnas.1116437108

Tscharntke T, Rand TA, Bianchi FJ (2005) The landscape context of trophic interactions: insect spillover across the crop-noncrop interface. Ann Zool Fenn 42:421–432

Tscharntke T, Bommarco R, Clough Y, Crist TO et al (2007) Conservation biological control and enemy diversity on a landscape scale. Biol Control 43:294–309. https://doi.org/10.1016/j.biocontrol.2007.08.006

Tscharntke T, Clough Y, Wanger TC, Jackson L et al (2012) Global food security, biodiversity conservation and the future of agricultural intensification. Biol Cons 151:53–59. https://doi.org/10.1016/j.biocon.2012.01.068

Tscharntke T, Karp DS, Chaplin-Kramer R, Batary P et al (2016) When natural habitat fails to enhance biological pest control - Five hypotheses. Biol Cons 204:449–458. https://doi.org/10.1016/j.biocon.2016.10.001

Tschumi M, Albrecht M, Entling MH, Jacot K (2015) High effectiveness of tailored flower strips in reducing pests and crop plant damage. Proc R Soc B Biol Sci 282:20151369. https://doi.org/10.1098/rspb.2015.1369

Turner KG, Anderson S, Gonzalez M, Costanza R et al (2014) Toward an integrated ecology and economics of land degradation and restoration: Methods, data, and models. Report to the ELD Project Data and Methodology Working Group, pp 1–60

Tylianakis JM, Binzer A (2014) Effects of global environmental changes on parasitoid-host food webs and biological control. Biol Control 75:77–86. https://doi.org/10.1016/j.biocontrol.2013.10.003

Tylianakis JM, Didham RK, Bascompte J, Wardle DA (2008) Global change and species interactions in terrestrial ecosystems. Ecol Lett 11(12):1351–1363. https://doi.org/10.1111/j.1461-0248.2008.01250.x

Tylianakis JM, Romo CM (2010) Natural enemy diversity and biological control: Making sense of the context-dependency. Basic Appl Ecol 11:657–668. https://doi.org/10.1016/j.baae.2010.08.005

United Nations (2017) Report of the special rapporteur on the right to food. A/HRC/34/48. Human Rights Council, United Nations General Assembly. https://documents-dds-ny.un.org/doc/UNDOC/GEN/G17/017/85/PDF/G1701785.pdf?OpenElement. Accessed 8 Nov 2017

Unruh TR, Pfannenstiel RS, Peters C, Brunner JF, Jones VP (2012) Parasitism of leafrollers in Washington fruit orchards is enhanced by perimeter plantings of rose and strawberry. Biol Control 62:162–172. https://doi.org/10.1016/j.biocontrol.2012.04.007

Van Dyck H, Bonte D, Puls R, Gotthard K, Maes D (2015) The lost generation hypothesis: Could climate change drive ectotherms into a developmental trap? Oikos 124:54–61. https://doi.org/10.1111/oik.0206

Vong J, Song I, Mandal P (2013) Application of ICT to improve rural livelihood in Vietnam. Int J Electron Financ 7(2):132–145. https://doi.org/10.1007/978-981-287-347-7_14

Waddington H, Snilstveit B, Hombrados JG, Vojtkova M, Anderson J, White H (2014) Farmer field schools for improving farming practices and farmer outcomes in low-and middle-income countries: a systematic review. Campbell Syst Rev 10(6), 335

War AR, Taggar GK, War MY, Hussain B (2016) Impact of climate change on insect pests, plant chemical ecology, tritrophic interactions and food production. Int J Clin Biol Sci 1(2):16–29

Warner KD (2007) Agroecology in action: extending alternative agriculture through social networks. MIT Press, Massachusetts, USA, p 269

Wasserberg G, White L, Bullard A, King J, Maxwell R (2013) Oviposition site selection in *Aedes albopictus* (Diptera: Culicidae): are the effects of predation risk and food level independent? J Med Entomol 50:1159–1164. https://doi.org/10.1603/ME12275

Wilson MR, Leather SR (2012) The effect of past natural enemy activity on host-plant preference of two aphid species. Entomol Exp Appl 144(2):216–222. https://doi.org/10.1111/j.1570-7458.2012.01282.x

Winding A, Binnerup SJ, Pritchard H (2004) Non-target effects of bacterial biological control agents suppressing root pathogenic fungi. FEMS Microbiol Ecol 47:129–141. https://doi.org/10.1016/S0168-6496(03)00261-7

Winkler K, Wäckers FL, Termorshuizen AJ, van Lenteren JC (2010) Assessing risks and benefits of floral supplements in conservation biological control. Biocontrol 55(6):719–727. https://doi.org/10.1007/s10526-010-9296-8

Wolfenbarger LL, Naranjo SE, Lundgren JG, Bitzer RJ, Watrud LS (2008) Bt crop effects on functional guilds of non-target arthropods: a meta-analysis. PLoS ONE, 3. https://doi.org/10.1371/journal.pone.0002118

Wollaeger HE, Getter KL, Behe BK (2015) Consumer preferences for traditional, neonicotinoid-free, bee-friendly, or biological control pest management practices on floriculture crops. HortScience 50(5):721–732

Woodward A, Porter JR (2016) Food, hunger, health, and climate change. The Lancet 387 (10031):1886–1887. https://doi.org/10.1016/S0140-6736(16)00349-4

Wratten SD (1976) Searching by *Adalia bipunctata* (L.) and escape behaviour of its aphid and cicadellid prey on lime (*Tilia* x *vulgaris* Hayne). Ecol Entomol 1:139–142. https://doi.org/10.1111/j.1365-2311.1976.tb01215.x

Wratten SD, Gonzalez-Chang M, Shields MW, Nboyine J et al (In press) Understanding the pathway from biodiversity to agro-ecological outcomes: a new, interactive approach. Ecol Appl

Wyckhuys KAG, Bentley JW, Lie R, Phuong LTN, Fredrix M (2017) Maximizing farm-level uptake and diffusion of biological control innovations in today's digital era. Biocontrol 62:1–16. https://doi.org/10.1007/s10526-017-9820-1

Zhang W, Ricketts TH, Kremen C, Carney K, Swinton SM (2007) Ecosystem services and dis-services to agriculture. Ecol Econ 64(2):253–260. https://doi.org/10.1016/j.ecolecon.2007.02.024

Zehnder G, Gurr GM, Kühne S, Wade MR, Wratten SD, Wyss E (2007) Arthropod pest management in organic crops. Annu Rev Entomol 52:57–80. https://doi.org/10.1146/annurev.ento.52.110405.091337

Zhou K, Huang J, Deng X, van der Werf W et al (2014) Effects of land use and insecticides on natural enemies of aphids in cotton: First evidence from smallholder agriculture in the North China Plain. Agr Ecosyst Environ 183:176–184. https://doi.org/10.1016/j.agee.2013.11.008

Chapter 4
Application of Conservation Agriculture Principles for the Management of Field Crops Pests

Morris Fanadzo, Mvuselelo Dalicuba and Ernest Dube

Abstract Worldwide, farmers are called upon to abandon harmful pesticides and adopt conservation agriculture for improving environmental sustainability, soil fertility, pest management and farm profits, among other benefits. Whereas the positive environmental benefits of conservation agriculture are non-questionable, pest management benefits are still a subject of debate. Abandonment of the plough and harmful pesticides towards conservation agriculture presented new challenges to farmers in terms of pest management. Pest problems are frequently reported as the main yield limiting factor for conservation agriculture in many production systems of the world, especially among the resource poor farmers. Here we first review the pest management benefits of conservation agriculture principles, with special focus on weeds and animal pests. In conservation agriculture, emphasis should be placed on use of different multiple and varied tactics incorporated into the cropping system design to avoid damaging levels of pests, thus minimizing the need for curative solutions. Conservation agriculture embraces integrated pest management, as it aims to incorporate reduced pesticide applications with cover crops, conservation tillage and crop rotation to strengthen natural pest control. We show that effective long term weed management in conservation agriculture systems is based on an integration of measures for limiting competitiveness of the weeds that are already in the field and growing with the crop, preventing the introduction of new weeds, and preventing the multiplication of the weeds that are already there. Although the abandonment of tillage towards no-till requires an initial investment on herbicides for weed control, herbicide requirement tends to decline over time with proper application of conservation agriculture. Proper selection of planting date, density and spatial arrangement of a crop can maximize the space it occupies early in the season and put competitive pressure on weeds. Crops can be rotated in sequences that are not only profitable, but highly effective at breaking animal pest

M. Fanadzo (✉) · M. Dalicuba
Department of Agriculture, Faculty of Applied Sciences, Cape Peninsula University of Technology, Private Bag X8, Wellington 7654, South Africa
e-mail: FanadzoM@cput.ac.za

E. Dube
ARC–Small Grain, Private Bag X29, Bethlehem, South Africa

S. Gaba et al. (eds.), *Sustainable Agriculture Reviews 28*, Ecology for Agriculture 28,
https://doi.org/10.1007/978-3-319-90309-5_4

cycles. Mixed cropping reduces pest populations by increasing environmental diversity and lowering the overall attractiveness of the environment. We then highlight some possible solutions to the major challenges for pest management through conservation agriculture practice. Promotion of integration of conservation agriculture principles with cultural measures is essential for pest management in conservation agriculture systems. In conservation agriculture, it is important for farmers to employ several strategies simultaneously so that if one strategy fails, then the other ones operate to prevent yield loss. The focus should not be just on how to fit various pest management tactics into the conservation agriculture production system, but also on how the system can be modified to accommodate various pest control tactics. We demonstrate that farmers practicing conservation agriculture have several cultural methods that they can put together to build up a good pest management strategy. Although cover crops and mulches are generally viewed as the first line of defense against weeds, the reduction in weeds is not enough to eliminate the need for chemical control. Cover crops can be used to reduce animal pest dispersal, colonization and reproduction on crops through maintenance of the cover crop as a sink for various pests, confusing the pests visually and by causing microclimate changes that reduce pest success. Fertilizer timing and placement strongly influences crop competition; and deep banding of fertilizer has the potential to enhance not only fertilizer use efficiency, but also crop resistance to animal pests and competitive ability against weeds. Proper selection of planting date, density and spatial arrangement of a crop can put competitive pressure on weeds and break animal pest cycles. Sanitation practices are important tools in conservation agriculture because of their ability to eliminate necessities that are important to the pests' survival. The development of pest resistance will likely be minimal if host plant resistance is integrated with other control measures through conservation agriculture practice. A more holistic, integration approach of control tactics in conservation agriculture, which goes beyond the three principles, is essential for effective pest management.

Keywords Crop production · Pest control · Pesticides · Sustainable farming

In conservation agriculture, it is important for farmers to employ several strategies simultaneously so that if one strategy fails, then the other ones operate to prevent yield loss.

4.1 Introduction

Pests of field crops are a major cause crop yield losses across the world. Broadly speaking, a pest is an organism that reduces the availability, quality, or value of a plant or animal grown for food, fiber, or recreation (Flint and Van den Bosch 1981).

Crop pests are grouped into three broad categories namely—animal pests such as insects, mites, other arthropods, and vertebrates such as birds and rodents, plant pathogens such as fungi, bacteria, viruses and nematodes and lastly, plant pests commonly known as weeds. Most species reach pest status when the losses they cause exceed 5–10% (Hill 2008). Economic losses that are caused by pests do not only include the direct action of these organisms as they damage crops, but also indirect economic losses related to the costs of controlling the pest and environmental damage caused by pesticides. These costs include the purchase and application of pesticides, expenses related to medical treatment for people poisoned by pesticides, and damage caused by environmental contamination (Oliveira et al. 2014).

In developing countries, excessive usage of harmful pesticides for crop production is common. It has been estimated that only about 0.3% of these pesticides come into direct contact with the target pests (Pimentel 1995). A wide range of studies attest negative impacts of pesticides (e.g. Eskenazi et al. 2007; Rashid et al. 2010). This can be the case even if official guidelines for "safe" pesticide usage are followed (Stehle and Schulz 2015). Harmful effects of pesticides that end up in the environment include air, water, soil and food contamination. Concerns about the adverse impacts of pesticides started being voiced in the early 1960s (Van der Werf 1996), mostly in developed countries. Consumption of raw fresh produce is an increasing trend among health conscious consumers in developed countries and as such, many markets have placed zero tolerance of certain pesticide residues on fresh vegetables. This has increased pressure on farmers to refine their pest management systems for maintaining fresh produce's aesthetic quality without using harmful pesticides. As resistance of pests to pesticides and consumer awareness of the health hazards of pesticides increases, there is ever-growing interest in sustainable, alternative pest management strategies.

Usage of chemical pesticides is generally limited in undeveloped countries because majority of farmers are resource poor. Here, crop pests are frequently the major cause of food shortages, malnutrition and poor quality of life (Oerke and Dehne 2004). Women and children in many resource poor smallholder farmer communities of Africa spend significant amounts of time removing weeds manually using hand hoes (Giller et al. 2009). Literacy rates are low, such that even where pesticides are made freely available, there is a general lack of technical knowledge on how to use them efficiently. Neither planting more productive cultivars nor increasing land area under production can meet the need for food in the underdeveloped world, if pest management is not improved. Meanwhile, resource poor farmers are being called upon to abandon conventional farming practices and adopt conservation agriculture for the purpose of improving environmental sustainability, soil fertility and farm profits, among other envisaged benefits. Whereas the soil fertility and environmental benefits of conservation agriculture are well established, pest management benefits of the practice are still a subject of much debate.

Conservation agriculture is a crop management system based on three principles: (a) minimum soil disturbance, (b) permanent soil surface cover with crop residues and cover crops, and (c) crop rotations that include diverse species. Conservation agriculture strives to achieve acceptable profits along with high and sustained production levels, while concurrently conserving the environment through minimizing the requirement of pesticides and external fertilizer inputs (Verhulst et al. 2010). Unlike conventional pest control methods which tend to be reactive, conservation agriculture emphasizes ecological methods and aims to keep the pest population density below the economic injury level. The ultimate aim is to make field conditions more favorable for crop growth and less favorable for pests. Whereas the concept of conservation agriculture is fairly new, the principles are simple, and probably the oldest methods of crop cultivation that were used by mankind before the advent of synthetic pesticides and tillage equipment.

Abandonment of the plough and harmful pesticides towards conservation agriculture presented a myriad of new challenges to farmers in terms of pest management. Conventional tillage was useful for controlling not just weeds, but animal pests as well through exposing the pests to their natural enemies or directly by physical damage inflicted during the tillage process. Tillage was also used to bury populations of weeds and volunteer crop plants that harbour plant pathogens into deeper layers of the soil where they cause less or no disease. The soil conservation benefits of conservation agriculture over conventional farming are non-questionable, but effects on yields and pest problems tend to be variable, especially in the short term. For example, Brainard et al. (2016) reported that conservation agriculture had variable effects on weeds and natural enemies in snap bean (*Phaseolus vulgaris* L.) production, with no consistent impacts relative to full tillage. From the same study, it was concluded that the adoption of conservation agriculture resulted in greater pest and cover crop management costs that outweighed savings due to reduced tillage, resulting in short-term financial losses. Giller et al. (2009) recommended the need for a critical assessment of the suitability of conservation agriculture for smallholder farmer systems in Africa because of the numerous constraints associated with the practice, which included an increased weeding requirement for crops. Pest problems are frequently reported as the main yield limiting factor for conservation agriculture in many production systems of the world. Table 4.1 summarizes some of the major pest problems that have been observed in various conservation agriculture systems around the world.

In this chapter, pest management benefits of conservation agriculture principles are reviewed as an important step towards identifying opportunities for refining conservation agriculture. Some possible solutions to the major challenges for pest management through conservation agriculture practice are highlighted, and recommendations for further research are provided. It is hoped that this information would be useful to conservation agriculture researchers, practitioners and policy makers in improving the sustainable management of field crop pests.

Table 4.1 Some problematic weeds and animal pests as reported from various conservation agriculture systems across the world

Pest problem	Damage	References
Cutworms (*Agrotis* spp.)	Serious damage from variegated cutworm [*Peridroma saucia* (Hubner)] in a cotton (*Gossypium hirsutum* L.) crop as a result of conservation tillage with crimson clover (*Trifolium incarnatum* L.) cover crop	Gaylor et al. (1984)
	Infestations of black cutworm (*Agrotis ipsilon* Hufnagel) were increased by reduced tillage and rotation of maize (*Zea mays* L.) with either wheat (*Triticum aestivum* L.) or soybean (*Glycine max* [L.] Merrill). Moldboard ploughed plots had very little damage from this pest under any rotation	Johnson et al. (1984)
Slugs (*Deroceras reticulatum*) Agriolimacidae and Arionidae: Mollusca	Wide scale adoption of conservation tillage elevated the status of slugs as an important pest of all crops around the world. Slugs eat all crops and they inflict most of their damage during crop establishment and early growth. They are mostly abundant where there is more crop residue cover	Douglas and Tooker (2012), Glen and Sysmondson (2003)
Armyworm (*Pseudaletia unipuncta*), cutworms (*Agrotis ipsilon*); stalk borer (*Papaipema nebris*), rootworms, white grubs, slugs, and rodents	In maize and soybean production systems of Kentucky USA, it was reported that there was heavier infestation of no-till fields with these pests than reduced tillage fields	Gregory and Musick (1976)
Amaranthus retroflexus, Chenopodium album, Portulaca oleracea, Digitaria sanguinalis and *Conyza canadensis*	The problem of these weeds in a maize crop was increased with no till. A rye cover crop mulch slightly decreased the weed problem, but had no effect on survival. Late emerging weeds survived to maturity better than earlier emerging weeds in the no till system	Mohler and Calloway (1995), Dube et al. (2012) Barberi and Carcio (2001)

4.2 Application of Conservation Agriculture Principles in Weed Management

Out of all the pests of field crops, weeds are the most damaging as they cause average yield losses of 20–50% in most places around the world (Oerke and Dehne 2004). Some characteristics of weeds that make them difficult to manage include the ability to germinate under many environments, persistent seed, rapid growth, ability to reproduce through both self and cross pollination, long distance dispersal of reproductive units and prolific seed production. The concept of effective weed management is based on limiting competitiveness of the weeds that are already in the field and growing with the crop, preventing the introduction of new weeds, and preventing the multiplication of the weeds that are already there. The weed seed bank is the reserve of weed seeds present on the soil surface and scattered in the soil profile (Menalled 2008). It is the resting place of weed seeds and is an important component of the life cycle of weeds. Weeds emerging during later stages of crop development are often considered by farmers as harmless, but seed production by these weeds contributes to future weed infestations. The sections below are a discussion of the application of conservation agriculture principles in managing weed populations of field crops.

4.2.1 Cover Crops Management for Weed Control in Conservation Agriculture

Cover crops are generally viewed as the first line of defense against weeds in conservation agriculture. An ideal cover crop species should have the following characteristics: (i) be easy to establish and adapted to the environment; (ii) have a rapid growth rate so as to provide ground coverage quickly; (iii) be economically viable; (iv) be disease resistant and not act as a host for plant diseases; (v) be easy to kill and most importantly; (vi) produce a sufficient quantity of biomass. The various benefits and shortcomings of some selected, popular cover crops are presented in Table 4.2.

Vigorously growing, high biomass yielding cover crops normally provide excellent weed control by competing with weeds for light, moisture, nutrients and space. While actively growing cover crops are efficient at outcompeting and suppressing weeds, residue remaining after cover crop death may be less reliable for suppressing weeds, particularly for the duration of a cash-cropping season. Adequate cover with cover crop residues can be expected to reduce weed emergence by up to 99%, but weed suppression will decline during the course of the season according to the rate of residue decomposition. Therefore, cover crop residues may not provide full-season weed control. Figure 4.1 illustrates the failure of a high biomass yielding vetch (*Vicia villosa* Roth), and oat (*Avena sativa*) cover crop to provide full season weed control in an irrigated maize (*Zea mays* L.) crop,

Table 4.2 The pest control benefits of various popular cover crop species

Cover crop species	Benefits and shortcomings
Oat (*Avena sativa*)	The mulch is persistent and very effective at suppressing broadleaf weeds (Dube et al. 2012)
Rye grass (*Secale cereale*)	Has excellent weed suppression ability through allelopathy. It also breaks nematodes cycles (Clark 2007)
Vetches (*Vicia* spp.)	Residues have fast decomposition but can provide early weed suppression. Host for soybean cyst nematode (*Heterodera glycines*) and root-knot nematode (Smith 2006)
Barley (*Hordeum vulgare*)	The mulch suppresses weeds. However, barley is host to the rootknot nematode
Sorghum (*Sorghum bicolor*)	Residues of sorghum have allelopathic effects on weeds. However, it harbours root knot nematode
Buckwheat (*Fagopyrum esculentum*)	The flowers are attractive to bees, wasps and parasitic flies. It is allelopathic to many weeds but harbours root lesion nematode (*Pratylynchus penetrans*) (Marks and Townsend 1973)
Canola (*Brassica napus*)	Its flowers are attractive to bees, wasps and parasitic flies. It is susceptible to *Sclerotinia sclerotiorum*
Cowpea (*Vigna unguiculata*)	Attracts beneficial insects such as wasps and honey bees. It can reduce rootknot nematodes (*Meloidogyne arenaria* and *M. incognita*) and soybean cyst nematode (*Heterodera glycines*) (Clark 2007)
Amaranth (*Amaranthus* spp.)	It is tolerant to many diseases and nematodes and can be used to break their cycles. However, it may be allelopathic on some crops and needs to be terminated early before planting
Sunhemp (*Crotalaria juncea*)	Claimed to reduce many types of nematodes (Sipes and Arakaki 1997)
Mustards (*Brassica* spp.)	It is used for soil sanitisation through fumigant effects. Used to clean up nematodes. Must have rain or irrigation after termination to help release glucosinilates (Henderson et al. 2009)
Broad bean (*Vicia faba*)	It reduces the incidence of take-all of wheat (*Gaeumannomyces graminis*) (Sattell et al. 1998)

based on conservation agriculture studies from maize-based smallholder farmer systems of South Africa (Dube et al. 2012).

For long season crops in hot and humid areas, farmers may not be able to rely solely on cover crop residue biomass for weed control. Insufficient crop reside cover may actually enhance the weed problem for the duration of the season by creating more favorable conditions for weed seed germination and growth (Brainard et al. 2016). Vetch and oat cover crop residues were better at suppressing broadleaf weeds such as purslane (*Portulaca oleracea*) and pigweed (*Amaranthus hybridus*), but failed to significantly suppress nutsedge emergence in summer maize (Dube et al. 2012). Narrow shaped leaves of grass weeds are in many instances able to emerge through mulched surfaces. Other researchers have found that in no till systems, cover crops reduced weeds, but not enough to eliminate the need for chemical control (Yenish et al. 1996; Teasdale 1996). In some instances, the

Fig. 4.1 Failure of cover crop residue mulch to provide full season weed control. Figure 4.1a shows a thick mat of mulch from high biomass yielding, irrigated oat and vetch cover crops in experimental plots immediately after termination of the cover crops. Figure 4.1b shows maize at 3 weeks after emergence, and much of the crop is relatively weed free. Figure 4.1c shows grass weeds (*Digitaria sanguinalis* and *Cyperus esculentus*) invading all plots at 7 weeks after emergence. Location of this experiment was the warm temperate region of the Eastern Cape, South Africa. Photographs by E. Dube

practice of conservation agriculture has been observed to gradually increase the herbicide requirement for weed control because of problematic grass weeds (Chiduza and Dube 2013).

4.2.2 Crop Rotation and Mixed Cropping to Reduce Weed Problems in Conservation Agriculture

Maximum potential yields of many crops are known to be compromised by monocropping. Altieri and Nicholls (2004) reported that over 90% of the world's 1.5 billion hectares of cropland was under monocultures of wheat ((*Triticum aestivum* L.), maize, cotton (*Gossypium hirsutum* L.) and soybeans (*Glycine max* [L.] Merrill), and in these monoculture systems, there was a heavy reliance on pesticide use for pest control. Conway and Pretty (1991) suggested that excessive use of agrochemicals in conjunction with expanding monocultures was the major reason for herbicide resistance in field crop production of the world. Weeds tend to thrive in the presence of crops whose growth requirements and characteristics are similar to their own, and rotating crops with different growth characteristics creates a changing environment which prevents the dominance of such weed species.

The inclusion within rotations of densely planted, high biomass yielding, fast-growing crop species and those exhibiting allelopathic potential provides further opportunities for weed suppression. Allelopathy refers to inhibition of the growth of one plant by chemical compounds released into the soil from neighbouring plants. For example, rye (*Secale cereale* L.), sorghum (*Sorghum bicolor* [L.] Moench), rice (*Oryza sativa* L.), sunflower (*Helianthus annuus* L.), rapeseed (*Brassica napus* L.), and wheat are crops known to have allelopathic effects on weeds (Jabran et al. 2015). Cool season crops should be rotated with warm season crops in order to interrupt cycles of weeds that are adapted to cropping cycles. Mixed cropping, sometimes called intercropping, is the agronomic practice of growing two or more crops in the same field at the same time. An inter-planted crop may provide weed control by allelopathy, smothering, or reducing the competitive ability of weeds (Strand 2000). Intercropping is widely practiced in smallholder farmer conservation agriculture systems across Latin America, Asia and Africa as a means of increasing crop production per unit land area with limited capital investment and minimal risk of total crop failure. It is not popular on commercial, large scale production farms because most modern field equipment and machinery is not designed for intercropping scenarios.

4.2.3 Reduced Tillage and Crop Residue Retention

Tillage of the soil is an old practice, dating back to 5,000–9,000 years ago (Lichtfouse et al. 2009) and it serves primarily for seedbed preparation and weed control. The harms of conventional tillage to both the soil and the environment are well documented (Sturz et al. 1997). Mechanical tillage implements destroy soil structure by reducing aggregate size and currently, conventional tillage is a major cause of soil erosion and desertification in many developing countries. Deep tillage accelerates loss of soil organic matter and soil organisms, which are important for soil tilth and water infiltration. The soil then becomes vulnerable to compaction, requiring even more deep tillage. Thus, the use of the plough has proven to be unsustainable. Conservation tillage is a generic term that describes any tillage practice that reduces loss of soil and water, and includes minimum tillage and no-tillage or zero tillage (Sturz et al. 1997). Conservation tillage may also be defined as those practices leaving more than 30% crop residue cover on the soil surface after planting (Dumanski et al. 2006). Reduced tillage causes a buildup of weed seeds on the soil surface (Swanton et al. 2000), where seed fate is strongly influenced by the presence of crop residues. In theory, weeding requirement under conservation agriculture is expected to decline over time as permanent soil cover prevents weed emergence and cause a decline of weed seedbanks (FAO 2008, Dube et al. 2012). Organic mulches also encourage biotic activity which increases loss of weed seeds through predation and non-recruitment of new seeds (Power et al. 1986).

Soils in many conventional tillage systems of resource poor farmers contain a large reservoir of seeds of problematic weeds and in severe cases; they result in abandonment of fields. High costs of production stemming from increased weeding requirements in the early phases of adoption are regarded as a hindrance to widespread conservation agriculture adoption in smallholder farmer systems of Africa (Giller et al. 2009). Invention of the cheap herbicide, glyphosate N-(phosphonomethyl) glycine, together with glyphosate resistant crops may not have solved this problem. Recent extensive studies of the potential toxicity of glyphosate to biological systems have revealed several direct and indirect harmful effects of this herbicide. Glyphosate, was deemed to be "probably cancerogenic to humans" by the World Health Organization's cancer research body in 2015 (Guyton et al. 2015), reflecting findings of various scientific studies that falsified the herbicide producer's claim that this substance was harmless to humans and to animals. Current estimates are that 237 weed species have developed resistance to herbicides, with resistance reported in 61 countries and 66 crops to 155 different herbicides (Heap 2014). In resource poor farming communities of Africa, there is a lack of technical knowhow on herbicide application. Even when the cash is available, the smallholder farmers may be reluctant to spend it on herbicides because of other priorities. Therefore, continued promotion of integration of reduced herbicide dosages with other non-chemical methods is essential for improving pest management in conservation agriculture systems.

4.3 Integration of Conservation Agriculture Principles in Weed Management

For the benefits of conservation agriculture to accrue, the general recommendation is that the three principles in terms of crop rotation, reduced tillage and permanent soil cover through cover crops must not be applied singly in the system, but holistically (Dumanski et al. 2006). For the purposes of adequate weed control, it may be important for farmers who adopt conservation agriculture to embrace integrated pest management. Integrated pest management is the optimization of pest control in an economically and ecologically sound manner, accomplished by the coordinated use of multiple tactics to assure stable crop production and to maintain pest damage below the economic injury level while minimizing hazards to humans, animals, plants, and the environment (Kogan 1998). Under integrated pest management, natural enemies, cultural practices, resistant crops, microbial agents, genetic manipulation and pesticides become mutually augment active (Flint and Van den Bosch 1981). In the following sections, a discussion of some of the integrated pest management practices which can complement conservation agriculture principles to enhance weed control is provided.

4.3.1 Managed Application of Fertilizer for Weed Management in Conservation Agriculture

One of the major objectives of conservation agriculture is to reduce dependency on external fertilizer inputs. The need to manage fertilizer correctly has been recently proposed to be a fourth principle of conservation agriculture (Vanlauwe et al. 2014). Fertilizer timing and placement strongly influences crop competition; and band application of fertilizer has the potential to enhance not only fertilizer use efficiency, but also crop competitive ability against weeds (Mohler 2001; Blackshaw et al. 2007). The optimal fertilizer type, application rate, and placement to favour the crop will depend upon crop type and growth stage, weed species and environmental conditions. In a situation where there is a high density of weeds in a crop, added nutrients favour weed growth, often reducing fertilizer use efficiency and yield benefits. Banding fertilizers within the crop row of bean (Otabbong et al. 1991), soybean and wheat (Cochran et al. 1990) not only lowered weed populations compared to broadcast applications, but also increased crop yield.

Otabbong et al. (1991) compared the effect of weeds on bean yield using three fertilization methods; broadcast application, surface banding (5 cm strip) on seed row, and deep banding within the seed row 7 cm below seed level. Surface banding in the crop row had little beneficial effect on bean yield and weed suppression, and even reduced bean yield in non-weeded plots. This was presumably due to increased access by weeds growing in the crop row to concentrated levels of nutrients. In contrast, deep banding of the fertilizer in the crop seed row significantly increased bean biomass and yield, particularly in non-weeded plots, while also suppressing weed biomass by 44%. Similar results were reported for deep placement of fertilizers in rice (Moody 1981). Hence, for weed control, the best results are obtained with deep banding of fertilizer rather than surface application. This implies that conservation agriculture should be promoted together with planting implements that allow deep banding of fertilizer, and the practice of surface application of fertilizer must be discouraged. It should be noted that reduced tillage tends to improve the fertility of the top soil, mostly because of the decomposition of crop residue mulches on the surface. If weeds are not managed well, this soil organic matter rich surface layer creates favorable nutrient conditions for weeds that emerge between the crop rows, thus worsening the competitive ability of weeds in the system.

4.3.2 Cultivar Choice, Planting Dates and Planting Patterns

In addition to cover crops, crop rotation and good fertilizer management, well-adapted varieties that are established at adequate plant populations are helpful in reducing weed problems. Vigorous cultivars are better competitors with weeds for resources, such as light, water and nutrients. As previously alluded to, weeds are highly competitive and successful organisms. Most weeds grow rapidly and tend to reproduce early, especially when they experience stress. Cultivars with an ability to grow relatively fast during early growth stages are particularly important for weed smothering. Selection of the largest seeds and seed priming are some of the means of providing crop plants a favourable starting position (Bastiaans et al. 2008). Tall cereal cultivars with planophile leaf inclination tend to increase light interception and weed suppression more efficiently than shorter erectophile ones (Drews et al. 2009). Tall grain crops, for example, are generally more competitive with weeds because they intercept light better (Finney and Creamer 2008). Cultivars that emerge quicker than weeds in the field also have a competitive advantage. In places where wild oat (*Avena fatua* L.) can be a serious problem in spring sown wheat, delayed planting has been shown to reduce wild oat densities and increase crop yields (Liebman and Dyck 1993). Proper selection of planting date, density and spatial arrangement of a crop can maximize the space it occupies early in the season and put competitive pressure on weeds.

Under conditions where water and nutrients are not limiting, increasing planting density can sometimes be used to improve crop yields. It is therefore important for farmers to adopt cultivars that can be planted at high density. This tactic is useful for weed management, although it can actually create conducive conditions for development of plant pathogens. Higher seed rates can compensate for crop losses that occur because of pest damage, or during weed control operations (Gunsolus et al. 2010). The establishment of a crop with a more uniform and dense plant distribution may result in better use of light, water and nutrients, and lead to greater crop competitive ability against weeds (Swanton and Weise 1991). Increasing plant density would lead to early canopy closure and thereby limit light penetration into the inter-row spaces and lead to the suppression of many dominant weeds.

Increased seed rates are beneficial mostly for the control of associated weeds, which are phenotypically similar to crops (Lemerle et al. 2001). For instance, annual ryegrass (*Lolium rigidum* Gaud.) has similar morphology and growth habit to wheat and is difficult to weed out by mechanical or chemical means. In a study from southern Queensland, sorghum planted at the density of 7.5 plants m^{-2} reduced weed seed production of Japanese millet (*Echinochloa esculenta* [A. Braun] H. Scholz) by 38% as compared with sorghum density of 4.0 plants m^{-2} (Wu et al. 2010). In multi-location trials conducted in Australia, higher wheat crop densities increased grain yield and reduced *L. rigidum* biomass while any reductions in grain size were negligible (Lemerle et al. 2004). The general recommendation from these trials was that high wheat plant densities suppressed *L. rigidum* and increased crop yield. Increasing crop density to suppress weeds through competition has also been found effective to reduce herbicide rates (Walker et al. 2002). High wheat densities up to 150 plants m^{-2} provided maximum grain yield and reduced the seed production of two important weeds, *Artemisia ludoviciana* Nutt. and *Phalaris paradoxa* L. whereas, the herbicide rate was reduced by 50% of the recommended rate (Walker et al. 2002). The adoption of increased seed rates has become an essential part of integrated weed management plans for grain growers in Australia, especially in areas where herbicide-resistant weeds are dominant (Bajwa et al. 2017).

In the US, increasing the seeding rate of wheat from 50 to 300 kg ha^{-1} reduced seed production of *Erodium cicutarium* by 95%, and increased wheat yields by 56–98% over the four years of the study (Blackshaw et al. 2000). Teasdale (1998) observed a 99% reduction in seed production of *Abutilon theophrasti* in maize planted at 128,000 compared to 64,000 plants ha^{-1}. Increasing maize planting density from 59,300 to 72,900 plants ha^{-1} reduced weed seed production by 50% and improved maize yields (Tharp and Kells 2001). Williams and Boydston (2013) also observed a decline from 72 to 27% in seed production of *Panicum miliaceum* L. with an increase in maize seeding rate from 35,000 to 105,000 seeds ha^{-1}. The need to increase plant densities may also be justified by the fact that under mulched surfaces in conservation agriculture systems, seed mortality tends to be high because of a proliferation of seedling pests such as slugs and cutworms.

4.3.3 *Sanitation*

Sanitation in crop production is an important practice whereby the food, water, shelter or other necessities that are important to the pest's survival are eliminated. Sanitation practices in weed management include the prevention of weed seed movement by mechanical or human means (McCarty and Murphy 1994). Many on-farm weed populations exist because of the natural movement of weed seeds and propagules from both neighbouring and distant populations by wind, animals, people, and other carriers (Finney and Creamer 2008). Practices such as use of clean seed and clean equipment are examples of good field sanitation (Svotwa and Jiyane 2006). Removing or destroying weeds in fields or near fields before they flower and release weed seed is also a good sanitation plan (Rasmussen 2004). Research shows that machinery and tools that were used in more than one location should be thoroughly cleaned before use in a different field, as they are the sources of weed seed dispersion. Animal manures should be properly exposed to high temperatures to destroy all viable weed seeds before they are used.

4.3.4 *Narrow Rows*

Row spacing manipulation can be a very useful weed management tool for integrated weed management in conservation agriculture systems. Research on use of narrow rows for weed management in field crops has largely shown superiority of weed control and in some cases, higher yields, when compared to wider rows. Decreasing the inter-row spacing can increase crop competitiveness with weeds (Teasdale 1995). Crops grown in narrower rows start competing with weeds at an earlier stage than those in wide rows because of more rapid canopy closure and probably better root distribution (Swanton and Weise 1991). The establishment of a crop with narrow rows, compensated for by wider intra-row spaces may result in better use of light, water, and nutrients and lead to greater crop competitive ability (Berkowitz and Rabin 1988; Minotti and Sweet 1981).

Reducing space between crop rows has also been discussed as a strategy to reduce the need for herbicides (Sankula et al. 2001). Teasdale (1995) concluded that growing maize in 38-cm rows and a double population may have potential for improving weed control in reduced-herbicide systems. Forcella et al. (1992) demonstrated that reduced-herbicide programs performed more consistently when maize was grown in 38-cm rather than 76-cm rows. A major limitation to the use of narrow rows maybe the need to accommodate field equipment, especially for large scale commercial farmers. They generally make the rows wide enough to accommodate tyre tracks, which in some cases may be up to 1 m wide for the large tractors. Conservation agriculture practitioners should also opt for planting equipment with adjustable row width, in order to be able to exploit the potential benefits of narrow rows. Table 4.3 presents information on superior weed suppression as reported in experiments with the major field crops through the use of narrow rows.

4.4 Animal Pest Management in Conservation Agriculture

The following sections outline the benefits and limitations of conservation agriculture in management of insects and other animal-type pests such as rodents, snails and slugs. Pest losses from insects across the world were estimated at 13% (Koul and Cuperus 2007). Animal type pests require a basic set of resources to live and reproduce, and production practices that deprive them of at least one needed element of life at a particular stage of the life cycle may eradicate or maintain the pest populations below economically damaging levels for extended periods (Linker et al. 2009).

4.4.1 Crop Rotations and Mixed Cropping to Manage Animal Pests in Conservation Agriculture Systems

Crops can be rotated in sequences that are not only profitable, but highly effective at breaking pest cycles. Crop rotation was reported as a highly effective tool to reduce maize pests such as rootworms (*Diabrotica virgifera virgifera*) and stalk borer

Table 4.3 Weed management and yield benefits due to narrow row spacing

Crop	Treatments compared	Major findings	Reference
Field peas (*Pisum sativum*)	Field peas planted in 36 and 18 cm rows	Control of annual ryegrass (*Lolium rigidum* Gaud.) improved substantially in 18 cm rows	Lemerle et al. (2006)
Soybean (*Glycine max*)	Soybean planted at 20 and 75 cm rows	Weed suppression in 20 cm rows was 60% higher than in 75 cm rows. Early planting in 20 cm rows was recommended for higher yield, effective weed suppression, and less lodging of the plant	Matsuo et al. (2015)
Sorghum (*Sorghum bicolor*) and sunflower (*Helianthus annuus* L.)	Sorghum and sunflower planted at 100 and 150 cm	Narrow rows (100 cm) reduced weed pressure and weed seed production in both crops. Yield penalty due to direct weed competition was reduced in 100 cm rows	Osten et al. (2006)
Cotton (*Gossypium hirsutum* L.)	Cotton planted in 19 cm twin rows (76 cm apart) was compared with the 76 cm single row planting pattern	Cotton planted in 19 cm twin rows had greater control of the weeds *Commelina benghalensis* L., *Senna obtusifolia* L. and *Jacquemontia tamnifolia* L. compared with the 76 cm single rows	Stephenson and Brecke (2010)
Wheat (*Triticum aestivum L.*)	Wheat planted in 11, 15 and 23 cm rows	Narrow spacing (11 cm) resulted in higher grain yield and reduced growth of the troublesome annual weeds *G. aparine* and *Lepidium sativum* L.	Fahad et al. (2015)
Sugarcane (*Saccharum officinarum*)	Sugarcane planted in 60, 90 and 120 cm rows	Lowest weed density was observed in 60 cm spaced crop rows compared to 90 and 120 cm rows. However cane yield decreased with reduced spacing between rows	Munsif et al. (2015)
Maize (*Zea mays L.*)	Maize planted in 15, 25 and 35 cm rows	Narrow rows (15 cm) resulted in 72 and 71% reduction in *Echinochloa colona* and *Trianthema portulacastrum* shoot biomass, respectively, and	Fahad et al. (2014)

(continued)

Table 4.3 (continued)

Crop	Treatments compared	Major findings	Reference
		showed 13% increase in maize grain yield compared to 35 cm rows	
Peanut (*Arachis hypogaea* L.)	Peanut planted in 81, 41 and 20 cm rows	Reducing peanut row spacing from 81 to 41 or 20 cm decreased weed biomass by 25–50%	Hauser and Buchanan (1982)
Snap beans (*Phaseolus vulgaris* L.)	Snap beans planted in 91, 36 and 15 cm rows	Row spacing of 15 and 36 cm suppressed weed growth by 18% relative to 91 cm spacing when weeds were allowed to emerge with snap beans	Teasdale and Frank (1983)
Mungbean (*Vigna radiata*)	Mungbean planted in 25, 50 and 75 cm rows over 2 seasons	Weed biomass in 25 and 50 cm rows was 35–46 g m^{-2} and 5–9 g m^{-2} compared to 117 g m^{-2} and 56 g m^{-2} in 75 cm rows for weeds allowed to grow beyond 3 and 6 weeks after planting, respectively	Chauhan et al. (2017)

(*Busseola fusca* Fuller) (Levine and Oloumi-Sadeghi 1991; Spencer et al. 2009). Maize rootworm eggs overwinter in maize fields and larvae are present to feed on maize roots the following year, and rotating to a different non-host crop such as soybeans denies food to hatching rootworm larvae (Spencer et al. 2009). For this strategy to be effective, it is generally recommended that rotations between susceptible crops should be at least three to seven years (Linker et al. 2009). This is especially important in conservation agriculture systems where farmers should retain crop residues on the soil surface as mulch.

Mixed crops also reduce pest populations by increasing environmental diversity. In some cases, intercropping lowers the overall attractiveness of the environment, as in the case where host and non-host plant species are mixed together in a single planting; while in other cases, intercropping may concentrate the pest in a smaller, more manageable area where it can be controlled by some other tactic. With intercropping, animal pests may feed preferentially on the second crop, or it may provide a more favourable habitat to increase natural enemies (Strand 2000). Intercropping is most effective against exogenous pests, such as locusts that enter the crop for only part of their life cycle (Hill 1983). A study by Jackai and Adalla (1997) showed that the population density of flower thrips was consistently lower in cowpea (*Vigna unguiculata* L. Walp.) intercropped with maize or sorghum.

With reference to mixed cropping patterns, trap crops are plant stands that are designed to attract, divert, intercept and/or retain targeted insects or the pathogens they vector in order to reduce damage to the main crop (Shelton and Badenes-Perez 2006). Plantings of highly-preferred host plants can arrest arriving pests and "trap" them, indirectly protecting less-attractive (to the pest), but economically-valuable, nearby crop species (Shelton and Badenes-Perez 2006). Prior to the introduction of modern synthetic insecticides, trap cropping was a common method of pest control for several cropping systems (Talekar and Shelton 1993). There is a resurgence of interest in trap cropping in conservation agriculture systems where one of the major objectives is to reduce reliance on pesticides. Conventional trap cropping is the most general practice of trap cropping, in which a trap crop planted next to a higher value crop is naturally more attractive to a pest as either a food source or oviposition site than is the main crop, thus preventing or making less likely the arrival of the pest to the main crop and/or concentrating it in the trap crop where it can be economically destroyed (Shelton and Badenes-Perez 2006). Dead-end trap cropping describes plants that are highly attractive to insects, but on which they or their offspring cannot survive (Shelton and Nault 2004). Dead-end trap crops serve as a sink for pests, preventing their movement from the trap crop to the main crop later in the season (Badenes-Perez et al. 2004). In sequential trap cropping, trap crops are planted earlier and/or later than the main crop to enhance the attractiveness of the trap crop to the targeted insect pest. Multiple trap cropping involves planting several plant species simultaneously as trap crops with the purpose of either managing several insect pests at the same time or enhancing the control of one insect pest by combining plants whose growth stages enhance attractiveness to the pest at different times (Shelton and Badenes-Perez 2006). Push-pull trapping (Khan et al. 2001) strategy is based on a combination of a trap crop with a repellent intercrop. In this case, the trap crop attracts the insect pest and, combined with the repellent intercrop, diverts the insect pest away from the main crop.

In most instances, trap cropping is focused on attracting and arresting the movement of adult insects, thus keeping them from moving to the cash crop (Caldwell et al. 2005). The trap crops can maintain the pest population to serve as a resource on which natural enemies can increase; natural enemies may suppress the pest population, preventing it from spilling over onto the cash crop, or the trap crop may serve as an initial source of natural enemies that move to the cash crop (Caldwell et al. 2005). One of the most widely cited examples of successful conventional trap cropping is alfalfa as a trap crop for lygus bugs (Hemiptera: Miridae) in cotton (Godfrey and Leigh 1994). This example is remarkable because it is still used today at the commercial level. Other examples of conventional trap cropping in commercial operation include the use of highly attractive varieties of squash (*Cucurbita* spp.) to manage insect pests in several cucurbitaceous crops (Pair 1997). Yellowrocket (*Barbarea vulgaris* var. *arcuata*) works as a dead-end trap crop for the diamond back moth (*Plutella xylostella*) (Badenes-Perez et al. 2005; Lu et al. 2004; Shelton and Nault 2004). Sunhemp (*Crotalaria juncea*) has also been suggested as a dead-end trap crop for the bean pod borer (*Maruca testulalis*) (Jackai and Singh 1983). Borders of early-planted potatoes (*Solanum tuberosum* L.) have been used as a trap crop for Colorado potato beetle, which moves to potato fields from overwintering sites next to the crop, becoming concentrated in the outer rows, where it can be treated with insecticides, cultural practices, or even propane flamers (Hoy et al. 2000). Similar success has been reportedly achieved in commercial fields with perimeter trap cropping for control of pepper maggot (*Zonosemata electa*), in bell peppers by using a trap crop of hot cherry peppers (Boucher et al. 2003).

Indian mustard is used as a trap crop for diamondback moth, which requires planting mustard two or three times through the cabbage season because Indian mustard has a shorter crop cycle than cabbage and other cole crops (Srinivasan and Krishna Moorthy 1991). An example of multiple trap cropping is a mixture of Chinese cabbage (*Brassica campestris* ssp. *pekinensis*), marigolds (*Tagetes* spp.), rapes, and sunflower that was successfully demonstrated to be an effective trap crop for the pollen beetle (*Melighetes aeneus*), in cauliflower (*Brassica oleracea* var. *botrytis*) fields in Finland (Hokkanen 1989). Other cases of multiple trap cropping are the use of a mixture of castor (*Ricinus communis*), millet, and soybean to control groundnut leafminer (*Aproarema medicella*) (Muthiah 2003) in India, and the use of maize and potato plants combined as a trap crop to control wireworms in sweet potato fields (Seal et al. 1992). Push-pull trap cropping based on using either napier grass (*Pennisetum purpureum* K. Schum) or sudan grass (*Sorghum vulgare* Pers.) as a trap crop planted around the main crop, and either *Desmodium* or molasses grass planted within the field as a repellent intercrop, greatly increased the effectiveness of trap cropping for stem borers in several countries in Africa (Khan et al. 2001). The use of molasses grass as a repellent intercrop enhances stem borer parasitoid abundance, thereby improving stem borer control (Khan et al. 2000). Stem borers are the most important biotic constraint to maize production in Africa, and the push-pull strategy was reported to be an important strategy which allowed small farmers to control them while managing various parasitic weed species in the genus *Striga* (Khan and Pickett 2004).

4.4.2 Cover Crops and Crop Residues Use in Animal Pest Management

Cover crops and crop residues are mostly beneficial for breaking pest cycles and enhancing the effectiveness of the natural enemies. Residues of various cover crop species were found to reduce insect pest abundance in cabbage fields (Bottenberg et al. 1997; Xu et al. 2011). Xu et al. (2011) found greater natural enemy densities in cabbage plots with plant residues present on the soil surface compared to bare soils. Mulched surfaces provide favorable microhabitats for beneficial insects such as carabids, staphylinids, and spiders (Altieri and Schmidt 1985) which can more effectively control pests. Several studies have also demonstrated that fire ants can be ecologically and economically important insect predators in a variety of cropping systems (Morrill 1977) and they are enhanced under mulched surfaces.

On the other hand, the combination of reduced tillage and crop residue mulch retention has, in many instances shown to increase the abundance of seedling pests such as slugs, cutworms and rodents (Gaylor et al. 1984, Douglas and Tooker 2012; Chiduza and Dube 2013; Fig. 4.2). Soil conditions under mulches tend to be cooler, and these slows down seedling emergence rate, allowing the seedling pests to inflict even more seedling damage. An option for overcoming this problem would be to breed for cultivars that have improved vigor and emergence under mulched surfaces. At present, there is no record of studies that have been carried out to evaluate the suitability of different crop cultivars for conservation agriculture with respect to emergence under mulched surfaces. Genetically modified cultivars, such as those with the Bt gene could also be useful in reducing seedling pest damage. Additionally, conservation agriculture farmers planting on mulched surface should be encouraged to increase seed rate in order to compensate for possible losses from seedling pests. Proponents of conservation agriculture have always argued that when it comes to crop residue biomass input, the more there is the better. However, issues relating to the lowering of soil temperatures especially in cooler areas and the proliferation of seedling pests brings up the question of 'adequate biomass for conservation agriculture'. Research is required in order to establish biomass input thresholds for different conservation agriculture systems at which the crop residue effect is diminished.

Fig. 4.2 Large patches of irrigated wheat seedlings (up to 30%) are damaged by rodents in conservation agriculture fields in KwaZulu-Natal, South Africa (2016). Burrows of these rodents were previously controlled by the plough, and these rodents have no natural enemies in this habitat. They have now emerged as a major pest of no-till wheat in the region. Photographs by E. Dube

4.5 Integration of Conservation Agriculture Principles with Other Measures for Improving Animal Pest Control in Field Crops

4.5.1 Managed Application of Fertiliser

Fertility practices that replenish and maintain high soil organic matter provide an environment that enhances plant health (McGuiness 1993). Soils with high organic matter and active soil biology generally exhibit good fertility as well as complex food webs and beneficial organisms (Magdoff and Van Es 2000). The practice of conservation agriculture, through increasing soil organic matter, is generally expected to increase plant resistance to pests. Plant fertility and water stress are known to play a major role in plant susceptibility to herbivore feeding, tolerance to insect injury, and insect population growth (Linker et al. 2009; Rebek et al. 2012). For example, excess nitrogen concentration within plants has a tendency to increase plant quality for herbivores (Rebek et al. 2012). Crops grown in more balanced, organic matter rich soils generally exhibit lower abundance of several insect herbivores, reductions that may be attributed to a lower nitrogen content in organically farmed crops (Altieri and Nicholls 2003). There is evidence which shows that the ability of a crop to resist or tolerate pests is tied to optimal physical, chemical and mainly biological properties of soils (Altieri and Nicholls 2003). More balanced soil fertility due to increased diversity of cropping systems under conservation agriculture should thus help reduce the incidence of insect pests. For potato crops, increasing nitrogen fertilization increases leaf consumption, reduces development time, and increases abundance of potato beetles (Rebek et al. 2012).

4.5.2 Cultivar Choice

Because most crops are susceptible to animal pests only during certain stages of growth and many pests are present only for a few days or weeks of the year, pest attack can be avoided by simply choosing the correct planting date (Pingali 1993). Farmers can choose varieties that can be planted or harvested early while still achieving a full yield in order to avoid pests. This cultivar choice depends upon the farmer knowing the emergence times and life cycles of the pests to be controlled (Sarwar 2015). Early maize cultivars are usually more exposed to stalk borer damage than later planted ones. Delayed planting of summer crops in some cases helps to avoid seed and root rots, and promote vigorous growth (Watson et al. 2015). Late-season or early-season pest problems may be avoided by planting shorter season varieties (Strand 2000). Early sowing is used as a management tactic against cotton lygus (*Taylorilygus vosseleri*) and sorghum midge (*Contarinia sorghicola*) in Africa (Hill 1983). Farmers can also choose early cultivars that mature before the pest is abundant, or synchronize the pest with its natural enemies and climatic conditions that would adversely affect the pest (Strand 2000). The shorter the time a crop is in the field, the less time animal pests have to damage it; combining early planting with early maturing varieties generally allows a crop to mature before animal pests reach damaging levels (Linker et al. 2009).

Host plant resistance is a major tactic for the control of animal pests of field crops. Crop cultivars can be made resistant to insect pests because of specific resistant genes that are incorporated during breeding. Plant resistance to insects was reported as the most effective component of management for the Hessian fly and wheat stem sawfly (US Congress 1979). Use of high yielding resistant varieties has been the major technique for improving productivity of many staple food crops such as wheat and maize across the world. Pest resistant plants provide a natural, economic, environmentally safe, self-generating system that is compatible with conservation agriculture. With a highly resistant variety, a crop pest can be managed with no health risks to the farmer, farm workers, or the environment. Nowadays, many of the most damaging insect pests are effectively controlled by genetic resistance, and cultivars of major crops are high yielding because they possess resistance to a particular pest common to their production environment. Soybean yields are improved by as much as 30–40% simply by selecting the proper resistant variety (Vincelli 1994).

If cultivar resistance is not integrated with other control measures, pests have an ability to overcome the resistance. A good example is the recently reported resistance of *Busseola fusca* (Lepidoptera: Noctuidae) to Bt maize in the Vaalharts area of South Africa (Kruger et al. 2009). This resistance is thought to have emerged because of poor crop production practices by the farmers such as continuous maize monocropping, which enabled the stalk bores to have continuous, uninterrupted breeding cycles. The key concepts in host-plant resistance to animal pests are: (1) escape, which entails avoidance of pest problems due to inter-varietal phonological differences; (2) non-preference, whereby the host-plant variety does not attract pests; and (3) antibiosis, whereby the host-plant variety reduces viability of pests (Bugg 1992). At a higher integrative level, the aim should not be to achieve merely host-plant resistance, but agroecosystem resistance.

4.6 Conclusion

Worldwide, the demand for clean, pesticide free food has expanded quickly in recent years, stimulated by increasing consumer education and awareness regarding the harms of pesticides. Conservation agriculture has been proven as the best approach for sustainable crop production on many production lands across the world. Among other factors, serious pest management challenges hamper wide scale adoption of conservation agriculture, especially in the transitional stage. In this chapter, we explored some tactics that can be used for pest management within the context of conservation agriculture principles. Farmers practicing conservation agriculture have several cultural methods that they can put together to build up a good pest management strategy such as proper application of cultivar choice, timely planting dates, increased planting density, and irrigation and fertilizer management. Well planned sequences of crops in which each crop differs radically from its predecessor can be beneficial in the control of weeds and animal pests. Intercropping components are better than monoculture in reducing damage by weeds and insect pests. Much research work is still needed in order to refine the integration of cover crops, crop rotation and reduced tillage with other pest control tactics such as narrow rows, fertilizer application methods and reduced pesticide dosages for effective weed control in conservation agriculture. A more inclusive, integration approach of control tactics in conservation agriculture, which goes beyond the three principles is essential for effective pest management.

References

Altieri MA, Nicholls CI (2004) Biodiversity and pest management in agroecosystems. Food Products Press. An imprint of Haworth Press Inc., Binghamton, NY

Altieri MA, Nicholls CI (2003) Soil fertility management and insect pests: harmonizing soil and plant health in agroecosystems. Soil Till Res 72(2):203–211. https://doi.org/10.1016/S0167-1987(03)00089-8

Altieri MA, Schmidt LL (1985) Cover crop manipulation in northern California orchards and vineyards: effects on arthropod communities. Biol Agric Hortic 3:1–24. https://doi.org/10.1080/01448765.1985.9754453

Badenes-Perez FR, Shelton AM, Nault BA (2004) Evaluating trap crops for diamondback moth, *Plutella xylostella* (Lepidoptera: Plutellidae). J Econ Entomol 97:1365–1372. https://doi.org/10.1093/jee/97.4.1365

Badenes-Perez FR, Shelton AM, Nault BA (2005) Using yellow rocket as a trap crop for the diamondback moth, Plutella xylostella (L.) (Lepidoptera: Plutellidae). J Econ Entomol 98:884–890. https://doi.org/10.1603/0022-0493-98.3.884

Bajwa AA, Walsh M, Chauhan BS (2017) Weed management using crop competition in Australia. Crop Prot 95:8–13. https://doi.org/10.1016/j.cropro.2016.08.021

Bastiaans L, Paolini R, Baumann DT (2008) Focus on ecological weed management: what is hindering adoption? Weed Res 48(6):481–491. https://doi.org/10.1111/j.1365-3180.2008.00662.x

Berkowitz GA, Rabin J (1988) Antitranspirant associated abscisic acid effects on the water relations and yield of transplanted bell peppers. Plant Physiol 86:329–331. https://doi.org/10.1104/pp.86.2.329

Blackshaw RE, Anderson RL, Lemerle D (2007) Cultural weed management. In: Upadhyaya MK, Blackshaw RE (Eds) Non-chemical weed management. principles, concepts and technology. CABI, Wallingford, UK, pp 35–47

Blackshaw RE, Semach G, O'Donovan JT (2000) Utilization of wheat seed rate to manage redstem filaree (*Erodium cicutarium*) in a zero tillage cropping system. Weed Technol 14:389–396. https://doi.org/10.1614/0890-037X(2000)014[0389:UOWSRT]2.0.CO;2

Bottenberg H, Masiunas J, Eastman C, Eastburn D (1997) Yield and quality constraints of cabbage planted in rye mulch. Biol Agric Hortic 14:323–342. https://doi.org/10.1080/01448765.1997.9755168

Boucher TJ, Ashley R, Durgy R, Sciabarrasi M, Calderwood W (2003) Managing the pepper maggot (Diptera: Tephritidae) using perimeter trap cropping. J Econ Entomol 96:420–432. https://doi.org/10.1603/0022-0493-96.2.420

Brainard DC, Bryant A, Noyesa DC, Haramoto ER, Szendrei Z (2016) Evaluating pest-regulating services under conservation agriculture: a case study in snap beans. Agric Ecosyst Environ 235:142–154. https://doi.org/10.1016/j.agee.2016.09.032
Bugg RL (1992) Using cover crops to manage anthropods on truck farms. HortScience 27: 741–745
Caldwell BA, Sideman E, Seaman A, Brown Rosen E, Shelton AM, Smart CD (2005) Resource guide to organic insect and disease management. New York State Agricultural Experiment Station. http://web.pppmb.cals.cornell.edu/resourceguide/pdf/resource-guide-for-organic-insect-and-disease-management.pdf. Accessed 28 December 2016
Chauhan BS, Florentine SK, Ferguson JC, Chechetto RG (2017) Implications of narrow crop row spacing in managing weeds in mungbean (*Vigna radiata*). Crop Prot 95:116–119. https://doi.org/10.1016/j.cropro.2016.07.004
Chiduza C, Dube E (2013) Maize production challenges in high biomass input smallholder farmer conservation agriculture systems: a practical research experience from South Africa. African Crop Sci Conf Proc 11:23–27
Clark A (2007) Managing cover crops profitably, 3rd edn. Sustainable Agriculture Network. United Book Press, Inc., p 247. http://www.sare.org/publications/covercrops/covercrops.pdf. Accessed 24 Feb 2016
Cochran VL, Morrow LA, Schirman RD (1990) The effect of N placement on grass weeds and winter wheat in three tillage systems. Soil Till Res 18:347–355. https://doi.org/10.1016/0167-1987(90)90119-X
Conway GR, Pretty J (1991) Unwelcome harvest: agriculture and pollution. Earthscan, London
Douglas MR, Tooker JF (2012) Slug (Mollusca: Agriolimacidae, Arionidae) ecology and management in no-till field crops, with an emphasis on the mid-Atlantic region. J Integr Pest Manag 3(1):C1–C9. https://doi.org/10.1603/IPM11023
Drews S, Neuhoff D, Köpke U (2009) Weed suppression ability of three winter wheat varieties at different row spacing under organic farming conditions. Weed Res 49(5):526–533. https://doi.org/10.1111/j.1365-3180.2009.00720.x
Dube E, Chiduza C, Muchaonyerwa P, Fanadzo M, Mthoko TS (2012) Winter cover crops and fertiliser effects on the weed seed bank in a low-input maize-based conservation agriculture system. S Afr J Plant Soil 29(3–4):195–197. https://doi.org/10.1080/02571862.2012.730637
Dumanski J, Peiretti R, Benetis J, McGarry D, Pieri C (2006) The paradigm of conservation tillage. In: Proceedings of world association of soil and water conservation, FAO, Rome, pp 58–64
Eskenazi B, Marks AR, Bradman A, Harley K, Barr DB, Johnson C, Morga N, Jewell NP (2007) Organophosphate pesticide exposure and neurodevelopment in young Mexican-American children. Environ Health Perspect 115(5):792–798. https://doi.org/10.1289/ehp.9828
Fahad S, Hussain S, Chauhan BS, Saud S, Wu C, Hassan S, Huang J (2015) Weed growth and crop yield loss in wheat as influenced by row spacing and weed emergence times. Crop Prot 71:101–108. https://doi.org/10.1016/j.cropro.2015.02.005
Fahad S, Hussain S, Saud S, Hassan S, Muhammad H, Shan D, Chen C, Wu C, Xiong D, Khan SB, Jan A, Cui K, Huang J, Zwerger P (2014) Consequences of narrow crop row spacing and delayed Echinochloa colona and Trianthema portulacastrum emergence for weed growth and crop yield loss in maize. Weed Res 54:475–483. https://doi.org/10.1111/wre.12104
FAO (2008) Investing in Sustainable Agricultural Intensification: The role of conservation agriculture. A Framework for Action. Food and Agriculture Organization of the United Nations, Rome
Finney DM, Creamer NG (2008) Weed management on organic farms. The Organic Production Publication Series, CEFS, p 1–34
Flint ML, Van den Bosch R (1981) Introduction to Integrated Pest Management. Springer, New York Inc. https://doi.org/10.1007/978-1-4615-9212-9
Forcella F, Westgate ME, Warnes DD (1992) Effect of row width on herbicide and cultivation requirements in row crops. Am J Alt Agric 7:161–167. https://doi.org/10.1017/S0889189300004756

Gaylor JG, Fleischer SJ, Muehleisen DP, Edelson JV (1984) Insect populations in cotton produced under conservation tillage. J Soil Water Conserv 39:61–64
Giller KE, Witter E, Corbeels M, Tittonell P (2009) Conservation agriculture and smallholder farming in Africa: the heretics' view. Field Crop Res 114:23–34. https://doi.org/10.1016/j.fcr.2009.06.017
Glen DM, Symondson WOC (2003) Influence of soil tillage on slugs and their natural enemies. In: El Titi A (ed) The role of soil tillage in agroecosystems. CRC Press, Boca Raton, USA, pp 207–227
Godfrey LD, Leigh TF (1994) Alfalfa harvest strategy effect on Lygus bug (Hemiptera: Miridae) and insect predator population density: implications for use as trap crop in cotton. Environ Entomol 23:1106–1118. https://doi.org/10.1093/ee/23.5.1106
Gregory WW, Musick GJ (1976) Insect management in reduced tillage systems. Bull Entomol Soc Am 22:302–304. https://doi.org/10.1093/besa/22.3.302
Gunsolus J, Wyse D, Moncada K, Fernholz C (2010) Weed management. In: Moncada KM (ed)
Guyton KZ, Loomis DY, El Ghissassi F, Benbrahim-Tallaa L, Guha N, Scoccianti C, Mattock H, Straif K, International Agency for Research on Cancer Monograph Working Group, IARC, Lyon, France (2015) Carcinogenicity of tetrachlorvinphos, parathion, malathion, diazinon, and glyphosate. Lancet Oncol 16(5):490–491
Hauser EW, Buchanan GA (1982) Production of peanuts as affected by weed competition and row spacing. Alabama Agric Exp Stn Bull 538:35
Heap I (2014) Global perspective of herbicide-resistant weeds. Pest Manag Sci 70(9):1306–1315. https://doi.org/10.1002/ps.3696
Henderson DR, Riga E, Ramirez RA, Wilson J, Snyder WE (2009) Mustard biofumigation disrupts biological control by Steinernema spp. nematodes in the soil. J Nematol 41(4):337–337. https://doi.org/10.1016/j.biocontrol.2008.12.004
Hill DS (1983) Agricultural insect pests of the tropics and their control. Cambridge University Press, London
Hill DS (2008) Pests of crops in warmer climates and their control. Springer, Dordrecht, The Netherlands
Hokkanen HMT (1989) Biological and agrotechnical control of the rape blossom beetle Meligethes aeneus (Coleoptera: Nitidulidae). Acta Entomol Fenn 53:25–30
Hoy CW, Vaughn TT, East DA (2000) Increasing the effectiveness of spring trap crops for Leptinotarsa decemlineata. Entomol Exp Appl 96:193–204. https://doi.org/10.1046/j.1570-7458.2000.00697.x
Jabran K, Mahajan G, Sardana V, Chauhan BS (2015) Allelopathy for weed control in agricultural systems. Crop Prot 72:57–65. https://doi.org/10.1016/j.cropro.2015.03.004
Jackai LEN, Adalla CB (1997) Pest management practices in cowpea: a review. In: Singh BB, Mohan Raj DR, Dashiell KE, Jackai LEN (eds) Advances in cowpea research. Copublication of International Institute of Tropical Agriculture (IITA) and Japan International Research Center for Agricultural Sciences (JIRCAS), pp 240–257
Jackai LEN, Singh SR (1983) Suitability of selected leguminous plants for development of *Maruca testulalis* larvae. Entomol Exp Appl 34:174–178. https://doi.org/10.1111/j.1570-7458.1983.tb03314.x
Johnson TB, Turpin FT, Schreiber MM, Griffith DR (1984) Effects of crop rotation, tillage, and weed management systems on black cutworm (Lepidoptera: Noctuidae) infestations in corn. J Econ Entomol 77(4):919–921. https://doi.org/10.1093/jee/77.4.919
Khan ZR, Pickett JA (2004) The 'push–pull' strategy for stemborer management: a case study in exploiting biodiversity and chemical ecology. In: Gurr GM, Wratten SD, Altieri MA (eds) Ecological engineering for pest management: advances in habitat manipulation for arthropods. CABI Publishing, Wallingford, Oxon, UK
Khan ZR, Pickett JA, Van den Berg J, Wadhams LJ, Woodcock CM (2000) Exploiting chemical ecology and species diversity: stem borer and striga control for maize and sorghum in Africa. Pest Manag Sci 56:957–962. https://doi.org/10.1002/1526-4998(200011)56:11<957:AID-PS236>3.0.CO;2-T

Khan ZR, Pickett JA, Wadhams L, Muyekho F (2001) Habitat management strategies for the control of cereal stemborers and striga in maize in Kenya. Insect Sci Appl 21:375–380. https://doi.org/10.1017/S1742758400008481

Kogan M (1998) Integrated pest management: historical perspectives and contemporary developments. Annu Rev Entomol 43(1):243–270. https://doi.org/10.1146/annurev.ento.43.1.243

Koul O, Cuperus GW (2007) Ecologically based integrated pest management. CABI Publishing, Wallingford

Kruger M, Van JBJ, Van den Berg J (2009) Perspective on the development of stem borer resistance to Bt maize and refuge compliance at the Vaalharts irrigation scheme in South Africa. Crop Prot 28:684–689. https://doi.org/10.1016/j.cropro.2009.04.001

Lemerle D, Cousens RD, Gill GS, Peltzer SJ, Moerkerk M, Murphy CE, Collins D, Cullis BR (2004) Reliability of higher seeding rates of wheat for increased competitiveness with weeds in low rainfall environments. J Agric Sci 142:395–409. https://doi.org/10.1017/S002185960400454X

Lemerle D, Gill GS, Murphy CE, Walker SR, Cousens RD, Mokhtari S, Peltzer SJ, Coleman R, Luckett DJ (2001) Genetic improvement and agronomy for enhanced wheat competitiveness with weeds. Crop Past Sci 52:527–548. https://doi.org/10.1071/AR00056

Lemerle D, Verbeek B, Diffey S (2006) Influences of field pea (*Pisum sativum*) density on grain yield and competitiveness with annual ryegrass (*Lolium rigidum*) in south-eastern Australia. Aust J Exp Agric 46:1465–1472. https://doi.org/10.1071/EA04233

Levine E, Oloumi-Sadeghi H (1991) Management of diabroticite rootworms in corn. Annu Rev Entomol 36:229–255. https://doi.org/10.1146/annurev.en.36.010191.001305

Lichtfouse E, Navarrete M, Debaeke P, Souchere V, Alberola C, Menassieu J (2009) Agronomy for sustainable agriculture: a review. Agron Sustain Dev 29:1–6. https://doi.org/10.1051/agro:2008054

Liebman M, Dyck E (1993) Crop rotation and intercropping strategies for weed management. Ecol Appl 3(1):92–122. https://doi.org/10.2307/1941795

Linker HM, Orr DB, Barbercheck ME (2009) Insect Management on Organic Farms. North Carolina Cooperative Extension Service. https://cefs.ncsu.edu/wp-content/uploads/insectmgmtfinaljan09.pdf?x47549. Accessed 27 December 2016

Lu J, Liu YB, Shelton AM (2004) Laboratory evaluations of a wild crucifer Barbarea vulgaris as a management tool for diamondback moth. Bull Entomol Res 94:509–516. https://doi.org/10.1079/BER2004328

Magdoff F, Van Es H (2000) Building soils for better crops. SARE, Washington, DC

Marks CF, Townshend JL (1973) Multiplication of the root lesion nematode *Pratylynchus penetrans* under orchard cover crops. Can J Plant Sci 53:187–188. https://doi.org/10.4141/cjps73-034

Matsuo N, Yamada T, Hajika M, Fukami K, Tsuchiya S (2015) Planting date and row width effects on soybean production in Southwestern Japan. Agron J 107:415–424. https://doi.org/10.2134/agronj14.0268

McCarty LB, Murphy TR (1994) Control of turfgrass weeds. In: Turgeon AJ, Kral DM, Viney MK (eds) Turfgrass weeds and their control. ASA and CSSA, Madison, Wisconsin

McGuiness H (1993) Living soils: sustainable alternatives to chemical fertilizers or developing countries. Consumers Policy Institute, New York

Menalled F (2008) Weed seedbank dynamics and integrated management of agricultural weeds. Bozeman: Extension Publication, Montana State University. http://www.msuextension.org/publications/AgandNaturalResources/MT200808AG.pdf. Accessed 9 Jan 2017

Minotti PL, Sweet RD (1981) Role of crop competition in limiting losses from weeds. In: Pimentel D (ed) CRC Handbook of pest management in agriculture, vol 2. CRC Press Inc, Boca Raton, Florida

Mohler CL (2001) Enhancing the competitive ability of crops. In: Liebman M, Mohler CL, Staver CP (eds) Ecological management of agricultural weeds. Cambridge University Press, Cambridge, UK

Mohler CL, Callaway MB (1995) Effects of tillage and mulch on weed seed production and seed banks in sweet corn. J Appl Ecol 32:627–639. https://doi.org/10.2307/2404658

Moody K (1981) Weed-fertilizer interactions in rice. IRRI research paper series, No, p 68

Morrill WL (1977) Red imported fire ant control with diazinon and chlorpyrifos drenches. J Georgia Entomol Soc 12:96–100

Munsif F, Ali K, Khalid S, Ali A, Ali M, Ahmad M, Ahmad W, Ahmad I, Basir A (2015) Influence of row spacing on weed density, biomass and yield of chip bud settling of sugarcane. Pak. J Weed Sci Res 21:137–144

Muthiah C (2003) Integrated management of leafminer (Aproaerema modicella) in groundnut (*Arachis hypogaea*). Ind J Agric Sci 73:466–468

Oerke EC, Dehne HW (2004) Safeguarding production: Losses in major crops and the role of crop protection. Crop Prot 23:275–285. https://doi.org/10.1016/j.cropro.2003.10.001

Oliveira CM, Auad AM, Mendes SM, Frizzas MR (2014) Crop losses and the economic impact of insect pests on Brazilian agriculture. Crop Prot 56:50–54. https://doi.org/10.1016/j.cropro.2013.10.022

Osten V, Wu H, Walker S, Wright G, Shields A (2006) Weeds and summer crop row spacing studies in Queensland. In: Preston C, Watts JH, Crossman ND (eds) Proceedings of the 15th Australian weeds conference, 24–28 Sep 2006. Adelaide, South Australia, pp 347–350

Otabbong E, Izquierdo MML, Talavera SFT, Geber UH, Ohlander LJR (1991) Response to P fertilizer of *Phaseolus vulgaris* L. growing with or without weeds in a highly P-fixing mollic Andosol. Trop Agric 68:339–343

Pair SD (1997) Evaluation of systemically treated squash trap plants and attracticidal baits for early-season control of striped and spotted cucumber beetles (Coleoptera: Chrysomelidae) and squash bug (Hemiptera: Coreidae) in cucurbit crops. J Econ Entomol 90:1307–1314. https://doi.org/10.1093/jee/90.5.1307

Pimentel D (1995) Amounts of pesticides reaching target pests: environmental impacts and ethics. J Agric Environ Ethics 8(1):17–29

Pingali PL (1993) Pesticides, rice productivity, and farmers' health: an economic assessment. Plant Dis 70:906–911

Power JF, Wilhelm WW, Doran JW (1986) Crop residue effects on soil environment and dryland maize and soya bean production. Soil Till Res 8:101–111. https://doi.org/10.1016/0167-1987(86)90326-0

Rashid A, Nawaz S, Barker H, Ahmad I, Ashraf M (2010) Development of a simple extraction and clean-up procedure for determination of organochlorine pesticides in soil using gas chromatography–tandem mass spectrometry. J Chromatogr A 1217:2933–2939. https://doi.org/10.1016/j.chroma.2010.02.060

Rasmussen IA (2004) The effect of sowing date, stale seedbed, row width and mechanical weed control on weeds and yields of organic winter wheat. Weed Res 44:12–20. https://doi.org/10.1046/j.1365-3180.2003.00367.x

Rebek EJ, Frank SD, Royer TA, Bográn CE (2012) Alternatives to chemical control of insect pests. Insecticides–basic and other applications, pp 171–196. http://cdn.intechopen.com/pdfs-wm/27804.pdf. Accessed 27 Dec 2016

Sankula S, VanGessel MJ, Kee WE, Beste CE, Everts KL (2001) Narrow row spacing does not affect lima bean yield or management of weeds and other pests. HortScience 36:884–888

Sarwar M (2015) Mechanical Control Prospectus to Aid in Management of Fruit Flies and Correlated Tephritid (Diptera: Tephritidae) Pests. Int J Anim Biol 1(5):190–195

Sattell R, Dick R, Mcgrath D (1998) Faba bean (*Vicia faba* L.). Oregon State University Extension Service, P 2

Seal DR, Chalfant RB, Hall MR (1992) Effects of cultural practices and rotational crops on abundance of wireworms (Coleoptera: Elateridae) affecting sweet potato in Georgia. Environ Entomol 21:969–974. https://doi.org/10.1093/ee/21.5.969

Shelton AM, Badenes-Perez FR (2006) Concepts and applications of trap cropping in pest management. Annu Rev Entomol 51:285–308. https://doi.org/10.1146/annurev.ento.51.110104.150959

Shelton AM, Nault BA (2004) Dead-end trap cropping: a technique to improve management of the diamondback moth, *Plutella xylostella* (Lepidoptera: *Plutellidae*). Crop Prot 23(6):497–503. https://doi.org/10.1016/j.cropro.2003.10.005

Sipes BS, Arakaki AS (1997) Root-knot nematode management in dryland taro with tropical cover crops. Suppl J Nematol 29:721–724

Smith B (2006) The farming handbook. University of KwaZulu-Natal Press, South Africa, p 431

Spencer JL, Hibbard BE, Moeser J, Onstad DW (2009) Behaviour and ecology of the Western corn rootworm (*Diabrotica virgifera* LeConte). Agr Forest Meteorol 11(1):9–27. https://doi.org/10.1111/j.1461-9563.2008.00399.x

Srinivasan K, Krishna Moorthy PN (1991) Indian mustard as a trap crop for management of major lepidopterous pests on cabbage. Trop Pest Manag 37:26–32. https://doi.org/10.1080/09670879109371532

Stehle S, Schulz R (2015) Agricultural insecticides threaten surface waters at the global scale. PNAS 112:5750–5755

Stephenson DO, Brecke BJ (2010) Weed management in single- versus twin-row cotton (*Gossypium hirsutum*). Weed Technol 24:275–280. https://doi.org/10.1614/WT-D-09-00056.1

Strand JF (2000) Some agrometeorological aspects of pest and disease management for the 21st century. Agr Forest Meteorol 103(1):73–82. https://doi.org/10.1016/S0168-1923(00)00119-2

Sturz AV, Carter MR, Johnston HW (1997) A review of plant disease, pathogen interactions and microbial antagonism under conservation tillage in temperate humid agriculture. Soil Till Res 41(3):169–189. https://doi.org/10.1016/S0167-1987(96)01095-1

Svotwa E, Jiyane J, Ndangana F (2006) Integrated weed management: a possible solution to weed problems in Zimbabwe. In: Muchabayiwa B, Trimble J, Dube S (eds.) Proceedings from the 2nd international conference on appropriate technology. National University of Science and Technology, Bulawayo, Zimbabwe, July 12–15, 2006

Swanton CJ, Weise SF (1991) Integrated weed management: the rationale and approach. Weed Technol:657–663. https://doi.org/10.1017/S0890037X00027512

Swanton CJ, Shrestha A, Knezevic SZ, Roy RC, Ball-Coelho BR (2000) Influence of tillage type on vertical seed bank distribution in a sandy soil. Can J Plant Sci 80:455–457. https://doi.org/10.4141/P99-020

Talekar NS, Shelton AM (1993) Biology, ecology, and management of the diamondback moth. Annu Rev Entomol 38:275–301. https://doi.org/10.1146/annurev.en.38.010193.001423

Teasdale JR (1996) Contribution of cover crops to weed management in sustainable agricultural systems. J Prod Agric 9:475–479. https://doi.org/10.2134/jpa1996.0475

Teasdale JR, Frank JR (1983) Effect of row spacing on weed competition with snap beans (*Phaseolus vulgaris*). Weed Sci 31:81–85. https://doi.org/10.1017/S0043174500068582

Teasdale J (1998) Influence of corn (Zea mays) population and row spacing on corn and velvetleaf (*Abutilon theophrasti*) yield. Weed Sci 46:447–453

Teasdale J (1995) Influence of narrow row/high population corn on weed control and light transmittance. Weed Technol 9:113–118

Tharp BE, Kells JJ (2001) Effect of glufosinate-resistant corn (*Zea mays*) population and row spacing on light interception, corn yield, and common lambsquarters (*Chenopodium album*) growth. Weed Technol 15:413–418. https://doi.org/10.1614/0890-037X(2001)015[0413:EOGRCZ]2.0.CO;2

US Congress (1979) Pest Management Strategies in Crop Production. Office of Technology Assessment, October 1979

Van der Werf HMG (1996) Assessing the impact of pesticides on the environment. Agric Ecosyst Environ 60:81–96. https://doi.org/10.1016/S0167-8809(96)01096-1

Vanlauwe B, Wendt J, Giller KE, Corbeels M, Gerard B, Nolte C (2014) A fourth principle is required to define conservation agriculture in sub-Saharan Africa: the appropriate use of fertiliser to enhance crop productivity. Field Crops Res 155:10–13. https://doi.org/10.1016/j.fcr.2013.10.002

Verhulst N, Govaerts B, Verachtert E, Castellanos-Navarrete A, Mezzalama M, Wall P, Deckers J, Sayre KD (2010) Conservation agriculture, improving soil quality for sustainable production systems? In: Lal R, Stewart BA (eds) Advances in soil science: food security and soil quality. CRC Press, Boca Raton, pp 137–208

Vincelli PC (1994) Fundamental principles of plant pathology for agricultural producers. Agriculture and Natural Resources Publications. Paper 77. http://uknowledge.uky.edu/anr_reports/77. Accessed 27 Dec 2016

Walker SR, Medd RW, Robinson GR, Cullis BR (2002) Improved management of *Avena ludoviciana* and *Phalaris paradoxa* with more densely sown wheat and less herbicide. Weed Res 42:257–270. https://doi.org/10.1046/j.1365-3180.2002.00283.x

Watson W, Orr D and Bambara S (2015) NC cooperative extension, North Carolina cooperative extension

Williams MM II, Boydston RA (2013) Crop seeding level: implications for weed management in sweet corn. Weed Sci 61:437–442. https://doi.org/10.1614/WS-D-12-00205.1

Wu H, Walker SR, Osten VA, Robinson G (2010) Competition of sorghum cultivars and densities with Japanese millet (*Echinochloa esculenta*). Weed Biol Manage 10:185–193. https://doi.org/10.1111/j.1445-6664.2010.00383.x

Xu QC, Fujiyama S, Xu HL (2011) Biological pest control by enhancing populations of natural enemies in organic farming systems. J Food Agric Environ 9:455–463

Yenish JP, Worsham AD, York AC (1996) Cover crops for herbicide replacement in no-tillage corn (*Zea mays*). Weed Technol 10:815–821. https://doi.org/10.1017/S0890037X00040859

Chapter 5
Population Ecology of Aphid Pests Infesting Potato

Mohd Abas Shah, Sridhar Jandrajupalli, Vallepu Venkateshwarlu, Kamlesh Malik, Anuj Bhatnagar and Sanjeev Sharma

Abstract Potato is one of the most important food crops contributing to nutritional and food security in the world. It is grown as a summer crop in temperate areas of the world and as a winter crop in the subtropics of India and China. Potato crop is damaged by numerous pests and diseases of which aphid transmitted viruses are the major concern for healthy seed potato production. Since potato is a vegetatively propagated crop, the viral diseases lead to an ongoing decline in health of the propagating material i.e., seed degeneration. Hence, the management of aphids and aphid transmitted viruses is the first and foremost requirement for seed potato production. Potatoes are infested by a large number of aphid species very few of them actually able to colonise the crop. Most of the aphids are non-specific to the crop and are cosmopolitan and polyphagous. The common colonising aphid species are *Myzus persicae*, *Macrosiphum euphorbiae*, *Aulcorthum solani*, *Aphis gossypii*, *A. fabae*, *A. spiraecola* etc.; while as more than 100 species of aphids are reported to transiently visit potato crop. Of these, more than 65 species are known to transmit one or more potato viruses. Aphids because of their cyclic parthenogenesis and short life cycle can assume epic proportions while on the other hand, their host

M. A. Shah (✉)
ICAR-Central Potato Research Station, Post Bag no. 01, Model Town PO, Jalandhar 144003, Punjab, India
e-mail: mabas.shah@icar.gov.in; khubaib20@gmail.com

S. Jandrajupalli · V. Venkateshwarlu · S. Sharma
ICAR-Central Potato Research Institute, Shimla 171001, Himachal Pardesh, India
e-mail: brosridhar@gmail.com

V. Venkateshwarlu
e-mail: venkiiari@gmail.com

S. Sharma
e-mail: sanjeevsharma.cpri@gmail.com

K. Malik · A. Bhatnagar
ICAR-Central Potato Research Station, Modipuram, Meerut 250110, Uttar Pradesh, India
e-mail: malikkamlesh7@gmail.com

A. Bhatnagar
e-mail: dr.anujbhatnagar@gmail.com

S. Gaba et al. (eds.), *Sustainable Agriculture Reviews 28*, Ecology for Agriculture 28,
https://doi.org/10.1007/978-3-319-90309-5_5

selection and feeding behaviour predisposes them to being the most effective virus vectors. The host rang and life cycle characteristics of aphids are also key in determining the rate of spread of the viruses. Keeping in view the importance of life cycle variation and population ecology of aphids for virus transmission in potato crop, information is compiled and analysed to identify the gaps in knowledge and help determine the direction of future research, with special emphasis on the subtropics.

Keywords Migration · *Myzus persicae* · Non-persistent transmission Parthenogenesis · Potato virus Y · Primary host · Seed potato

5.1 Introduction

Potato (*Solanum tuberosum* L.) is a major world food crop, exceeded only by maize, rice and wheat in world production. Over three-fourth of the potato area is in Europe and Asia and the remainder is in Africa, North-Central America, South America and Oceania. Major areas are located between 45°N and 57°N that represent potato production zones in the temperate climate where potato is a summer crop (Fig. 5.1). Another potato concentration area is between 23°N and 34°N that mainly represents potato area in Gangetic plains, southern China and Egypt, where potato is grown as a winter crop. This belt that goes from south-west to south China and continues into the plains of Ganges River dominates potato production in Asia (Khurana and Naik 2003).

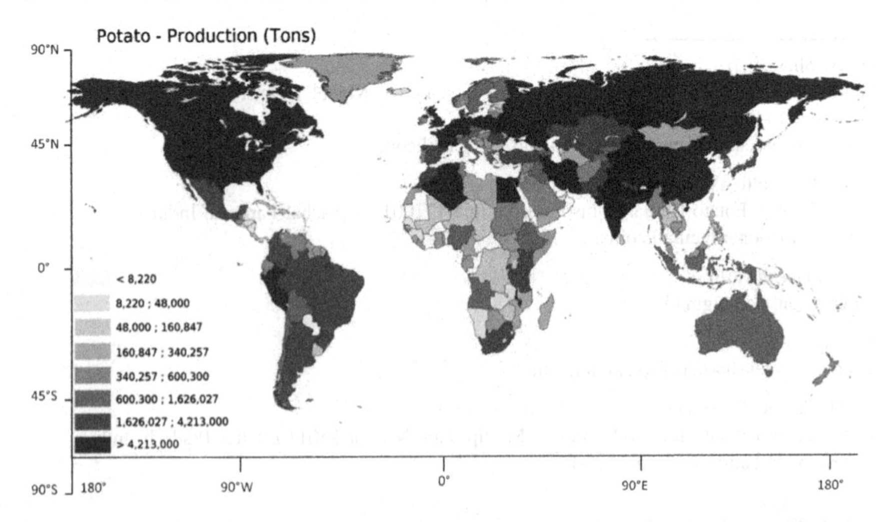

Fig. 5.1 Worldwide distribution of potato producing areas, Source: FOA, 2014. (Modified after Actualitix.com)

In India, potato is cultivated in almost all states under very diverse agro-climatic conditions (Pandey and Kang 2003). More than 85% of India's potatoes are grown in the vast Indo-Gangetic plains of north India (subtropics) during short winter days from October to March. The states of Punjab, Uttar Pradesh, West Bengal, and Bihar account for more than 75% of the potato-growing area in India and for about 80% of total production. Hilly areas, where the crop is grown during the summer from April to September (temperate), account for less than 5% of production. In the plateau regions of south-eastern, central, and peninsular India (tropics), which constitute about 6% of the potato-growing area, potato is mainly a rain-fed crop or is irrigated as winter crop. In the Nilgiri and Palni hills of Tamil Nadu, the crop is grown year-round under both irrigated and rain-fed conditions (Chandel et al. 2013).

Among the various biotic and abiotic stresses that constrain potato production, viruses are of highest importance for quality seed production. A series of mosaic causing potyviruses like potato virus Y (PVY) and the potato leaf roll virus (PLRV) are of utmost concern (Fig. 5.2). These viruses are transmitted by aphids in non-persistent and persistent manner, respectively. Aphids rarely cause a direct damage as a pest by feeding on the plants but they are important for potato being the natural vectors that transmit a large number of viruses responsible for progressive degeneration of the seed stock. Among the aphids, *Myzus persicae* (Sulzer) is the most efficient and most common vector of potato viruses. The main aphid species associated with potatoes worldwide are non-specific to this crop. Most of them are cosmopolitan and polyphagous. Apart from the aphid species that colonise potato crop, hundreds of non-colonising species are important to various extents. In this chapter, we first present attributes common to most aphid species and then we discuss characteristics that are specific to the most important aphid species of potato. In this chapter, emphasis is laid on the population ecology of aphids, mainly their life cycle and host range that are of potential importance in subtropical and tropical areas of potential production mainly so in India. An attempt has been made to bring out the knowledge gaps so that investigations could be taken up for the

Fig. 5.2 **a** Healthy potato plant. **b** Potato plant with viral mosaic symptoms

better understanding of the biology and ecology of the concerned species for the purpose of modelling, forecasting and hence better management.

5.2 Aphids Infesting Potato; Fundamentals of Population Ecology

Aphids have complex life cycles, and their classification depends on host alternation and on their mode of reproduction (Blackman and Eastop 2000; Williams and Dixon 2007). Different morphs are associated with these life cycles. Heteroecy and monoecy refer to the status of aphids regarding their host plant. Aphids that practice host alternation are heteroecious. They live on a primary host during the winter and colonize secondary hosts during the rest of the year before coming back to their primary host. Heteroecy occurs in only 10% of aphid species that generally colonize herbaceous plants, including economically important crop species such as potatoes. In contrast, a majority of aphid species live on the same plant throughout the year, do not have host plant alternation, and are classified as monoecious species. Some of these species are monophagous. Other species are oligo- or polyphagous and may migrate between plant species. However, they do not have regular alteration of primary and secondary hosts in their life cycles (Williams and Dixon 2007).

Holocycly and anholocycly refer to the ability of aphids to reproduce using parthenogenesis alone or in combination with sexual reproduction (Blackman and Eastop 2000). Most aphid species alternate parthenogenesis and sexual reproduction and are holocyclic. In this case, parthenogenesis occurs from the first generation in spring to the appearance of sexual morphs in autumn. The appearance of sexual individuals is triggered by seasonal changes in temperature and photoperiod. In contrast, some species are anholocyclic; they do not produce sexual morphs or eggs, and only reproduce by parthenogenesis (Williams and Dixon 2007). Anholocyclic life cycles may occur when climatic conditions are favourable for aphids to maintain populations on various plants during winter. Depending on the region, certain populations in some holocyclic aphid species can lose their sexual phase and become anholocyclic or generate only male populations (androcycly) during winter (Blackman 1971; Fenton et al. 1998; Williams and Dixon 2007). Aphids that colonize potato are mainly heteroecious and holocyclic. Their life cycles include an overwintering phase, during which fertilized eggs constitute the resistant form during periods of cold temperatures. Because potato is an annual plant, its colonizing aphids are heteroecious.

The viviparous mode of reproduction in aphids confers a rapid reproduction rate with short developmental times, resulting in population growth that is atypically high, even for insects. For instance, Dixon (1971) estimated that aphid populations in potato fields can reach densities of 2×10^9 individuals per hectare. Douglas (2003) suggested that such rates of population increase reflect nutrient allocation to the reproductive system. Energy is preferentially invested in embryo biomass and larval development rather than in maternal tissues. Aphids have telescoping generations—i.e., ovarian development and embryo formation start at the same time in embryonic mothers (Powell et al. 2006). Parthenogenetic reproduction results in clonal aphid colonies that have the same genotype. With this reproduction mode, an atypical characteristic can be amplified and become predominant in a given population after several generations. This can explain why aphids are able to adapt quickly to disturbances in their environment. Aphid populations may crash depending on the weather (Barlow and Dixon 1980), deteriorating resources, or pesticide treatments. However, parthenogenesis rapidly generates new populations that are adapted to their environment and, in some cases, resistant to pesticides. Parthenogenesis generally occurs during the warmer months of the year and maximizes offspring production. In fall, it is interrupted and followed by sexual reproduction that produces overwintering eggs.

Aphids produce both apterous (wingless) and alate (winged) morphs. Production of alate morphs is energetically costly (Dixon et al. 1993). Alates appear at different times during the year. They are considered to be colonizers, and use winds to disperse and locate new hosts. Wingless fundatrices emerge from eggs laid on the primary host. Their alate progeny are the spring migrants. Alate production is completed within a 2-week period (Radcliffe 1982). These individuals fly to secondary hosts (e.g., potato) and, when conditions are favourable, generate apterous and parthenogenetic populations. During summer, overpopulation of aphids, degradation of host plant nutritional suitability, or variations in light intensity, temperature, and precipitation induce the decline in aphid populations and the appearance of winged morphs that move to more suitable host habitats.

In autumn, as day length and temperature decrease, the quality of secondary host plants is altered. These factors generate the appearance of a new generation of virginoparous alates that migrate to the primary host. After the second generation on the primary host, oviparous females appear and are fertilized by winged males (Radcliffe 1982). After reproduction, oviparous females lay their eggs on the primary host for overwintering (Powell et al. 2006). Timing of flight and the number of migrants are important for colonization, clonal fitness, and overwintering success.

5.3 Biodiversity of Economically Important Aphid Species on Potato

Aphids belong to the Stenorrhyncha (Hemiptera). The most important genera found on potato crops, i.e. *Myzus* spp., *Aulacorthum* spp., *Aphis* spp. and *Macrosiphum* spp. belong to the family Aphididae. Aphids are characterized by high polymorphism. The colour of the individuals can also be highly variable within a given population and can be influenced by the symbiotic bacteria they host (Tsuchida et al. 2010). Morphology is influenced by several factors, such as environmental, climatic, and seasonal conditions; quality of the host plants and population densities. Blackman and Eastop (1994, 2000) listed more than 22 species of aphids that commonly infest potato plants (Table 5.1), and provided an identification key using morpho-anatomic criteria such as body colour, length, shape, and segmentation; antennal tubercles, head, siphunculi, legs and femurs, cauda and anal plate; and hairs on these structures. Verma and Chandla (1990) described five major species ingesting potato under Indian conditions viz., *Myzus persicae*, *Aphis gossypii*,

Table 5.1 List of aphid species colonising potato (After Blackman and Eastop 1994, 2000, 2006)

S. No	Aphid species	Subfamily	Life cycle
1.	*Acyrthosiphon malvae*	Aphidinae	Autoecious Holocyclic
2.	*Aphis craccivora*	Aphidinae	Anholocyclic, sexual morphs recorded from India and Germany
3.	*Aphis fabae*	Aphidinae	Heteroecious Holocyclic
4.	*Aphis frangulae* ssp. *beccabungae*	Aphidinae	Heteroecious Holocyclic
5.	*Aphis gossypii*	Aphidinae	Anholocyclic/Holocyclic
6.	*Aphis nasturtii*	Aphidinae	Heteroecious Holocyclic
7.	*Aphis spiraecola*	Aphidinae	Anholocyclic/Holocyclic
8.	*Aulacorthum solani*	Aphidinae	Anholocyclic/Holocyclic
9.	*Brachycaudus helichrysi*	Aphidinae	Heteroecious Holocyclic/Anholocyclic
10.	*Macrosiphum euphorbiae*	Aphidinae	Heteroecious Holocyclic/Anholocyclic
11.	*Myzus antirrhinii*	Aphidinae	Anholocyclic
12.	*Myzus ascalonicus*	Aphidinae	Anholocyclic
13.	*Myzus ornatus*	Aphidinae	Anholocyclic, Males recorded from India
14.	*Myzus persicae*	Aphidinae	Heteroecious Holocyclic/Anholocyclic
15.	*Neomyzus circumflexus*	Aphidinae	Anholocyclic
16.	*Pemphigus* sp.	Eriosomatinae	Not clear
17.	*Pseudomegoura magnoliae*	Aphidinae	Mainly Anholocyclic
18.	*Rhopalosiphoninus latysiphon*	Aphidinae	Anholocyclic
19.	*Rhopalosiphum rufiabdominale*	Aphidinae	Heteroecious Holocyclic/Anholocyclic
20.	*Smynthurodes betae*	Eriosomatinae	Heteroecious Holocyclic/Anholocyclic
21.	*Uroleucon compositae*	Aphidinae	Anholocyclic

A. fabae, *Rhopalosiphoninus latysiphon* and *Rhopalosiphum rufiabdominalis* with notes on biology, life cycle, migration and management, in addition to two minor species *Rhopalosiphum nymphaeae* and *Tetraneura nigriabdominalis*. Later, Shridhar et al. (2015) compiled information on 13 species of aphids recorded on potato viz., *M. persicae*, *A. gossypii*, *A. fabae*, *A. phis spiraecola*, *A. nerii*, *A. craciivora*, *Macrosiphum euphorbiae*, *Brevicoryne brassicae*, *Aulacorthum solani*, *Lipaphis erysimi*, *Hydaphis coriandari*, *Rhopalosiphum rufiabdominalis* and *Rhopalosiphum maidis*.

Although green peach aphid is the most efficient vector of potato virus Y (PVY) in potato (van Hoof 1980; Sigvald 1984; Singh et al. 1996; Fernandez-Calvino et al. 2006), several non-colonizing species also contribute to PVY prevalence (Ragsdale et al. 2001). Non-colonizing species do not reproduce on potato, but may transiently visit potato plants. Such species can occur in very large numbers, making their effect on virus spread large despite their lower virus transmission efficiency (Halbert et al. 2003). Hundreds of non-colonising aphid species have been reported from potato fields and tested for virus transmission ability. In Table 5.2, a consensus list of non-colonising aphid species is given most of which have been implicated with virus transmission albeit with low efficiency. In addition, notes on common hosts are also given which are indicative of the probable host plants from which they originate or are oriented to while crossing through potato fields.

Next, we will briefly introduce the main species of aphids infesting potato from ecological point of view and the variation they exhibit round the world.

5.3.1 Acyrthosiphon pisum *(Harris) (Pea Aphid)*

Acyrthosiphon pisum is a complex of races and subspecies with different host ranges and preferences. This species feeds mostly on Leguminosae. It is worldwide in distribution. Biology is holocyclic on various leguminous hosts in temperate regions; in warmer climates there is presumably facultative anholocycly (Blackman and Eastop 2000).

Not much is known about the biology of the aphid in India where both adult apterous and alate viviparous females reproduce parthenogenetically. One generation apparently takes about a week for completion. The colony of green or pink aphids is found round the stems, young shoots and the underside of leaves, they drop to the ground at the slight disturbance. It is sporadically serious pest of peas and causes cupping and distortion of leaves. It's distributed throughout the country and has wide host range. It happens to be potential vector of more than 30 viruses, including Potato virus Y (Misra 2002).

Table 5.2 List of aphid species associated with potato but not colonising the potato crop[a]

S. No	Aphid Species	Major host plants
1.	*Acrythosiphon pisum*	Fabaceae, important pest of peas and alfalfa
2.	*Acyrthosiphon kondoi*	Fabaceae such as *Medicago*, *Melilotus*, *Trifolium*, *Dorycnium*, *Lotus*; also *Pisum*, *Vicia* and *Lens*
3.	*Acyrthosiphon lactucae*	*Lactuca* spp.
4.	*Aphis glycines*	Fabaceae, particularly *Glycine* spp., a major pest of soybean
5.	*Aphis helianthi (= Aphis carduella)*	Asteraceae and Apiaceae, sexual phase on *Cornus stolonifera*
6.	*Aphis pomi*	Rosaceae including *Chaenomeles*, *Cydonia*, *Malus* and *Pyracantha*
7.	*Aphis sambuci*	*Sambucus* spp., Host alternation occurs to roots of plant such as *Cerastium*, *Dianthus*, *Silene*, *Melandrium*, *Moehringia*, *Spergula*), also often on *Rumex*, *Capsella*, *Oenothera* and *Saxifraga*
8.	*Brachycaudus cardui*	Compositae e.g. *Arctium*, *Carduus*, *Cirsium*, *Cynara*, *Chrysanthemum*, *Tanacetum*, *Matricaria*) and Boraginaceae e.g. *Borago*, *Cynoglossum*, *Echium*, *Symphytum*
9.	*Capitophorous elaeagni*	*Elaeagnus* spp., and sometimes on *Hippophae*, migrate to Compositae (*Arctium, Carduus, Cirsium, Cynara, Gerbera, Silybum*)
10.	*Capitophorous hippophaes*	Elaeagnaceae (*Elaeagnus* spp., *Hippophae* spp.), migrate to Polygonacease such as *Polygonum* and *Persicaria* spp.
11.	*Caveriella aegopodii*	Numerous genera and species of Umbelliferae, sexual phase on various *Salix* spp.
12.	*Cryptomyzus galeopsidis*	*Ribes* spp., migrating to *Lamium* and *Galeopsis*
13.	*Cryptomyzus ribis*	On *Ribes* spp., migrating to *Stachys* spp.
14.	*Diuraphis noxia*	On grasses and cereals *Agropyron*, *Anisantha*, *Andropogon*, *Bromus*, *Elymus*, *Hordeum*, *Phleum*, *Triticum*
15.	*Dysaphis aucuparie*	On *Sorbus terminalis* migrating to *Plantago* spp.
16..	*Hyalopterus pruni*	On *Prunus domesticus*, migrating to *Phragmites*, or sometimes to *Arundo donax*
17.	*Hydaphis foeniculi*	On *Lonicera* spp., migrating to various Umbelliferae
18.	*Hyperomyzus lactucae*	On *Ribes* spp. migrating to *Sonchus* spp., and occasionally other Asteraceae
19.	*Lipaphis erysimi*	On various Brassicaceae (*Arabis*, *Capsella*, *Coronopus*, *Erysimum*, *Isatis*, *Lepidium*, *Matthiola*, *Sinapis*, *Sisymbrium*, *Thlaspi*, etc.), but not usually on field *Brassica* crops
20.	*Macrosiphum rosae*	On *Rosa* spp. in spring, migrating to Dipsacaceae (*Dipsacus,Knautia*, *Succisa*) and Valerianaceae (*Centranthus*, *Valeriana*)
21.	*Metopolophium albidum*	On grasses such as *Arrhenatherum elatius*

(continued)

Table 5.2 (continued)

S. No	Aphid Species	Major host plants
22.	*Metopolophium dirhodum*	On *Rosa* spp. in spring, migrating to numerous species of Poaceae and Cyperaceae
23.	*Myzus cerasi*	On *Prunus* spp., migrating to secondary hosts in Rubiaceae (*Asperula, Galium*), Orobanchaceae (*Euphrasia, Rhinanthus*), Plantaginaceae (*Veronica*), and certain Brassicaceae (*Capsella, Cardamine, Coronopus, Lepidium*)
24.	*Myzus certus*	On Caryophyllaceae (*Cerastium, Dianthus, Stellaria*)
25.	*Nasonovia ribis-nigri*	On *Ribes* spp., migrating to liguliferous compositae (*Cichorium, Crepis, Lactuca, Lampsana*)
26.	*Phorodon humuli*	On *Prunus* spp., migrating to *Humulus lupulus*
27.	*Prociphilus americanus*	Host-alternating between *Fraxinus* spp. and roots of *Abies* (*balsamea, grandis, procera*)
28.	*Rhopalosiphum maidis*	On *Avena, Hordeum, Oryza, Saccharum, Secale, Sorghum, Triticum* and *Zea*, migrating to *Prunus* spp.
29.	*Rhopalosiphum nymphaeae*	Variety of water plants (*Alisma, Butomus, Callitriche, Echinodorus, Juncus, Nelumbo, Nuphar, Nymphaea, Potamogeton, Sagittaria, Sparganium, Triglochin, Typha*), sexual phase on *Prunus* spp.
30.	*Rhopalosiphum oxyacanthae* (= *R. insertum*)	On Pyroideae (*Cotoneaster, Crataegus, Malus, Pyrus, Sorbus*) migrating to Poaceae (*Agropyron, Agrostis, Alopecurus, Dactylis, Festuca, Glyceria, Phalaris, Poa, Triticum*)
31.	*Rhopalosiphum padi*	On *Prunus* spp., migrating to to numerous grasses and cereals,
32.	*Rhopolomyzys poae*	Grasses *Agrostis, Dactylis, Festuca, Glyceria, Phalaris, Poa*), migrating to *Lonicera* spp.
33.	*Schizaphis graminum*	Various species of Poaceae
34.	*Sipha elegans*	Various grasses and cereals (*Aegilops, Agropyron, Agrostis, Ammophila, Arrhenatherum, Bromus, Elymus, Festuca, Hordeum, Phleum, Puccinellia, Setaria, Triticum*)
35.	*Sitobion avenae*	On numerous species of Poaceae, including all the cereals and pasture grasses
36.	*Sitobion fragariae*	Apterae on *Rubus* and other Rosaceae, migrating to Poaceae
37.	*Therioaphis riehmi*	Commonly on *Melilotus* spp., also recorded from *Medicago, Trigonella* and *Trifolium*
38.	*Therioaphis trifolii*	On many plants of Leguminosae/Fabaceae in genera *Astragalus, Lotus, Medicago, Melilotus, Onobrychis, Ononis* and *Trifolium*
39.	*Uroleucon sonchi*	Mainly on *Sonchus* spp. and other genera in the tribe Cichoriaceae (*Lactuca, Cichorium, Hieracium, Ixeridium, Picris, Reichardia*)

[a]Blackman and Eastop 2000; De Bokx and Pirion 1990; DiFonzon et al. 1997; Halbert et al. 2003; Harrington and Gibson 1989; Harrington et al. 1986; Katis and Gibson 1985; Mondal et al 2016a, b; Piron 1986; Sigvald 1984, 1987, 1989, 1990; van Hoof 1977, 1980

5.3.2 Aphis craccivora *Koch (Cowpea Aphid; Groundnut Aphid; Black Legume Aphid) (Fig. 5.3)*

Aphis craccivora Koch, commonly known as black aphid, attacks many plants and most often leguminous crops. It is a major pest of bean and cowpea. The aphid species is known to infest at least 90 plant species belonging to 23 plant families in India (Raychaudhuri 1983; Chakrabarti and Sarkar 2001). The species acts as a vector of numerous viral diseases.

Both apterous and alate viviparous females reproduce asexually and both breed throughout the year (Talati and Butani 1980). Reproduction is the highest in February and lowest in June. The species completes 31 overlapping generations in a year (Bakhetia and Sidhu 1977). This species is anholocyclic almost everywhere, but monoecious holocyclic populations are recorded from Germany and India; sexuales on leguminous plants in Germany (Falk 1957/58) and on *Tinospora cordifolia* (Family Leguminosae) in Calcutta. Verma and Khurana (1974) reported males and oviparous females on green gram during December, 1973 at Haryana (India). Sometimes alatae of *A. craccivora* and *A. gossypii* are confused particularly if small, lightly coloured specimens are involved. However, both can be separated by the relative lengths of u.r.s. and h.t.2. Also, it differs from *gossypii* in the transverse sclerites on the individual tergites being much longer, particularly on the posterior segments, when they often extend along the entire width of the tergites, uniting with the postsiphuncular and marginal sclerites (Cottier 1953).

Fig. 5.3 *Aphis craccivora* apterae. Courtesy: Andy Jensen (aphidtrek.org)

5.3.3 Aphis fabae *Scopoli (Black Bean Aphid) (Fig. 5.4)*

Aphis fabae causes curling of the leaves of *Euonymus europaeus* in spring and migrates to a wide range of secondary hosts, including young growth of some trees, and many crops. It has one of the broadest host ranges, having been recorded from nearly 120 plant families. It is particularly important on beans, peas, beets, crucifers, cucurbits, potato, tobacco, tomato, and tulip (Blackman and Eastop 2007). Anholocyclic populations of aphids of the *A. fabae* group occur on secondary hosts in southern Europe, south-west Asia, Africa, Indian subcontinent, Korea (Kim et al. 2006), South America, Hawaii and Auckland Isles. Over much of Europe, *A. fabae* is heteroecious holocyclic, alternating between *Euonymous* and a wide range of secondary host plants.

Aphis fabae is a bewildering complex of species, at least some of which also have wide host ranges. Favret (2014) has recognised five subspecies of *Aphis fabae* viz. *cirsiiacanthoidis* Scopoli, 1763, *eryngii* Blanchard, 1923, *evonymi* Fabricius, 1775, *fabae* Scopoli, 1763, *mordvilkoi* Mordvilko, 1923, in addition to two species viz. *Aphis solanella* Theobold, 1914 and *Aphis euonymi* Gmelin in the context of the species complex.

In India, adults of both apterae and alatae reproduce parthenogenetically throughout the year. The life cycle is rather very complicated. The female deposits about 100 nymphs which become adults in about 10 days, passing through four moults. When the population is overcrowded and food source is hampered, the number of adult winged viviparous females increases so that they migrate from one plant to another. However, sexual forms of both apterous oviparous females as well as alate males are on record (Raychaudhuri et al. 1980a) from India. This suggests that the pest species enjoys both anhlolocyclic and holocyclic life cycle in the Indian conditions.

Fig. 5.4 *Aphis fabae* alate adult and nymphs Courtesy: Bob Dransfield (influentialpoints.com)

5.3.4 Aphis gossypii *Glover (Melon Aphid; Cotton Aphid) (Fig. 5.5)*

Aphis gossypii occurs on a very wide range of host plants, its polyphagy being particularly evident during the dry season in hot countries. It's a major pest of cotton and cucurbits, and in glasshouses in cold temperate regions (Blackman and Eastop 2000). Distributed almost worldwide, *A. gossypii* is particularly abundant and well-distributed in the tropics. Life cycle is anholocyclic in warm climates, and mainly so in Europe. Host alternation and a sexual phase occur more regularly in parts of east Asia (e.g. Japan, Takada 1988; China, Zhang and Zhong 1982) and also in North America (Kring 1959), with several unrelated plants utilised as primary hosts (such as *Catalpa bignonioides*, *Hibiscus syriacus*, *Celastrus orbiculatus*, *Rhamnus* spp. and *Punica granatum*). Apterae of spring populations on primary hosts are usually greenish. Males are always alate.

In India, *A. gossypii* is highly polyphagous and infests more than 500 species of plants in 76 families both cultivated as well as wild. On cotton it inflicts appreciable damage to the crop and hence it is commonly known as cotton aphid. Both adults and nymphs suck plant sap of many ornamental plants like *Hibiscus rosa-sinensis* Linn., *Cassia glauca* Linn., *Tecoma capensis* Lindl. and *Rosa* spp. from September to April in northern India. The maximum population is observed in *H. rosa-sinensis* during March-April. On *Rosa* spp., it is also observed in September to October and on *C. glauca* in March-April. The aphid is present throughout the year in the plains. In Punjab it remains very active from 4th week of July to third week of October on okra, from first week of August to 4th week of November on brinjal and from second week of August to fourth week of December on chilli (Jamwal and Kandoria 1990). In western Uttar Pradesh it is found in early, main and late potato crops while in Bihar its population is heavy on the seed crop (Verma and Parihar 1991). In Bangalore, the population of *A. gossypii* remains high from 3rd week of October to 1st week of January, and from last week of June to the last week of July,

Fig. 5.5 *Aphis gossypii* **a** alate **b** apterae. Courtesy: (a) Bob Dransfield (influentialpoints.com) (b) Andy Jensen (aphidtrek.org)

respectively. Aphids are observed in more numbers in August planted cotton crop than the March planted crop (Jalali et al. 2000).

The life cycle of this cotton aphid is very complicated. It is polymorphic and adults of both apterae and alatae are viviparous and reproduce parthenogenetically. The female deposits 80–100 nymphs which become adults in 7–9 days on cotton after passing through four moults. When the population is overcrowded, the number of winged adults increases so that they migrate from one plant to others. Both males and oviparous females have been reported from India (Ghosh 1970; Basu and Raychaudhuri 1980). The complete life-cycle is still unknown under Indian conditions.

5.3.5 Aphis nasturtii *Kaltenbach (Buckthorn Aphid; Buckthorn-Potato Aphid)*

It is found on a wide range of herbaceous plants in summer, including *Nasturtium officinale*, *Solanum tuberosum*, *Veronica beccabunga*, *Drosera rotundifolia* (Müller 1978) and *Rumex* spp. In India, the species infests nearly 74 species belonging to 64 genera under 37 plant families (Ghosh 1990). Severe infestation results in curling of leaves, started growth and gradual drying and death of young plants. Adults of both apterae and alatae are viviparous and reproduce by parthenogenesis. However, sexual morphs of both apterous oviparous females and alate males are known to occur in India. This hints at the possibility of holocyclic life cycle of the species in the Indian conditions. Life cycle is heteroecious holocyclic worldwide, with a sexual phase on *Rhamnus cathartica (*Blackman and Eastop 2006).

5.3.6 Aphis spiraecola *Patch (Spiraea Aphid; Green Citrus Aphid) (Fig. 5.6)*

Highly polyphagous species, usually colonises the under surface of leaves and tender buds of the host plants chiefly *Citrus* group of fruit trees. The species is however, not host specific to *Citrus* plants; instead it has wide preference of alternate hosts. Emigrant alatae normally initiate colonies on the tender leaves and shoots during late monsoon and sizeable population is produced on apical twigs, branches and sometimes even on young fruits. Increase in population is accompanied by the production of alatae and a large part of the aphid infestation migrates to other plants. *Citrus* plants support moderate to heavy aphid population for 2–3 months and it is preceded and succeeded by poor colonization for 15–20 days respectively. The species usually reproduces anholocyclically. However, Ghosh et al. (1972) reported sexual female of the species for the first time in India. Life cycle is holocyclic in North America and Brazil where *Spiraea* is the primary host.

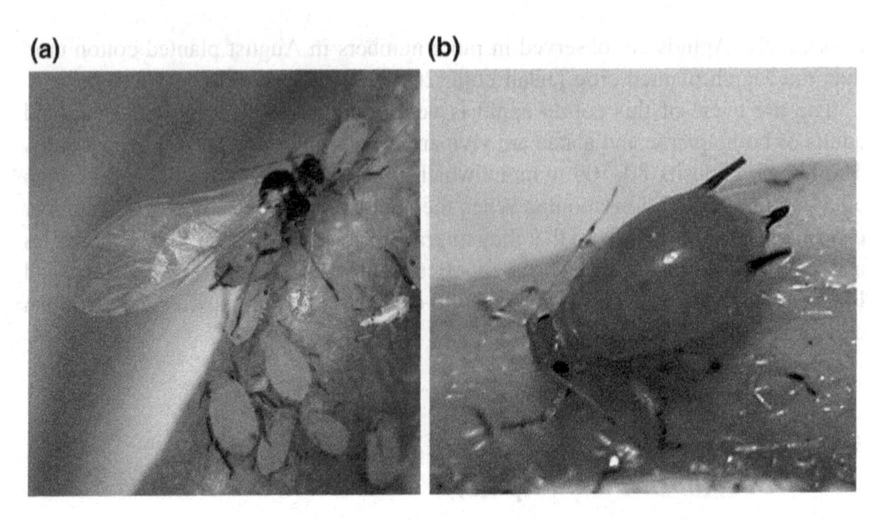

Fig. 5.6 *Aphis spiraecola* **a** alate **b** apterous. Courtesy: Andy Jensen (aphidtrek.org)

In Japan both *Spiraea* and *Citrus* serve as primary hosts; differences are indicative of either separate host races or species occur between the forms colonising these two plants (Komakazi et al. 1979).

5.3.7 **Aulacorthum solani** *(Kaltenbach) (Glasshouse-Potato Aphid; Foxglove Aphid) (Fig. 5.7)*

Aulacorthum solani is extremely polyphagous, colonising nearly 40 species belonging to 21 plant families. The important families are Asteraceae, Brassicaceae,

Fig. 5.7 *Aulacorthum solani* **a** alate **b** apterous. Courtesy: (a) Bob Dransfield (influentialpoints.com) (b) Andy Jensen (aphidtrek.org)

Leguminasae, Poaceae, Polygonaceae, Solanaceae etc. Probably of European origin, now almost world-wide in distribution. Life cycle is monoecious holocyclic with apterous or alate males, and with the unusual ability to go through the sexual phase on many different plant species. Also, anholocycly is commonly reported in mild climates and glasshouses (Blackman and Eastop 2006).

Aulacorthum solani was reported by David (1958) for the first time from India. Apterous and alate viviparous females were found on the leaves of *Digitalis* in the Nilgiris. David (1958) believed that the aphid was introduced around the mid-twentieth century into the country. Basu (1967) recorded apterous and alate viviparous females on potato in West Bengal. In Indian Conditions, this aphid reproduces parthenogenetically throughout the year. Several generations are completed in a year. One generation takes about two weeks in favourable conditions. The biology of the species is not properly known.

The brownish aphids usually infest young shoots and the undersides of young leaves of the host plants. As a result, leaves are cupped or otherwise distorted and become yellowish. Drops of sticky honey dew or patches of sooty mould on the upper sides of leaves. According to Misra (2002) this aphid is a vector of more than 30 viruses and is known to be present throughout the country.

5.3.8 **Brachycaudus helichrysi** *(Kaltenbach) (Leaf-Curling Plum Aphid)*

Brachycaudus helichrysi is extremely polyphagous, infesting nearly 185 species belonging to 115 genera under 49 plant families. Primary host plants are various *Prunus* species, notably *domestica, insititia, spinosa*. Secondary hosts are numerous species of Compositae (e.g. *Achillea, Chrysenthemum, Erigeron, Ageratum*) and Boraginaceae (*Myosotis, Cyanoglossum*) and sometimes other plants e.g., *Cucurbita, Rumex, Veronica* etc. life cycle is heteroecious holocyclic, but with widespread anholocycly in mild climates and in glasshouses (Blackman and Eastop 2000). The damage is caused by nymphs and females (apterous and alate) which are confined to the growing shoots and leaves, from where they suck the cell sap. It also sucks the sap from buds, blossoms, petioles, tender fruits and leaves. Excess feeding causes ventral curling of peach leaves which adversely affects the yield. On sprouting the leaves emerge curled while the fruits either do not set or falloff prematurely. The infestation starts with the commencement of bud swelling and continues during and after flowering.

In India, the pest species is active from February to March on temperate fruits. During the winter, it is found only in the egg stage at the base of the buds. During spring the eggs hatch and nymphs move on to the young leaves. Here they start sucking the plant sap. In about 4 weeks the nymphs become apterous viviparous female adults. The viviparous females give birth to young ones. 3 or 4 generations are completed on the fruit plants. With the warming up winged females and probably males are also produced. They migrate to other alternative host plants such as golden-rod and start reproducing parthenogenetically again. 4 or 5 generations are completed on the said plant from June to October. Early in November, the alate viviparous females are produced again. They migrate back to peach, plum and other fruit trees. Later, the females lay eggs at the base of the buds. Egg laying is completed by the middle of December. In Kumaon hills, this peach leaf curl aphid infests peach, plum and almond from November to May and from May to October it spends on an alternate host *Erigeron Canadensis* and other Asteraceae. On both the fruit trees and *E. canadenses* reproduction is normally through parthenogenesis. Basu et al. (1970) reported ovipara from *Prunus* sp. and alate males at Shillong. Alate males were also recorded (Ghosh et al. 1971) from *Prunus* sp. at Kalimpong (N-E Himalaya). Ghosh (1986) reported both oviparous female and alate males from N.W. Himalaya. Thus, it is a holocyclic and heteroecious species alternating between *Prunus* spp. (primary host) and a number of heterogenous secondary host plants including *Anaphalis* sp., *Ageratus conyzydes, Chrysanthemum* sp., *Eupatorium* sp., *Erigeron canadensis* (Raychaudhuri 1983)

It is widely distributed in North India, South India including Nilgiris and Coimbatore and also Eastern India including Assam, Meghalaya, Manipur, Sikkim and West Bengal.

5.3.9 **Brevicoryne brassicae** *(L.) (Cabbage Aphid; Mealy Cabbage Aphid)*

Brevicoryne brassicae is virtually restricted to members of the Cruciferae and is common in all temperate and warm temperate parts of the world. Life cycle is monoecious holocyclic in colder regions, anholocyclic where winter climate is mild (Blackman and Eastop 2000).

It forms colony of soft mealy-grey nymphs and apterous viviparous females that are found feeding in clusters on leaves, stems and flowers. White cast skins and drops of sticky honeydew and sooty mould growing on the honeydew are often noticed on the leaves. In India, both apterous and alate viviparous females are reported besides the sexuales; apterous oviparous females and alate males. In most of the cases, the aphid reproduces parthenogenetically both in the plains and altitudes (Debraj et al. 1995). Although anholocycly takes place in most of the cases of this aphid, sexual reproduction plays significant role in the biology of the aphid occurring at the higher elevations where cold climate prevails and day length is short. David (1958) reported the occurrence of only apterous oviparous females on *Brassica oleracea* from Shimla. Banerjee et al. (1969) reported males (alate) from Kuti valley (Uttar Pradesh) and Ghosh et al. (1969) reported alate male on *B. oleracea* in colony with viviparae at Shimla (Himachal Pradesh). All the above records of the aphid are from the plants of Brassicaceae growing at an elevation of above c 5,000 feet in the Western Himalaya. This suggests that *Brevicoryne brassicae* (L.) reproduces both anholocyclically and holocyclically at higher elevations. It is known that the species is more common above ca. 3,000 feet in Indian conditions where there is a chance of overlapping of mustard aphid *Lipaphis erysini* (Kaltenbach) and cabbage aphid *Brevicoryne brassicae* (L.)

5.3.10 **Macrosiphum euphorbiae** *(Thomas) (Potato Aphid) (Fig. 5.8)*

The potato aphid is a common and highly polyphagous species. Primary host plant is Rosa spp., and the species is hoghly polyphagous on secondary hosts feeding on over 200 plant species. It is often a pest on various crops such as potato (*Solanum tuberosum*), lettuce (*Lactuca sativa*) and beets (*Beta vulgaris*) as well as on numerous garden ornamentals. *Macrosiphum euphorbiae* is a vector of about one hundred plant viruses. The species originates from the north-eastern USA where it produces sexual forms and host alternates with rose (*Rosa*) as its primary host. Elsewhere it usually overwinters as viviparae. Aphid numbers increase rapidly from early spring, and alatae spread infestations to other plants. It is an especial problem in unheated greenhouses (Blackman and Eastop 2000).

Life cycle is heteroecious holocyclic with a sexual phase on *Rosa* in north-eastern USA, but elsewhere probably mainly or entirely anholocyclic on secondary hosts in more than 20 different plant families.

5.3.11 **Myzus antirrhinii** *(Macchiati)*

These aphids occur on leaves and young growth of numerous plants, on which it may be confused with *M. persicae* (Blackman and Paterson 1986). It often forms

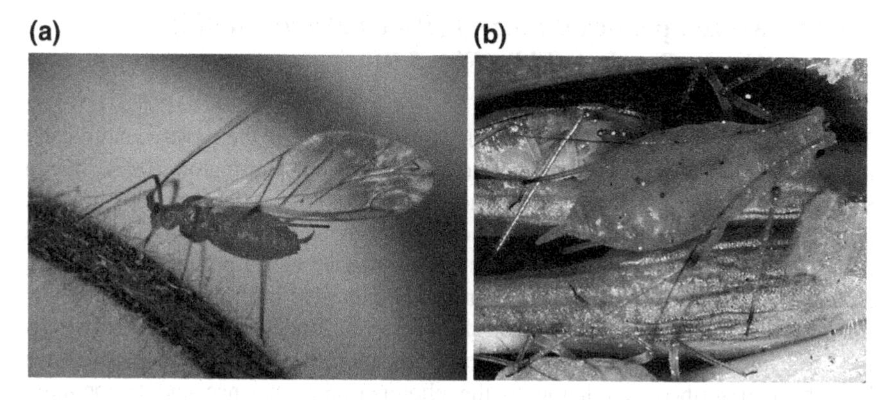

Fig. 5.8 *Macrosiphum euphorbiae* **a** alate **b** apterous. Courtesy: Andy Jensen (aphidtrek.org)

large, dense colonies, and only produces alatae rather sporadically. This species enjoys anholocycly almost everywhere, and produces alatae only sporadically, so it is most often found on perennial plants. However, there is now evidence of a possible sexual phase in Japan (Shigehara et al. 2006). Separation from *M. persicae* is difficult except using enzyme or molecular analysis (Hales et al. 2000).

5.3.12 **Myzus ascalonicus** *Doncaster (Shallot Aphid)*

Myzus ascalonicus is extremely polyphagous, with hosts in more than 20 families, but particularly Alliaceae, Caryophyllaceae, Compositae, Brassicaceae, Liliaceae and Rosaceae (Blackman and Eastop 2000). This species of aphids is apparently completely anholocyclic everywhere. During winter, it is frequently found on stored bulbs, potatoes or root vegetables, in glasshouses and on potted plants (Müller and Möller 1968).

5.3.13 **Myzus ornatus** *Laing Violet Aphid*

Myzus ornatus is extremely polyphagous, infesting nearly 180 species of plants. They live singly on the leaves of host plants in many different plant families including especially Bignonaceae, Compositae, Lamiaceae, Polygonaceae, Primulaceae, Rosaceae, and Violaceae. Anholocyclic populations occur throughout the world, and in colder climates they probably overwinter in glasshouses, on pot plants, or in sheltered situations. Both alate males and oviparaous females (Maity and Chakrabarti 1981) are known from India. This suggests that the species enjoys holocycly besides usual anholocyclic life cycle in the Indian conditions.

5.3.14 **Myzus persicae** *(Sulzer) (Peach Potato Aphid; Green Peach Aphid) (Fig. 5.9)*

The peach potato or green peach aphid, *Myzus persicae* is the most economically important aphid crop pest worldwide (van Emden and Harrington 2007). It is a notable example of a heteroecious aphid species. As the day length drops below a critical level in the autumn, apterous holocyclic viviparae produce gynoparae and males on secondary (herbaceous) hosts which migrate to the primary host, peach, *Prunus persica* L. (Rosaceae). The gynoparae then give birth to oviparae that lay the overwintering eggs after mating with males (van Emden et al. 1969). However, the life cycle of *M. persicae* appears to be polymorphic. The life cycle categories have been described in relation to the photoperiodic response and temperature

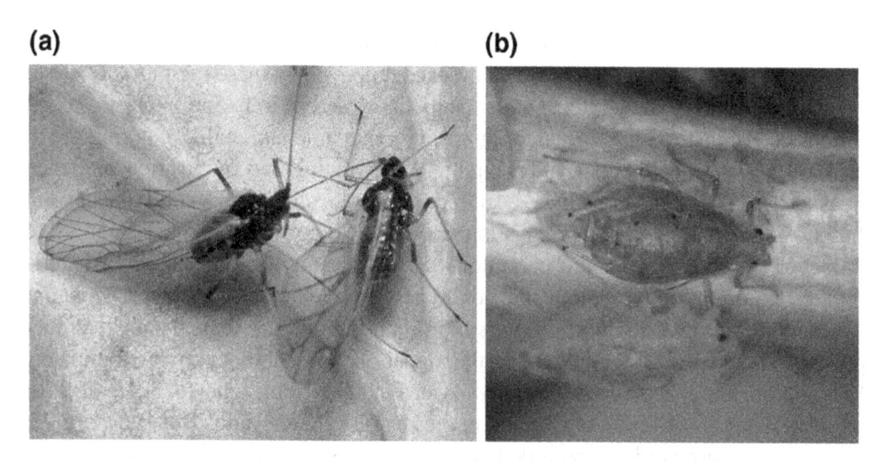

Fig. 5.9 *Myzus persicae* **a** alate **b** apterous. Courtesy: (a) Bob Dransfield (influentialpoints.com) (b) Andy Jensen (aphidtrek.org)

regime, i.e. holocyclic, anholocyclic, androcyclic and intermediate. In temperate regions clones with different reproductive strategies could coexist. The relative frequencies of holocyclic and anholocyclic populations in the spring depend on the severity of the previous winter. Anholocyclic clones are unable to produce any sexual morph and overwinter as parthenogenetic females on weeds or winter crops. Other genotypes are able to invest in both reproductive and overwintering strategies. Androcyclic clones, under short day conditions, produce parthenogenetic morphs and males, which are able to mate with oviparae of holocyclic or intermediate clones. Intermediate genotypes produce many apterous and alate virginoparae, some males and alate females which give birth both to virginoparae and oviparae (Blackman 1971, 1972).

Blackman (1974) reviewed the life cycle variation of *M. persicae* in different parts of the world and propounded that in tropics, where some varieties of peach are grown the mean monthly temperature never falls below 20 °C except at high altitudes, so that sexual morph production may be directly inhibited by high temperature all the year round. Parthenogenesis may continue uninterrupted for a long period of time, and although the holocyclic character could not be selected against, because it is not phenotypically expressed, one might expect loss of sexual viability due to the accumulation of chromosomal and genetic mutations. Apparently distinct anholocyclic biotypes of *M. persicae* such as that on tobacco may have originated in the tropics in this way. The extent to which splitting of the species occurs in tropical conditions must be largely governed by the frequency and degree of mixing between populations and integration of genotypes due to migrations between host-plants (Shaposhnikov 1966). The tropics cannot however be considered in isolation from the rest of the world, as the extent of frequently of long-range displacements into and out of the tropics of *M. persicae* from other latitudes is not known. In subtropical zone temperature is not low enough to permit the production of sexual morphs by October north of the equator, and by April south of the equator, then it will be too late for migration to *Prunus* and maturation of the oviparae before leaf fall. Therefore an induced holocycle in these regions is likely to be abortive. During the winter months temperature ceases to be inhibitory to sexual morph determination, and in winter or spring gynoparae and males may migrate to primary hosts (Bodenheimer and Swirski 1957). Mating and oviposition on peach trees in February and March have been recorded at localities in Egypt, Pakistan and India. As far as is known, any eggs laid in spring do not hatch. Where an abortive holocycle persists in climatic conditions which strongly favour anholocycly this implies immigration of holocyclic genotypes from other regions. This situation warrants further investigation. Winters are so mild in this subtropical zone that they present no obstacle to continuous parthenogenesis, and anholocycly predominates. It is significant that male *M. persicae* were caught from June to September in all-year-round traps for flying aphids in the region of Sao Paulo, Brazil, where the holocycle has not yet been recorded (Costa 1970). Males have also been collected from Solanaceae in Hong Kong in January and in Taiwan in March (Takahashi 1923). It is likely that the life cycle is androcyclic under such conditions.

Myzus persicae is a notorious polyphagous pest infesting nearly 250 species belonging to 77 genera in India, inflicting heavy losses to variety of crops and is also an important vector of many plant virus diseases (Raychaudhuri 1983; Chakrabarti and Sarkar 2001). It is universally distributed present in all ecological conditions prevailing in the country. The pest species usually appears on potato crop in the field from mid November onwards in most parts of Indo-Gangetic plains and does not migrate to the primary host plant for egg laying as in other temperate countries. Its population goes on increasing up to the end of February and early March (Chauhaf et al. 1975). However, by the end of March many alatae are formed and migrate to mid and then to higher hill where the climate is mild and suitable and a number of secondary host plants are available. The aphids keep on multiplying till November-December on high hills and thereafter, its return migration starts from hills to plains and vice versa. It is, thus, clear that *M. persicae* is present on the secondary host plants throughout the year either in the plains or hills (Nirula and Pushkarnath 1970). It can also overwinter in the hills, in green houses, sprouts of stored tuber and even in the fields (Lal and Verma 1987).

Ghosh and Raychaudhuri (1962) recorded sexual male from Delhi while sexual female was reported by Verma and Ghosh (1990) from northern part of the country such as Nainital (Uttarakhand) and Shillong (Meghalaya), but also from plains like Modipuram and Meerut (Uttar Pradesh). Verma and Chauhan (1993) reported that in the plains a few gynoparae (alatae) which produce oviparae, start arriving on peach trees by the end of January and February, the nymphs laid by these alatae mature into oviparous females by the middle of February. The males also start arriving during this period and mating takes place which lasts for about 2–5 min. The eggs are laid in the crevices of auxillary buds in clusters. Some eggs are also laid on the twigs. The eggs are first greenish in colour which later turn shining black. During this period, the temperature and day length go on rising and most of the eggs die and are also preyed on by the predators. It seems that these conditions are not suitable for egg hatching. Based on these observations, Singh and Ghosh

(2012) presumed that it enjoys both asexual and sexual life cycle in northern India which however, does not seem to be the case as this type of abortive life-cycle is known in many subtropical conditions and is referred to as androcycly (Margaritopoulos et al. 2002). The oviparae and eggs do not contribute to and nor continue the life cycle instead are a dead end to it. The life cycle continues through migration of alate virginoparae out of the subtropics to mild climate areas and then back as the temperature becomes suitable.

5.3.15 Neomyzus circumflexus *(Buckton) Mottled Arum Aphid*

It is extremely polyphagous, feeding on numerous species of both monocots and dicots, and even ferns and gymnosperms. In temperate climates *N. circumflexus* is found especially in glasshouses and on house plants (e.g. *Cineraria*, *Cyclamen*, *Fuschia*, *Zantedeschia*). Distribution is virtually world-wide. Apparently it is entirely anholocyclic; no sexual morphs have been recorded.

The crescent-marked lily aphid is entirely parthenogenetic with no sexual stage in the life cycle. In temperate climates it is primarily a pest of glasshouse crops where it attacks *Asparagus*, *Begonia*, *Fuchsia* and many others. Heavy infestations cause direct harm to many ornamental plants, and the aphids also transmit viruses (Blackman and Eastop 2000).

5.3.16 Pseudomegoura magnoliae (= Aulacorthum magnoliae) *(Essig and Kuwana)*

Polyphagous, feeding mainly on leaves of plants in over 20 different families, including Citrus and many ornamental shrubs and trees. Indian records are mainly from Cucurbitaceae (Raychaudhuri et al. 1980a). Life cycle is mainly anholocyclic, with a "relict" holocycle on *Sambucus* in Japan (Blackman and Eastop 2000). Matsuka and Imanishi (1982) studied its life cycle on *Sambucus sieboldiana* near Tokyo, where populations overwinter both as viviparae and, less commonly, as eggs. Clones descended from fundatrices produced males and viviparous females, but very few oviparae, so the sexual phase was almost non-existent in that population. Males are also recorded from India (Raychaudhuri et al. 1980b).

5.3.17 **Rhopalosiphoninus latysiphon** *(Davidson) Bulb and Potato Aphid*

It colonizes bulbs (*Tulipa, Gladiolus*) and potato tubers in store, and the roots of many plants, especially in clay soils (e.g. potato crops), or on etiolated stems or runners growing in darkness under stones (e.g. *Bromus sterilis, Convolvulus arvensis, Potentilla anserina, Vinca major, Urtica* spp.). It is a recorded vector of PVY and also has the ability to tranamit potato leaf roll virus (Bell 1988). Anholocyclic, overwintering on stored bulbs and potatoes in cold temperate regions; sexual morphs are not recorded. However, it is possible that *Rhopalosiphoninus deutzifoliae* (q.v.) on Hydrangaceae in Japan and east Siberia is the primary host form. *Rhopalosiphoninus latysiphon* ssp. *panaxis* Zhang, described from *Panax quinquefolium* in China (Qiao and Zhang 1999) appears to be a synonym (Blackman and Eastop 2006).

5.3.18 **Rhopalosiphum padi** *(L.) Bird Cherry-Oat Aphid*

Common primary hosts are *Prunus padus* in Europe and *P. virginiana* in North America. Secondary hosts are numerous species of Graminae, including all major cereal and pasture grasses. Also it has been found overwintering on dicotyledonous weeds (*Capsella, Stellaria*). Life cycle is heteroecious holocyclic between *Prunus padus* and Graminae in Europe, or anholocyclic on Graminae where winter conditions permit and in many part of the world where *P. padus* or alternative primary hosts are not available.

In India, *Rhopalosiphum padi* is known to infest about 30 species of plants belonging to many families Severe infestation is observed in the poaceous plants (November–March). From centres of infestation, it spreads in ever-widening circles. The aphid is responsible for Barley yellow dwarf (Nagaich and Vashisth 1963) and Wheat streak mosaic (Raychaudhuri and Ganguli 1968).

The species apparently reproduces parthenogenetically throughout the year in India. However, oviparae of this species are recorded from India (Raychaudhuri 1980a, b). This hints at the possibility that it may have sexual life cycle in the altitudinal areas where day length is short and temperature is low which may initiate the production of the sexuales. According to Richards (1960) the number of generations of alienicolae produced is unknown, but certainly several are produced. Fall migrant alate viviparous females resemble spring migrants but are produced by alienicolae and sexuales occur on the winter host from the middle of September to the end of October.

5.3.19 **Rhopalosiphum rufiabdominale** *(Sasaki) Rice Root Aphid*

Apterae in spring colonise on young leaves, stems and suckers of *Prunus* spp. (e.g. *mume, salicina, yedoensis*) (Moritsu 1983). *Rhodotypos scandens* may also be used as a primary host (Torikura 1991). Heteroecious holocyclic (Tanaka 1961); alatae migrate in May-June to form colonies on underground parts of numerous species of Poaceae, Cyperaceae and some dicots, particularly Solanaceae (potato, tomato, capsicums). Alatae on secondary hosts normally have 5-segmented antennae. It is a major pest of rice (Yano et al. 1983; Blackman and Eastop 2000). Anholocyclic populations occur throughout most of the world on secondary hosts, particularly in warmer climates and in glasshouses. However, early spring populations have recently been found on *Prunus* spp. (*armeniaca, domestica*) in Italy (Rakauskas et al. 2015), indicating that a holocycle may now have been established in Europe.

In India, *R. rufiabdominalis* has been found to attack roots and aerial parts of potato. Heavy infestations have been recorded on potato roots (Chahal et al. 1974) but the exact role the species plays is still unknown. The adults reproduce parthenogenetically and produce nymphs throughout the year. The alate forms that are produced after 2–3 generations of the apterous forms are responsible for dispersal of this aphid. They are carried to the potato fields by wind and subsequently nymphs are produced on the leaves, which move towards the roots at soil level. In India, alate male was reported and described for the first time from snow at Naini peak *(Ca.* 8,563 ft) in Uttar Pradesh by Ghosh (1969). Later, Verma (1988) reported apterous oviparous females collected on Peach. The finding of both sexual male and female from the same geographical range hints at the possibility that the species breeds holocyclically at least in the northern part of India.

Young et al. (1971) reported the aphid species from roots and underground stems of barley in Delhi. The aphid colonies were present on the crop from the 1st week of January to the end of February.

5.3.20 **Smynthurodes betae** *Westwood (Bean Root Aphid)*

Primary hosts are *Pistacia* spp. (*atlantica, mutica* and, rarely, *vera*). The galls on *Pistacia* spp. are yellow-green or red, spindle-shaped, about 20 mm long, formed by rolling of the edge of the leaflet near its base. These are secondary galls, produced by the progeny of the fundatrix, which lives in a small red mid-rib gall (Burstein and Wool 1991). *Smynthurodes betae* is heteroecious holocyclic with a two-year cycle; alatae emerge in September-November and migrate to roots of numerous, mostly dicotyledonous, plants. Secondary hosts are particularly Compositae/Asteraceae (*Artemisia, Arctium*), Leguminosae/Fabaceae (*Phaseolus, Vicia, Trifolium*), and Solanaceae (*Solanum tuberosum, S. nigrum, Lycopersicon esculentum*); also sometimes on *Beta, Brassica, Capsella, Gossypium,*

Heliotropum, *Rumex*, etc. Only rarely is it found on monocots (Poaceae, Cyperaceae). The holocycle is recorded throughout the range of the primary hosts; Algeria, Morocco, Israel, Syria, Iran, southern Crimea, Transcaucasus and Pakistan. Anholocyclic populations occur commonly on secondary hosts in other parts of the world (Blackman and Eastop 2000).

5.3.21 Uroleucon compositae *(Theobald)*

On flower stems, and in low numbers along the mid-ribs of the leaves, of a wide range of Compositae/Asteraceae in tropical and subtropical climates, particularly plants growing in moist or shady situations at the end of the dry season. It is a pest of *Carthamus tinctoria* (safflower) in India (Blackman and Eastop 2000), and is common on herbaceous *Vernonia* after the rains in Africa (Eastop 1958). Sometimes it is found on non-composite plants such as *Malva* and *Morus*. The aphid species is widely distributed in Africa, South America and on the Indian subcontinent. Apparently it is anholocyclic everywhere; no sexual morphs have been recorded. *U. compositae* is difficult to distinguish from the East Asian species *U. gobonis*, and it could even possibly be an anholocyclic form of that species. Early records of *U. jaceae* or *U. solidaginis* on safflower in Africa or India probably all refer to *U. compositae* (Blackman and Eastop 2000).

5.4 Summary

Various aphid species such as *M. persicae*, *A. gossypii* etc. exhibit huge variation in life cycle. It appears that in tropical and subtropical zones the aphid life cycle gets modified given the exorbitantly high temperature in summer and round the year availability of favourable temperature in tropics and absence of preferred primary hosts. The survival and life cycle variation of aphids in areas other than temperate zone needs thorough exploration.

A section of the population of *M. persicae* migrates from temperate to tropical and subtropical areas as was identified by Blackman (1974). However the factors inducing production of such migratory clones and their subsequent routes and behaviour in immigration zone needs further exploration. It is of paramount importance keeping in view the impact of *M. persicae* for the healthy seed production in subtropics like India.

It has been identified that there is influx of various species of aphids from various Poaceous plants into potato bringing viruses along. In this connection the importance of cropping pattern *vis-a-vis* healthy seed production needs exploration. Since the occasional visitor aphids can contribute to virus transmission and spread in an appreciable manner, the possibility of feeding repellents and physical barriers needs to be re-explored.

References

Bakhetia DRC, Sidhu AC (1977) Biology and seasonal activity of the groundnut aphid *Aphis craccivora* Koch. J Res Punjab Agric Univ 14(3):299–303. AGRIS record ID = US201302431594

Banerjee H, Ghosh AK, Raychaudhuri DN (1969) On a collection of aphids (Homoptera) from Kutivalley West Himalaya. Orient Ins 3(3):255–264. https://doi.org/10.1080/00305316.1969.10433914

Barlow ND, Dixon AFG (1980) Simulation of lime aphid population dynamics. Pudoc, Wageningen, 165 pp. ISBN: 9022007065

Basu AN (1967) One new genus and seven new species of aphids from Darjeeling district, West Bengal (Homoptera: Aphididae). Bull Ent 8(2):143–157

Basu RC, Raychaudhuri DN (1980) A study on the sexuales of aphids (Homoptera: Aphididae) in India. Records of Zoological Survey of India, Occasional Paper No. 18, 54 pp. ISSN: 0375-1511

Basu RC, Ghosh AK, Raychaudhuri DN (1970) A new genus and records of some sexual forms from Assam. Proc Zool Soc Calcutta 23:83–91

Bell AC (1988) The efficiency of the bulb and potato aphid *Rhopalosiphoninus latysiphon* (Davidson) as a vector of potato virus V. Potato Res 31(4):691–694. https://doi.org/10.1007/BF02361862

Blackman RL (1971) Variation in the photoperiodic response within natural populations of *Myzus persicae* (Sulz.). Bull Ent Res 60:533–546. https://doi.org/10.1017/S0007485300042292

Blackman RL (1972) The inheritance of life-cycle differences in *Myzus persicae* (Sulz.) (Hem., Aphididae). Bull Ent Res 62:281–294. https://doi.org/10.1017/S0007485300047726

Blackman RL (1974) Life-cycle variation of *Myzus persicae* (Sulz.) (Horn., Aphididae) in different parts of the world, in relation to genotype and environment. Bull Ent Res 63:595–607. https://doi.org/10.1017/S0007485300047830

Blackman RL, Eastop VF (1994) Aphids on the World's trees. CAB International, Wallingford, 987 pp. ISBN: 0851988776

Blackman RL, Eastop VF (2000) Aphids on the world's crops: An identification and information guide, 2nd edition. Wiley, Chichester, UK. 466 pp. ISBN: 978-0-471-85191-2

Blackman RL, Eastop VF (2006) Aphids on the world's herbaceous plants and shrubs. (2 vols) Wiley, Chichester, 1439 pp. ISBN: 978-0-471-48973-3

Blackman RL, Eastop VF (2007) Taxonomic Issues. In: van Emden, HF, Harrington R (eds), Aphids as Crop Pests. CABI, Wallingford, UK., pp 1–29. https://doi.org/10.1079/9780851998190.0001

Blackman RL, Paterson AJC (1986) Separation of *Myzus (Nectarosiphon) antirrhinii* from *M. (N.) persicae* and related species in Europe. Syst Ent 11:267–276. https://doi.org/10.1111/j.1365-3113.1986.tb00181.x

Bodenheimer FS, Swirski E (1957) The Aphidoidea of the Middle East. Weizmann Science Press, Jerusalem, p 378

Burstein M, Wool D (1991) A galling aphid with extra life-cycle complexity: population ecology and evolutionary considerations. Res Pop Ecol 33:307–322. https://doi.org/10.1007/BF02513556

Chahal BS, Sekhon SS, Sandhu MS (1974) Rhopalosiphum rufiabdominalis recorded on roots of potato in the Punjab. In: Symposium on problems in potato production. CPRI, Shimla, India

Chakrabarti S, Sarkar A (2001) A supplement to the' food plant catalogue of Indian Aphididae. J Aphidol 15:9–62

Chandel RS, Chandla VK, Verma KS, Pathania M (2013) Insect pests of potato in india: biology and management In: Giordanengo P, Vincent C, Alyokhin A (eds) Insect pests of potato. Global Perspectives on Biology and Management Elsevier Inc, USA, pp 227–270. http://dx.doi.org/10.1016/B978-0-12-386895-4.00008-9

Chauhaf BS, Sekhon SS, Bindra OS (1975) The incidence and the time of appearence of *Myzus persicae* in autumn potato crop under different agroclimatic conditions in the Punjab. Indian J Ecol 2:155–162. AGRIS record ID = IN19760083833

Costa CL (1970) Variacoes sazonais da migracao de *Myzus persicae* em Campinas nos anos de 1967 a 1969. Bragantia 29:347–359. http://repositorio.unb.br/handle/10482/15811

Cottier W (1953) Aphids of New Zealand. Bull N Z Dept Sci Ind Res 106:1–368

David SK (1958) Some rare Indian aphids. J Bombay Nat Hist Soc 55(1):110–116

De Bokx JA, Piron PGM (1990) Relative efficiency of a number of aphid species in the transmission of potato virus Y^N in the Netherlands. Ned J Pl Path 96(4):237–246. https://doi.org/10.1007/BF01974261

Debraj Y, Singh SL, Shantibala K, Singh TK (1995) Comparative biology of the cabbage aphid, *Breuicoryne brassicae* (L.) on six cruciferous hosts. J Aphidol 9:30–35

DiFonzo CD, Ragsdale DW, Radcliffe EB, Gudmestad NC, Secor GA (1997) Seasonal abundance of aphid vectors of potato virus Y in the Red River Valley of Minnesota and North Dakota. J Econ Ento 90:824–831. https://doi.org/10.1093/jee/90.3.824

Dixon AFG (1971) The role of intra-specific mechanisms and predation in regulating the numbers of the lime aphid *Eucallipterus tiliae* L. Oecol 8:179–193. https://doi.org/10.1007/BF00345812

Dixon AFG, Horth S, Kindlmann P (1993) Migration in Insects: cost and strategies. J Animal Ecol 62:182–190. https://doi.org/10.2307/5492

Douglas AE (2003) Nutritional physiology of aphids. Adv Insect Physiol 31:73–140. https://doi.org/10.1016/S0065-2806(03)31002-1

Eastop VF (1958) A study of the Aphididae of East Africa. Colonial Research Publication, H.M.S.O, London, 126 pp

Falk U (1957/58) Biologie and Taxonomic der Schwarzen Blattlause der Leguminosen. *Wiss.* Z. Uniu Rostock 7(4):616–634

Favret C (2014) Aphid Species File. Version 5.0/5.0. [30-12-2014]. http://Aphid.SpeciesFile.org

Fenton B, Woodford JAT, Malloch G (1998) Analysis of clonal diversity of the peach–potato aphid, *Myzus persicae* (Sulzer), in Scotland, UK and evidence for the existence of a predominant clone. Mol Ecol 7:1475–1487. http://doi.org/:10.1046/j.1365-294x.1998.00479.x

Fernandez-Calvino L, Lopez-Abella D, Lopez-Moya JJ, Fereres A (2006) Comparison of Potato virus Y and Plum pox virus transmission by two aphid species in relation to their probing behavior. Phytoparasitica 34:315–324. https://doi.org/10.1007/BF02980959

Ghosh LK (1969) Notes on the male of *Rhopolosiphum rufiabdominalis* (Sasaki) (Homoptera: Aphididae) from Uttar Pradesh, India. Indian J Sci Indust 3(4):215–217

Ghosh LK (1970) On a collection of aphids (Homoptera: Aphididae) from Rajasthan, India. Indian J Sci Indust (B) 4(2):85–89

Ghosh LK (1986) A conspectus of Aphididae (Homoptera) of Himachal Pradesh in Northwest Himalaya, India. Technical Monograph No. 16. Zoological Survey of India, 282 pp

Ghosh LK (1990) A taxonomic review of the genus *Aphis* Linnaeus (Homoptera: Aphididae) in India. Mem Zool Surv India 17(3):45–48

Ghosh AK, Chakrabarti S, Chowdhuri AN, Raychaudhuri DN (1969) Aphids (Homoptera) of Himachal Pradesh, India-II. Orient Ins 3(4):327–334

Ghosh AK, Raychaudhuri DN (1962) A preliminary account of bionomics and taxonomy of aphids from Assam. J Bombay Nat Hist Soc 59:238–253

Ghosh AK, Ghosh MR, Raychaudhuri DN (1971) Studies on the aphids (Homoptera: Aphididae) from Eastern India. VII. New species and new records from West Bengal. Oriental Ins 5 (2):209–221. http://dx.doi.org/10.1080/00305316.1971.10434009

Ghosh AK, Ghosh MR, Raychaudhuri DN (1972) Studies on the aphids (Homoptera: Aphididae) from eastern India XI: Descriptions of hitherto unknown or newly recorded sexual morphs of some species from West Bengal. Oriental Ins 6(3):333–341. http://dx.doi.org/10.1080/00305316.1972.10434083

Halbert SE, Corsini DL, Wiebe MA (2003) Potato Virus Y transmission efficiency for some common aphids in Idaho. Am J Potato Res 80:87–91. http://doi.org/10.1007/BF02870-207

Hales DH, Wilson ACC, Spence JM, Blackman RL (2000) Confirmation that *Myzus antirrhinii* (Macchiati) (Hemiptera: Aphididae) occurs in Australia, using morphometrics, microsatellite typing and analysis of novel karyotypes by fluorescence in situ hybridisation. Austr J Entomol 39:123–129. https://doi.org/10.1046/j.1440-6055.2000.00160.x

Harrington R, Gibson RW (1989) Transmission of potato virus Y by aphids trapped in potato crops in southern England. Potato Res 32(2):167–174

Harrington R, Katis N, Gibson RW (1986) Field assessment of the relative importance of different aphid species in the transmission of potato virus Y. Potato Res 29(1):67–76. https://doi.org/10.1007/BF02361982

Jalali SK, Singh SR, Biswas SR (2000) Population dynamics of *Aphis gossypii* Glover (Homoptera: Aphididae) as its natural enemies in the cotton ecosystem. J Aphidol 14:25–32

Jamwal R, Kandoria JL (1990) Appearance and build up of *Aphis gossypii* (Homoptera: Aphididae) on chilli, brinjal and okra in Punjab. J Aphidol 4:49–52

Katis N, Gibson RW (1985) Transmission of potato virus Y by cereal aphids. Potato Res 28:65–70. https://doi.org/10.1007/BF02357571

Khurana SMP, Naik PS (2003) The potato: an overview. In: Khurana SMP, Minhas JS, Pandey SK (eds) The potato—production and utilization in subtropics. Mehta Publishers, New Delhi, India, pp 1–14

Kim H, Lee W, Lee S (2006) Three new records of the genus *Aphis* (Hemiptera: Aphididae) from Korea. J Asia-Pacific Ent 9:301–312. https://doi.org/10.1016/S1226-8615(08)60307-6

Komakazi S, Sakagami Y, Korenaga R (1979) Overwintering of aphids on citrus trees. Jap J Appl Ent Zool 23: 246–250. https://doi.org/10.1303/jjaez.23.246

Kring JB (1959) The life cycle of the melon aphid, *Aphis gossypii* Glover, an example of facultative migration. Ann Ent Soc Am 52:284–286. https://doi.org/10.1093/aesa/52.3.284

Lal L, Verma KD (1987) Seasonal incidence and over seasoning of *Myzus persicae* (Sulzer) on potato in Meghalaya. Indian J Hill Frmg 1:35–39

Maity SP, Chakrabarti S (1981) On poplar inhabiting aphids (Homoptera: Aphididae) of India and adjoining countries with notes on some species. Entomon 6(4):297–305. AGRIS record no. 19820591768

Margaritopoulos JT, Tsitsipis JA, Goudoudaki S, Blackman RL (2002) Life cycle variation of *Myzus persicae* (Hemiptera: Aphididae) in Greece. Bull Entomol Res 92:309–319. https://doi.org/10.1079/BER2002167

Matsuka M, Imanishi M (1982) Life cycle of an aphid, *Acyrthosiphon magnoliae* (Essig et Kuwana) observed near Tokyo and in a laboratory under controlled condition. Bull Fac Agric Tamagawa Univ 22:56–66 (in Japanese)

Misra SS (2002) Aphids as Vector of plant diseases and their chemical control. Abs.: Nat. Seminar on Ecology and Diversity of aphids and aphidophaga complex, Tripura University May 4–5, 2002, p 26

Mondal S, Wenninger EJ, Hutchinson PJS, Weibe MA, Eigenbrode SD, Bosque-Pérez NA (2016a) Contribution of noncolonizing aphids to *potato virus y* prevalence in potato in Idaho. Env Ent pii: nvw131. http://doi.org/10.1093/ee/nvw131

Mondal S, Wenninger EJ, Hutchinson PJS, Whitworth JL, Shrestha D, Eigenbrode SD, Bosque-Pérez NA (2016b) Comparison of transmission efficiency of various isolates of Potato virus Y among three aphid vectors. Ent Exp et App 158(3):258–268. https://doi.org/10.1111/eea.12404

Moritsu M (1983) Aphids of Japan in colours. Zenkoku Noson, Tokyo, 545 pp (ISBN 4-88137-017-0)

Müller FP (1978) Untersuchungen über Blattläuse mecklenburgischer Hochmoore. Arch Freunde Naturg Mecklenb 18:31–41

Müller FP, Möller FW (1968) Ein bermerkenswerters Massenauftreten von Myzus ascalonicus Doncaster (Homoptera: Aphididae) in Freiland. Archiv der Freunde der Naturgeschichte in Mecklenburg 14:44–55

Nagaich BB, Vashisth KS (1963) Barley yellow dwarf, a new virus disease for India. Indian Phytopath 16:318–319

Pande SK, Kang GS (2003) Ecological and varietal improvement. In: Khurana SMP, Minhas JS, Pandey SK (eds) The Potato—Production and Utilization in Subtropics. Mehta Publishers, New Delhi, India, pp 48–60

Piron PGM (1986) New aphid vectors of potato virus Y^N. Ned J Pl Path 92(5):223–229. https://doi.org/10.1007/BF01977688

Powell G, Tosh CR, Hardie J (2006) Host plant selection by aphids: behavioral, evolutionary, and applied perspectives. Ann Rev Entomol 51:309–330. https://doi.org/10.1146/annurev.ento.51.110104.151107

Pushkarnath Nirula KK (1970) Aphid-warning for production of seed potato in subtropical plains of India. India J Agr. Sci 40(12):1061–1070

Qiao G, Zhang G (1999) Five new species and one new subspecies of Macrosiphinae from Fujian province, China. Acta Zootaxonomica Sinica 24:304–314

Radcliffe EB (1982) Insect pests of potato. Ann Rev Entomol 127:173–204. https://doi.org/10.1146/annurev.en.27.010182.001133

Ragsdale D, Radcliffe E, diFonzo CD (2001) Epidemiology and field control of PVY and PLRV. In: Lobenstein G, Berger, PH, Brunt AA, Lawson R (eds), Virus and virus-like diseases of potatoes and production of seed potatoes. Kluwar Academic Publishers, Dordrecht, Netherlands. pp 237–70. https://doi.org/10.1007/978-94-007-0842-6_22

Rakauskas R, Bašilova J, Bernotienė R (2015) *Aphis pomi* and *Aphis spiraecola* (Hemiptera: Sternorrhyncha: Aphididae) in Europe—new information on their distribution, molecular and morphological peculiarities. Eur J Ent 112:270–280. https://doi.org/10.14411/eje.2015.043

Raychaudhuri DN (1983) Food plant catalogue of Indian Aphididae. Grafic Print All, Calcutta, p 181

Raychaudhuri SP, Ganguly B (1968) A mosaic streak of wheat. Phytopathologische Zeitschrift 59:385–389

Raychaudhuri DN, Pal PK, Ghosh AK (1980a) Subfamily Pemphiginae. In: Raychaudhuri DN (ed) Aphids of North East India and Bhutan. The Zoological Society, Calcutta, pp 409–433

Raychaudhuri DN, Ghosh MR, Basu, RC (1980b) Subfamily Aphidinae. In: Raychaudhuri DN (ed) Aphids of North East India and Bhutan. The Zoological Society, Calcutta, pp 47–275

Richards WR (1960) A synopsis of the genus *Rhopalosiphum* in Canada. Memoirs Ent Soc Canada 92(S13):5–51. https://doi.org/10.4039/entm9213fv

Shaposhnikov GK (1966) Origin and breakdown of reproductive isolation and the criterion of the species. Ent Rev 45:1–18

Shigehara T, Komazaki S, Takada H (2006) Detection and characterization of new genotypes of *Myzus antirrhinii* in Japan, with evidence for production of sexual morphs. Bull Ent Res 96:605–611. https://doi.org/10.1079/BER2006460

Shridhar J, Venkateshwarlu, Nagesh M (2015) Aphids. In: Singh BP, Nagesh M, Sharma S, Sagar, Jeevalatha A, Shridhar J (eds) A manual on diseases and pests of potato. Technical Bulletin No. 101. ICAR-central Potato Research Institute Shimla, pp 56–61

Sigvald R (1984) The relative efficiency of some aphid species as vectors of potato virus Y^o (PVY^o). Potato Res 27(3):285–290. https://doi.org/10.1007/BF02357636

Sigvald R (1987) Aphid migration and the importance of some aphid species as vectors of potato virus Y^O (PVY^O) in Sweden. Potato Res 30:267–283. https://doi.org/10.1007/BF02357668

Sigvald R (1989) Relationship between aphid occurrence and spread of potato virus Y^o (PVY^O) in field experiments in southern Sweden. J App Ent 108:35–43. https://doi.org/10.1111/j.1439-0418.1989.tb00430.x

Sigvald R (1990) Aphids on potato foliage in Sweden and their importance as vectors of potato virus Y^o. Acta Agric Scand 40(1):53–58. https://doi.org/10.1080/00015129009438547

Singh R and Ghosh S. (2012) Sexuales of Aphids (Insecta: Homoptera: Aphididae) in India. Lambert Academic Publishing Gmbh & CO KG Germany, pp 402

Singh RP, Kurz J, Boiteau G (1996) Detection of stylet-borne and circulative potato viruses in aphids by duplex reverse transcription polymerase chain reaction. J Virol Methods 59:189–196. https://doi.org/10.1016/0166-0934(96)02043-5

Takada H (1988) Interclonal variability in the photoperiodic response for sexual morph production of Japanese *Aphis gossypii* Glover (Hom., Aphididae). J Appl Ent 106:188–197. https://doi.org/10.1111/j.1439-0418.1988.tb00582.x

Takahashi R (1923) Aphididae of Formosa Part 2. Rep. Govt Res. lnst. Dep. Agric. Formosa No. 4, 173 pp

Talati GM, Bhutami PG (1980) Reproduction and population dynamics of groundnut aphids. Gujarat Agric Univ J Res 5:54–56

Tanaka T (1961) The rice root aphids, their ecology and control. Spec Bull Coll Agric Utsunomiya 10:1–83

Torikura H (1991) Revisional notes on Japanese *Rhopalosiphum*, with keys to species based on the morphs on the primary host. Jpn J Ent 59:257–273

Tsuchida T, Koga R, Horikawa M, Tsunoda T, Maoka T, Matsumoto S, Simon JC, Fukatsu T (2010) Symbiotic bacterium modifies aphid body color. Science 330:1102–1104. https://doi.org/10.1126/science.1195463

van Emden HF, Harrington R (2007) Aphids as crop pests, 1st edn. CAB International, Willingford, United Kingdom

van Emden HF, Eastop VF, Hughes RD, Way MJ (1969) The Ecology of *Myzus persicae*. Ann Rev Entomol 14:197–270. https://doi.org/10.1146/annurev.en.14.010169.001213

van Hoof HA (1977) Determination of the infection pressure of potato virus Y^N. Ned J Pl Path 83:123–127. https://doi.org/10.1007/BF01981557

van Hoof HA (1980) Aphid vectors of potato virus Y^N. Eur J Pl Path 86(3):159–162. https://doi.org/10.1007/BF01989708

Verma KD (1988) First record of apterous oviparous females of potato root aphid from India. J Indian Potato Assoc 15:192

Verma KD, Chandla VK (1990) Potato aphids and their management. Technical Bulletin No. 26 (Revised). ICAR-Central Potato Research Institute, Shimla, 34 pp

Verma KD, Chauhan RS (1993) The life cycle of potato vector, *Myzus persieae* (Sulzer). Curr Sci 65:488–489

Verma KD, Ghosh LK (1990) Discovery of sexual female of Myzus persicae (Sulzer) with redescription of its alate male from India. J Aphidol 4:30–35

Verma KD, Khurana AD (1974) Sexuals of *Aphis craccivora* Koch, on green gram in India. Entomol News l 4:53

Verma KD, Parihar SBS (1991) Build up of the vector *Aphis gossypii* (Glover) on potato. J Aphidol 5:16–18

Williams IS, Dixon AFG (2007) Life cycles and polymorphism. In: van Emden H, Harrington R (eds) Aphids as crop pests. CAB International, Wallingford, UK, pp 69–86

Yano K, Miyake T, Eastop VF (1983) The biology and economic importance of rice aphids (Hemiptera: Aphididae): a review. Bull Ent Res 73:539–566. https://doi.org/10.1017/S0007485300009160

Young WR, Bhatia SK, Phadke KG (1971) Rice Root Aphid observed on Barley at Delhi. Entomol Newsl 1:53

Zhang G, Zhong T (1982) Experimental studies on some aphid life-cycle patterns. Sinozoologia 2:7–17

Takada H (1988) Interclonal variability in the photoperiodic response for sexual morph production of Japanese *Aphis gossypii* Glover (Hom., Aphididae). J Appl Ent 106:188–197. https://doi.org/10.1111/j.1439-0418.1988.tb00582.x
Takahashi R (1923) Aphididae of Formosa Part 2. Rep. Govt Res. Inst. Dep. Agric. Formosa No. 4, [illegible] pp
[illegible] GM, [illegible] PC (1990) Reproduction and population dynamics of groundnut aphids. [illegible] J Res 2:54–56
Tanaka T (1961) The rice root aphids: their ecology and control. Spec Bull Coll Agric Utsunomiya 10:1–83
Torikura H (1991) Revisional notes on Japanese *Rhopalosiphum*, with keys to species based on the morphs on the primary host. Jpn J Ent 59:257–273
Tsuchida T, Koga R, Horikawa M, Tsunoda T, Maoka T, Matsumoto S, Simon JC, Fukatsu T (2010) Symbiotic bacterium modifies aphid body color. Science 330:1102–1104. https://doi.org/10.1126/science.1195463
van Emden HF, Harrington R (2007) Aphids as crop pests. 1st edn. CAB International, Wallingford, United Kingdom
van Emden HF, Eastop VF, Hughes RD, Way MJ (1969) The ecology of *Myzus persicae*. Annu Rev Entomol 14:197–270. https://doi.org/10.1146/annurev.en.14.010169.001213
van Hoof HA (1977) Determination of the infection pressure of potato virus Y[N]. Neth J Pl Path 83:123–127. https://doi.org/10.1007/BF01981357
van Hoof HA (1980) Aphid vectors of potato virus Y[N]. Neth J Pl Path 86:159–162. https://doi.org/10.1007/BF01989708
Verma KD (1988) First record of [illegible] of [illegible] females of potato root aphid from India. [illegible] Assoc 12:[illegible]
Verma KD, Chandla VK (1990) Potato aphids and their management. Technical Bulletin No. 20 (Revised), ICAR-Central Potato Research Institute, Shimla, 34 pp
Verma KD, Chandla VK (1993) The life cycle of potato aphid, *Myzus persicae* (Sulzer). Curr Sci [illegible]
Verma KD, Ghosh LK (1990) Discovery of sexual morphs of [illegible] with description of its alate male from India. J Aphidol 4:30–35
Verma KD, [illegible] (1978) Sexuals of aphids [illegible]
Verma KD, [illegible] (1993) [illegible] *Aphis gossypii* (Glover) on potato. [illegible] Aphidol [illegible]
Williams IS, Dixon AFG (2007) Life cycles and polymorphism. In: van Emden HF, Harrington R (eds) Aphids as crop pests. CAB International, Wallingford, UK, pp 69–86
Yano K, Miyake T, Eastop VF (1983) The biology and economic importance of rice aphids (Hemiptera: Aphididae): a review. Bull Ent Res 73:539–566. https://doi.org/10.1017/S0007485300009160
Young WR, [illegible] (1971) Rice Root Aphid observed on barley at Delhi. Entomol Newsl 1:53
Zhang G, Zhong T (1983) Experimental studies on some aphid life-cycle patterns. Sinozoologia 3:7–17

Chapter 6
Organic Carbon and Ecosystem Services in Agricultural Soils of the Mediterranean Basin

Rosa Francaviglia, Luigi Ledda and Roberta Farina

Abstract Soil organic carbon (SOC), the major component of soil organic matter (SOM), is extremely important in all soil processes. Organic material in the soil is essentially derived from plant and animal residues, synthesized by microbes and decomposed under the influence of temperature, moisture and soil conditions. The problem of soil organic carbon depletion is of particular concern in the Mediterranean basin, with mild or moderately cold humid winters and warm dry summers, since high temperatures and reduced soil moisture conditions accelerate decomposition processes. This depletion is often in combination with non-conservative agronomic practices such as deep tillage and the low inputs of organic matter to soils, as well as other soil degradation processes, e.g. soil erosion by water. Typically, soils developed in the Mediterranean basin exhibit a high spatial variability of soil properties, are prone to drought, have low water holding capacity, and are shallow particularly on slopes or stony on the soil surface. They are also relatively fragile, and vulnerable to different human activities arising from changes in land cover and land use such as deforestation, urban development and deep soil tillage, and as a result of unsustainable agricultural and forestry practices. In this situation many ecosystem services (ES) are severely threatened. Here we describe the main ecosystems services including provisional, regulating, aesthetic and supporting services, with a focus on the provision of services from soil carbon and crop sustainable management in the Mediterranean basin, including the threats derived from soil erosion and floods. We highlight the specific measures for a sustainable cropland management that can decrease soil organic carbon (SOC) losses, increase the external organic matter (OM) input, and how to efficiently combine both. We reviewed different measures adopting external organic input addition to soil, conservation agriculture by no-tillage, residues retention,

R. Francaviglia (✉) · R. Farina
Consiglio per la ricerca in agricoltura e l'analisi dell'economia agraria,
Centro di ricerca Agricoltura e Ambiente, 00184 Rome, Italy
e-mail: rosa.francaviglia@crea.gov.it

L. Ledda
Dipartimento di Agraria, Sezione di Agronomia, Coltivazioni erbacee e Genetica,
Università di Sassari, 07100 Sassari, Italy

S. Gaba et al. (eds.), *Sustainable Agriculture Reviews 28*, Ecology for Agriculture 28,
https://doi.org/10.1007/978-3-319-90309-5_6

cover crops, organic farming compared to conventional agriculture and sustainable crop management by irrigation. In arable cropping systems, we reported an increase in C sequestration rate ranging from 1.3 to 5.3 Mg C ha^{-1} yr^{-1} with the addition of organic external inputs, and equal to 0.27 Mg C ha^{-1} yr^{-1} with the adoption of cover crops. No tillage and reduced tillage can increase C sequestration rate by 0.44 and 0.32 Mg C ha^{-1} yr^{-1} respectively. The adoption of combined management practices, where organic matter inputs and conservation tillage practices are simultaneously applied, increase C sequestration rate by 1.11 Mg C ha^{-1} yr^{-1}. Organic farming management increase C sequestration rate by 0.97 Mg C ha^{-1} yr^{-1} as average, ranging from 0.62 to 1.32 Mg C ha^{-1} yr^{-1} with compost application and manure combined with cover crops respectively. Organic farming is also effective in increasing soil organic carbon stocks by about 70% compared with conventional management, and depending on soil type in permanent crops such as olive groves. Regulated deficit irrigation in summer crops is able to decrease CO_2 emissions by about 10%, and consequently soil organic carbon losses without any negative effect on crop yields such as tomato. Soil erosion by water in permanent crops can be decreased by more than 70% with the use of cover crops, and by more than 40% with the adoption of temporary ditches on sloping soils in arable crops.

Keywords Soil organic carbon · Ecosystem services · Agricultural soils
Soil erosion · Crop management · Mediterranean basin

List of Abbreviations

Carbon dioxide (CO_2)
Carbon to Nitrogen Ratio (C/N)
Combined Management Practices (CMPs)
Conservation Agriculture (CA)
Conventional Tillage (CT)
Cover Crop (CC)
Crop Residues (CR)
European Union (EU)
Methane (CH_4)
Millennium Ecosystem Assessment (MEA)
Nitrous oxide (N_2O)
No Tillage (NT)
Other Nitrogen Oxide Compounds (NO_x)
Recommended Management Practices (RMPs)
Reduced Tillage (RT)
Soil Inorganic Carbon (SIC)
Soil Inorganic Matter (SOM)
Soil Organic Carbon (SOC)
United Nations Environment Programme (UNEP)
Unmanned Aerial Vehicles to gather GIS data (UAV-GIS)

6.1 Introduction

Organic material in the soil is essentially derived from residual plant and animal material, synthesized by microbes and decomposed under the influence of temperature, moisture and soil conditions. Soil organic carbon (SOC), the major component of soil organic matter (SOM), is extremely important in all soil processes. In general, SOC content is positively correlated with a fertile soil status, and plays a mostly beneficial role in determining the physical, chemical and biological qualities of a soil, the ecosystem functioning, and the magnitude of the different processes.

SOC varies among environments and management systems, and generally increases with higher mean annual rainfall (Burke et al. 1989) and lower mean annual temperatures (Jenny 1980); with higher clay content (Nichols 1984); with an intermediate grazing intensity (Parton et al. 1987; Schnabel et al. 2001); with higher crop residue inputs (Franzluebbers et al. 1998); with native vegetation compared with cultivated management (Burke et al. 1989; Francaviglia et al. 2014); with conservative tillage compared with plough tillage (Rasmussen and Collins 1991; Farina et al. 2011). In addition, SOC changes with land use, compaction, landscape position and slope (Cerdá et al. 2014; Fernández-Romero et al. 2014; Francaviglia et al. 2014; Lozano-García and Parras-Alcántara 2014; Parras-Alcántara et al. 2015a; Fissore et al. 2017).

SOC is of special interest in the Mediterranean basin, where rainfed cropping systems are prevalent, inputs of organic matter in soils are low and mostly rely on crop residue availability, while losses are high due to climatic and anthropogenic factors such as intensive farming practices that enhance SOC mineralization. Estimates indicate that 74% of the land in Southern Europe is covered by soil containing less than 2% of organic carbon, i.e. 3.4% organic matter in the top-soil (Zdruli et al. 2004).

6.1.1 *Soil Organic Carbon and Ecosystem Services*

The Millennium Ecosystem Assessment (MEA) investigated the consequences of ecosystem change for human well-being through an appraisal of ecosystem services (MEA 2005). The MEA categorizes ecosystem services into four different classes:

i. *Provisioning services*, the products obtained from ecosystems, including food, fiber, fuel, genetic resources, ornamental resources, freshwater, biochemical, natural medicines and pharmaceuticals.
ii. *Regulating Services*, the benefits obtained from the regulation of ecosystem processes including air quality, climate, water, erosion, water purification and waste treatment, disease, pest, pollination and natural hazard.

iii. *Cultural Services*, the non-material benefits people obtain from ecosystems including cultural diversity, spiritual and religious values, knowledge systems, educational values, inspiration, aesthetic values, social relations, sense of place, cultural heritage values, recreation and ecotourism.
iv. *Supporting services*, necessary for the production of all other ecosystem services. Some services, like erosion regulation, can be categorized as both a supporting and a regulating service. These services include soil formation, photosynthesis, primary production, and nutrient and water cycling.

The organic matter content of soils and more specifically soil organic carbon (SOC) is essential for the majority of these services. The more relevant services provided by agro-ecosystems and soils are shown in Table 6.1 and Fig. 6.1.

6.1.2 The Role of Soil and Soil Organic Matter in Providing Ecosystems Services

Despite the currency of the ecosystem services concept, there are few clear and comprehensive definitions of soil-based ecosystem services (Bennett et al. 2010). Following a recommendation by Fisher et al. (2008), the focus should be on final services because these are directly utilized by humans, and are supported by single

Table 6.1 Ecosystem services provided jointly by agro-ecosystems and soils according to the Millennium Ecosystem Assessment (Adapted from MEA 2005). They include the products obtained from ecosystems, the benefits obtained from regulation of ecosystem processes, the non-material benefits obtained from ecosystems, and the services necessary for the production of all other ecosystem services

Class	Ecosystem service
Provisioning services	Food, fiber and fuel
	Genetic resources
Regulatory services	Seed dispersal
	Pest regulation
	Disease regulation
	Climate regulation
	Water regulation
	Erosion regulation
	Natural hazard regulation
Cultural services	Recreation and ecotourism
	Cultural heritage
	Aesthetic values
Supporting services	Primary production
	Provision of habitat
	Soil formation and retention
	Nutrient cycling
	Water cycling

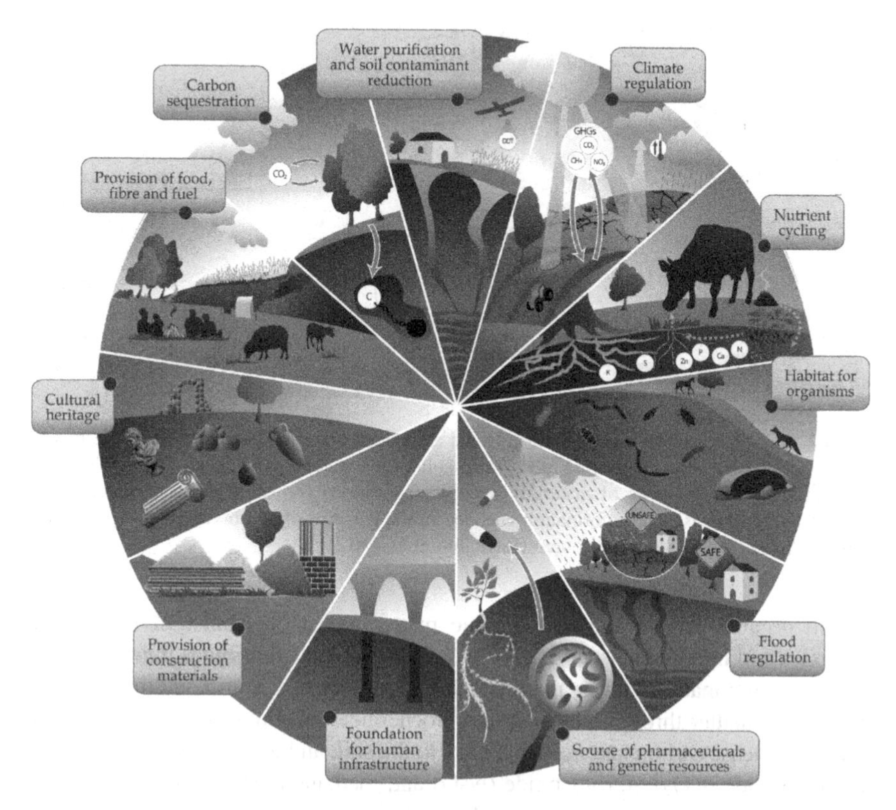

Fig. 6.1 Soils deliver provisioning, regulatory, cultural and supporting ecosystem services that enable life on earth. With permission of Philippe Baveye (Baveye et al. 2016)

or multiple intermediate services, which maintain the soil capital, but are not directly utilized.

Final services and examples of associated goods/benefits are: productive capacity of land for food or fiber, net gas emissions from land aiming to equitable climate, water runoff from land and flood control, groundwater and surface quality in particular for drinking water, ecological condition of rivers and lakes for present and future values of water quality, conditions suitable for recreation, habitat provision for landscapes and biodiversity.

Intermediate services and examples of associated processes are: soil structure maintenance such as aggregation, bioturbation, cheluviation, organic matter cycling, e.g. litter break up, decomposition, and humification, nutrient cycling linked with mineral weathering, mineralization, nitrification, ion retention and exchange, water cycling, gas cycling, and soil biological life cycles such as changes in biotic richness and composition.

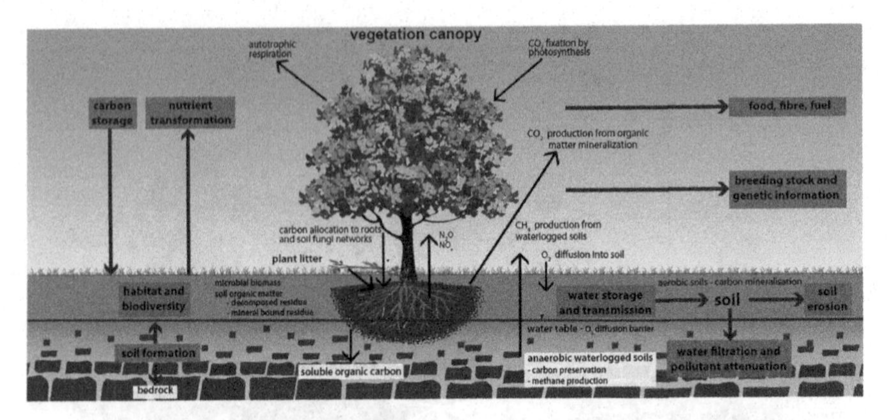

Fig. 6.2 Soil-plant carbon interrelationships and associated ecosystem services. Soils, formed by the action of biota and infiltrating water and solutes on parent rock material, provide many ecosystem services as flows of materials, such as sequestered carbon, water and solutes, plant nutrients, crop biomass, and information encoded in the genetics of soil organisms. Figure freely available from UNEP citing the source (Reynaldo et al. 2012)

Soil organic carbon and soil organic matter are at the center of soil-related processes and intermediate services (Fig. 6.2). Soil biota contributes living biomass to soil organic matter with its microbes and earthworms, that mix and break down the organic matter through physical and biochemical reactions. These biochemical reactions release carbon and nutrients back to the soil, and greenhouse gases such as carbon dioxide (CO_2), nitrous oxide (N_2O) and methane (CH_4) to the atmosphere.

Soil management can affect the relative balance of these processes and their environmental impacts. As soil organic matter is broken down, part of the carbon is mineralized rather rapidly to CO_2 and is lost from the soil. Soil organic matter may also be lost through physical erosion. Organic nitrogen contained in biodegrading soil organic matter is transformed to N_2O and other nitrogen oxide (NO_x) compounds. However, some fractions of soil organic matter are not readily degraded. Soil organic carbon content therefore tends to increase as soil is left undisturbed over time. In water-saturated soils, soil organic matter may even accumulate as thick layers of peat (Beer and Blodau 2007). Organic matter binds to minerals, particularly clay particles, a process that further protects carbon (Von Lützow et al. 2006). Organic matter also provides cohesive strength to soil and improves soil fertility, water movement, and resistance to erosion.

Ways to estimate soil carbon stocks and fluxes from field to global scales continue to be developed (Bernoux et al. 2010; Hillier et al. 2011). Actually, the lack of adequate methodologies and approaches has been one of the main constraints to accounting for the significant mitigation effects that land management planning actions can have, and there is a relevant need to develop universally agreed and reproducible field and laboratory methods for measuring, reporting and verifying changes in soil carbon over time. In fact, a number of studies points out the need to make a joint effort to homogenize the terminology, the methods to calculate the soil organic carbon storage, and even the soil sampling methods (Powlson et al. 2011; Parras-Alcántara et al. 2015b; Lorenz and Lal 2016; Francaviglia et al. 2017). Thus, the choice of sampling methods is important in providing results that are reliable, comparable and can be extrapolated (Lal 2005).

Soil carbon stocks are highly vulnerable to human activities. They decrease significantly and often rapidly in response to changes in land cover and land use such as deforestation, urban development and increased tillage, and as a result of unsustainable agricultural and forestry practices. Even deep soil carbon pools, previously considered resilient to degradation, have been shown to be vulnerable to environmental or anthropogenic change, contributing to net CO_2 exchange between the land and the atmosphere (Bernal et al. 2016).

Soil organic carbon may also be increased, although much more slowly by afforestation, retention of crops residues, direct addition of external organic carbon, and other activities that decrease the breakdown of soil organic matter, e.g. minimum or zero tillage, perennial pastures, designation of protected areas.

The incorporation of crop residues, such as cereal straw or cover crops, is an important measure to maintain or increase soil organic matter levels of cropland in particular under rainfed cropping system (Lugato et al. 2014).

External organic carbon can be applied to the soil as organic manure, compost, digestate, black carbon or biochar, and will be subjected to degradation, with CO_2 release (by heterotrophic respiration), and humification processes. Generally, mineralization of organic products ranks inversely with respect to their C/N ratio. Nowadays digestates tend to originate from biogas production of livestock biodegradable wastes, and are increasingly applied to soils as organic fertilizers, with positive environmental and agricultural benefits. However, anaerobic digestates generally have a lower C/N ratio than their aerobic compost counterparts (Canali et al. 2011), thus could represent a higher risk to the environment and human health than undigested animal manures and slurries, and they should be used cautiously (Nkoa 2014). The application of biochar or charcoal has been suggested to increase soil carbon stocks and to improve soil fertility especially of soils poor in carbon. Unfortunately, the impact of biochar addition to soils on heterotrophic respiration is not fully understood and inconsistent findings are reported (Mukherjee et al. 2014, and the references therein).

Climate change is expected to have significant impacts on soil carbon dynamics. Rising atmospheric CO_2 levels could increase biomass production and inputs of organic materials into soils. However, increasing temperatures could reduce soil organic carbon by accelerating the microbial decomposition and the oxidation of soil organic matter. Current scientific knowledge of how local soil properties and climatic conditions affect soil carbon stock changes and carbon fluxes is insufficient and conflicting (Tuomi et al. 2008; Conant et al. 2011; Falloon et al. 2011). Further study is needed to enable more accurate predictions of the impacts of climate change on soils, soil carbon and associated ecosystem services at scales relevant to local management, as well as to national carbon inventories.

6.2 Mediterranean Climate and Soils

Mediterranean climate is present on all continents apart from Antarctica, mainly at latitudes between 35° and 42°, and occurs in just five regions of the world: the Mediterranean Basin, the Cape Region of South Africa, Southwestern and South Australia, California, and Central Chile.

The Mediterranean basin is characterized by a specific climate, with mild or moderately cold humid winters and warm dry summers (Cowling et al. 2005), thus soil temperatures are higher than in Northern Europe, with an expected negative impact on soil organic carbon content. Moreover, Mediterranean and dryland ecosystems are more prone to land degradation due to soil organic carbon degradation and depletion, often coupled with erosive processes (Muñoz-Rojas et al. 2015). Erosion is further influenced by the incomplete coverage of the soil by the vegetation as a consequence of drought or land uses with bare soils such as vineyards and olive groves.

The issue of low soil organic carbon is of particular concern in perennial systems such as orchards and vineyards (Meersmans et al. 2012), which play an important role in Southern Europe. Grasslands and pastures can be subjected locally to overgrazing with a potential threat on soil organic matter. In addition, wildfires can also have a negative impact on soil organic matter, but they normally affect forests and rangelands and are of limited concern in cultivated agro-ecosystems. However, in the abandoned agricultural lands and rangeland, the consequent increase in dead plant biomass is expected to increase fire hazard.

Irrigation is widely utilized in intensive cropping systems in semi-arid environments, but this practice can induce an overall decrease in soil organic matter (Costantini and Lorenzetti 2013), unless combined with specific soil management techniques and rational use of irrigation for the summer crops (Boulal and Gómez-Macpherson 2010; Boulal et al. 2011; Di Bene et al. 2016; Farina et al. 2017). Typically soils developed in the Mediterranean basin exhibit a high spatial variability of soil properties, have unfavorable physical conditions due to drought and decreased water holding capacity, are shallow particularly on slopes, and stony on the soil surface (Rodeghiero et al. 2011). Calcareous soils with neutral to slightly alkaline pH values are more abundant than in Northern Europe, offering conditions that favor a rapid decay of SOM (Romanyà and Rovira 2011). The prevailing soil types are Cambisols, followed by Fluvisols, Luvisols and Leptosols. Cambisols are common in wide areas of the Iberian Peninsula and in most of the central and western Mediterranean islands, as well as in most parts of the Italian Peninsula. Fluvisols, Leptosols and Luvisols are also present in many Mediterranean areas. For instance, Leptosols are predominant in many areas of the countries influenced by the Aegean Sea, as well as some other parts of the eastern Iberian Peninsula. Luvisols are predominant in some areas of the regions influenced by the Marmara Sea, regions affected by the eastern part of the Ionian Sea or some areas of the central and western Iberian Peninsula (Rodeghiero et al. 2011).

6.3 Agro-Ecosystems Services from Soil Carbon in the Mediterranean Basin

6.3.1 Provisional Ecosystem Services: Food, Fiber and Fuel, Energy

Agriculture in the Mediterranean basin is mainly based on three cropping systems (Mørch 1999): (i) Rainfed annual crops based on the winter rain. (ii) Permanent crops, e.g. olive groves and vineyards, able to survive during the dry summer. (iii) Irrigated cropping systems, where irrigation is able to compensate for the lack of precipitation in summer, and enabling to grow more than one annual crop, e.g. horticultural crops, and fruits such as citrus, almonds, etc.

Winter crops are sown in the autumn, harvested in late spring-early summer, and thus utilize the winter rain and avoid the summer drought. Of the arable crops, half of them are cereals, of which 50% is wheat. Due to demand and European Union subsidies, wheat has gained importance over barley, the other classical Mediterranean cereal crop, and oats. More than two thirds of the wheat area is bread wheat, the rest is durum wheat located in the dryer parts of southern Mediterranean Europe and in North Africa. Barley is increasingly confined to semi-arid areas, e.g. in South-Eastern Spain, North Africa, and Near East because of its higher tolerance to drought. Maize is sometimes mentioned as a typical Mediterranean crop, but accounts for just one eight of the cereal area and is restricted to irrigated areas. Legume crops have a significance generally in rotation with cereals as fodder crops such as alfalfa, vetch, etc., and for human consumption since broad bean, chickpea, and lentils are important sources of protein in the traditional Mediterranean diet. Also vegetables are cultivated as field crops, and some, like tomato, peas, etc., are even cultivated for the agro-industrial production. The arable farming is predominantly located on plain land and terrain with moderate slopes. In rainfed agriculture, cereals are grown in rotation with fallow and possibly with legumes crops. Presently there is a trend to introduce cover crops after the winter cereals, to contribute to the storage of soil water for the following crop, to be used as green manure, and to uptake soil nitrogen before it can leach; in addition, legume cover crops fix nitrogen from the atmosphere that is available to subsequent crops (Dabney et al. 2001). But more complex rotations exist, depending on the agricultural system, soils, duration of drought period, and local tradition.

Due to extended root systems permanent crops such as fruit orchards, vineyards and olive tree plantations can better utilize soils which are too shallow, stony and steep for arable agriculture, and are grown in almost any type of terrain. In the more intensive systems, the soil is held free from weeds, thus the surface is exposed to surface run off and erosion. But cover crops can be established below the trees protecting the soil from erosion, contributing to soil carbon sequestration (González-Sánchez et al. 2012; Aguilera et al. 2013) and potentially serving as animal feed (Ramos et al. 2011). Olives and grapes are by far the most important crops, figs and almonds and locally carob, pistachio, apricot, and others are also widespread. Fruit tree orchards also produce large amounts of residual biomass in the form of pruning residues, which can be used for soil conditioning, animal feeding, or for energetic purposes (Kroodsma and Field 2006; Infante-Amate and González de Molina 2013).

The solution to the summer drought is irrigation, by which it is not only possible to utilize the arable land for a longer period, but also to take advantage of the high potential plant production in the summer period. However, irrigation depends on soil morphology, available water and technological level. The irrigated areas generally represent a very small share of the agricultural land, mainly on plain soils, and are under very intensive cultivation of citrus and other fruit crops, vegetables such as tomato, sweet pepper, aubergine, artichoke, fennel, cucumbers, melons, and many common industrial crops including tobacco, sugar beets, and cotton.

Growing energy crops has received increased interest as a non-traditional land use option able to convert solar energy into stored biomass relatively efficiently, and leading to positive input/output energy balances for the overall system (Sims et al. 2006). Combining production and environmental sustainability, they help to reduce soil degradation due to the protection that the dense canopy affords against the erosion caused by the intense precipitations that occur in Mediterranean areas (Grammelis et al. 2008; Lag-Brotons et al. 2014). Bioenergy crops produce both lignocellulosic biomass for solid biofuel production, and oil seeds for human consumption or biodiesel production (Acquadro et al. 2013; Ledda et al. 2013; Francaviglia et al. 2016), and have been recognized as promising energy crops for rainfed farmlands in Mediterranean climates under low external management supplies in marginal lands.

6.3.2 *Regulating Ecosystem Services: Climate and Gas Regulation, Carbon Sequestration, Erosion and Flood Control*

Agriculture is a major source of greenhouse gases and is also indirectly responsible for a large share of the greenhouse gases emitted in deforestation and other land use changes. Arable agriculture produces direct and indirect nitrous oxide (N_2O) emissions from nitrogen (N) application to soils, methane (CH_4) from rice paddies, carbon dioxide (CO_2) from direct fossil energy use, and N_2O and CH_4 from open biomass burning. Further emissions occur due to fossil energy use in the production of agricultural inputs, particularly N fertilizers (Aguilera et al. 2015). A recent meta-analysis of the N_2O emissions from Mediterranean cropping systems distinguished the effects of water management, crop type, and fertilizer management. Increasing the N fertilizer rate led to higher emissions, rainfed crops have lower emissions than irrigated crops, drip irrigation systems lower than sprinkler irrigation methods, extensive crops, such as winter cereals (wheat, oat and barley), had lower emissions than intensive crops such as maize. For flooded rice, anaerobic conditions likely led to complete denitrification and low emissions (Cayuela et al. 2017).

The CO_2-C emissions in a 20 years simulation in Southern Italy were higher in arable irrigated crops, land use change from pasture to arable and vineyards, with emissions equal to 41.6, 40.8, and 40.1 Mg CO_2-C ha^{-1} respectively. The emissions from other land uses, e.g. rainfed arable crops and olive groves were on average less than 39 Mg CO_2-C ha^{-1} (Farina et al. 2017). Di Bene et al. (2016) indicated that deficit irrigation in summer crops decrease CO_2 emissions by about 10% without affecting tomato yield.

Lastly, the other major link between agriculture and climate change is soil carbon (C) balance. Historical C losses from agricultural soils have contributed to the increase in both the global greenhouse gas budget and the vulnerability of agriculture to climate change (Lal et al. 2011), suggesting the need of management decisions to reduce soil erosion, increase carbon sequestration and contribute to the resilience of soils and cropping systems, and aimed to respond to climate change and related challenges such as food security.

Increasing soil organic carbon (SOC) is a key process in both mitigation and adaptation strategies to climate change. SOC is highly sensitive to changes in agricultural management under Mediterranean conditions, and Recommended Management Practices significantly increase soil organic carbon when compared to conventional management (Aguilera et al. 2013). Relevant increases are achieved by the application of external organic inputs to the soil, suggesting a high potential for carbon storage in the soil. Soil organic carbon is also increased by internal organic inputs and the reduction in soil disturbance by introducing cover crops, and adopting reduced tillage and no tillage. The combination of organic inputs and reduced disturbance is the most promising strategy to maximize soil organic carbon levels. Slurry applications only maintained soil organic carbon at the same level as mineral fertilization, while in unfertilized treatments a slight decrease was observed. Moreover, C sequestration is effectively promoted by organic farming practices and the relative increase of SOC sequestration over conventional practices is more marked in more intensive cropping systems, where differences in C inputs are higher (Aguilera et al. 2013).

Erosion has been recognized as one of the most significant environmental issue worldwide (Bakker et al. 2007), particularly in areas having seasonally contrasted climate and a long history of human pressure. This is the case of the Mediterranean basin, where concerns about soil loss and its consequences emerged mainly with the cultivation of steep slopes (López-Bermúdez 2008). A recent review of European sediment yields demonstrated that Mediterranean rivers have higher yields than those in the rest of Europe, which has been attributed to climate, topography, lithology and land use (Vanmaercke et al. 2011). Intense erosion processes are widespread, particularly sheet erosion, rilling, gullying, shallow land sliding and the development of active badlands in sub–humid and semi–arid areas (Figs. 6.3 and 6.4).

The complex mosaic that characterizes the Mediterranean landscape is a product of the occurrence of intense rainstorms and prolonged droughts, the presence of

(a)

(b)

Fig. 6.3 Olive groves in the province of Jaén (Spain). Soil with high erosion rates, due to the lack of vegetation from high herbicide use, leads to a heavy soil organic carbon decrease (**a**). Olive groves with spontaneous resident vegetation are adopted to protect soils from erosion and increase soil organic carbon content (**b**). With permission of José Luis Vicente-Vicente (University of Jaén, Spain)

Fig. 6.4 Arable fields in Rome province (Italy). Heavy rill erosion and sediment accumulation in a sloping arable field (**a**). Temporary ditches can be adopted to decrease water runoff and soil erosion (**b**). With permission of Paolo Bazzoffi and Ulderico Neri (Consiglio per la ricerca in agricoltura e l'analisi dell'economia agraria, Italy)

steep slopes, topographic diversity, high evapotranspiration, recent tectonic activity, and the long history of human activity reflected in the recurrent use of fire, overgrazing and farming (García-Ruiz et al. 2013). Such complexity explains why some areas are well preserved, including those with dense plant cover and deep soils, whereas in other areas most of the soil and plant cover has been lost, and they are now degraded, covered by stones and affected by rilling and gullying. Wealth and poverty, and land management intensification and extensification co-exist, resulting in a variety of hydrological and geomorphological processes. Thus, widespread land abandonment has occurred in all Mediterranean countries, particularly on the northern side of the basin. Other areas are affected by increasing erosion because of land use intensification arising from expansion of cultivation areas in response to population growth (North Africa), poorly targeted subsidies (vineyards, almond and olive orchards), attempts to increase profits through irrigation, and the expansion of urban areas into rural zones.

Many Mediterranean landscapes can be considered man-made, and to be preserved they require the presence of a rural population. This is the case of bench-terraced fields, which have a history of construction over several centuries; these enabled cultivation on slopes and avoided soil erosion. Their recent evolution has been dominated by landslides between the terraces and by gullying, processes that have led to the degradation of these impressive cultural relics. In other cases, population migration and the European Union Agricultural Policy have encouraged land use changes that have not been considered in the context of their impact on infiltration rates, water storage and the connectivity between hillslopes and channels. Many areas of the Mediterranean appear to be degraded, while others have maintained their geoecological functions. Efforts must be invested in understanding the hydro morphological functioning of the former, even if they are poorly productive. Both types of areas need to be considered as part of a delicate equilibrium between nature and the long history of human activities.

Leaving aside droughts, floods are the most dangerous meteorological hazards affecting the Mediterranean countries, followed by windstorms and hail (Llasat et al. 2010). This is due not only to high flooding frequency, but also to the vulnerability caused by various human activities, e.g. soil sealing. Indeed, in Mediterranean regions such as eastern Spain, southern France, Italy and the west of the Balkan Peninsula, floods are frequent enough to be considered as a component of the local climate. These regions have widespread and intense economic activity and high population densities, with significant economic losses following flood events. Although floods affect the entire Mediterranean region, their frequency and impact is not homogeneous over the entire area. Their greater frequency and social impact in the north-western part, together with major preventive measures such as emergency plans, environmental laws, participation in international projects, contrasts with the scant information available on floods in some southern and eastern countries. This fact points to a clear need for better coverage of such hazards in countries such as Spain, France and Italy. The spatial distribution of the different kinds of floods is neither homogeneous in the region nor stationary over time, and shows a clear difference between the western and eastern Mediterranean, with a major concentration in the former region. Flood events in the western part usually occur during autumn, while in the eastern part the major contribution is during the winter months. Spain and Italy show the highest floods frequency, though the material damage is worse in the latter country.

6.3.3 Cultural Ecosystem Services: Aesthetic Values, Recreation, Cultural Heritage

Cultural ecosystem services are commonly defined as the nonmaterial benefits people obtain from ecosystems through spiritual enrichment, cognitive development, reflection, recreation, and aesthetic experience, including, e.g. knowledge systems, social relations, and aesthetic values. In general, cultural ecosystem services incentivize the multifunctionality of landscapes. However, depending on the context, cultural ecosystem services can either encourage the maintenance of valuable landscapes or act as barriers to necessary innovation and transformation. Hence, cultural ecosystems services influence landowner decision-making, community engagement, and landscape planning (Plieninger et al. 2015).

Many people enjoy the beauty of natural or human-made landscapes and are fascinated by animals, plants and ecosystems. Landscapes have also been the source of inspiration for much of our art and culture as well as for technological innovations. Changes in land use and degradation caused by unsustainable land use reduce the attractiveness and scenic beauty of a landscape. They also compromise the environmental conditions that are crucial for all cultural ecosystem services. Cultural landscapes are especially vulnerable to social and economic changes and loss of traditional knowledge. Cultural services from landscapes are the expression of historical integration among social, economic and environmental factors, influencing all aspects of development. Many world heritage sites, e.g. the Amalfi coast in Italy (Fig. 6.5a), reflect a harmonious relationship between humankind and the natural environment and are of great aesthetic appeal. This area shows the versatility of the inhabitants in adapting their use of the land to the diverse nature of the terrain, which ranges from terraced vineyards and citrus orchards on the lower slopes to wide upland pastures. The Chianti wine region in Tuscany (Italy) can be considered as a well known example providing both provisional and cultural ecosystem services, linked to the aesthetic value of the landscape (Fig. 6.5b).

Very few studies on soil ecosystem services cover or identify “cultural services”. This is a curious omission as soils alone, as part of landscapes that supports vegetation, have been a source of aesthetic experiences, spiritual enrichment, and recreation (Dominati et al. 2010). At the non-physical level, ecosystems provide aesthetics, spiritual and cultural benefits through cultural services, thereby fulfilling self-actualization needs. The cultural and historic function of soil is mentioned by Weber (2007), the provision of recreational services by Swinton et al. (2007), the

Fig. 6.5 Terraced citrus orchards in the Amalfi coast (Italy) are an example of adapting the use of the land to the nature of the terrain (**a**). Provisional and aesthetic services in the Chianti region, Italy (**b**). Pictures are freely available from the WEB

aesthetic value of managed agro-ecosystems by Zhang et al. (2007), the aesthetic services by Sandhu et al. (2008) and Porter et al. (2009).

6.3.4 Supporting Ecosystem Services: Weathering/Soil Formation and Retention, Nutrient Cycling, Water Cycling

Soils are complex dynamic systems consisting of soil components (abiotic and biotic) interconnected by biological, physical and chemical processes. Soil processes support soil formation, which is the development of soil properties and soil natural capital stocks. Soil processes also form the core of soil functioning and allow the establishment of equilibria and the maintenance of natural capital stocks. The following supporting processes are included in the conceptual framework (Dominati et al. 2010):

i. Nutrient cycling, which refers to the processes by which a chemical element moves through both the biotic and abiotic compartments of soils. Nutrient cycles are a way to conceptualize the transformations of elements in a soil. The transformation, or cycling, of nutrients into different forms in soils is what maintain equilibria between forms, e.g. soil solution concentrations of nitrate drive many processes such as plant uptake, exchange reactions with clay surfaces or microbial immobilization.

ii. Water cycling, which refers to the physical processes enabling water to enter soils, be stored and released. Soil moisture is the driver of many chemical and biological processes and is therefore essential in soil development and functioning. The continuous movements of water through soils carrying nutrients disturb chemical equilibria, and thereby drive transformations.

iii. Soil biological activity: soils provide habitat to a great diversity of species, enabling them to function and develop. In return, the activity and diversity of soil biota are essential to soil structure, nutrient cycling, and detoxification. Biological processes include predation, excretion and primary production among others.

These processes are at the core of soil formation (pedogenesis), building up the physical, biological and chemical stocks of soils. Pedogenesis is the combined effect of physical, chemical, biological, and anthropogenic processes on soil parent material. Soils are formed from the rock materials that make up the earth's crust. Soils can be formed from the underlying bedrock, from material moved at relatively small distances e.g. down slope, or even at considerable distances from where the bedrock was originally exposed to the environment by heavy erosion events or landslides. The formation of a soil in these mineral deposits is a complex process. It may take centuries for a developing soil to acquire distinct profile characteristics. Minerals derived from weathered rocks undergo chemical weathering creating secondary minerals and other compounds that vary in water solubility. These constituents are translocated through the soil profile by water and biota. In addition to chemical weathering, physical weathering also takes place. It refers to the disintegration of mineral matter into increasingly smaller fragments or particles. Pedogenic processes, driven by nutrients and water cycles and biological activity, include the accumulation of organic matter, leaching, the accumulation of soluble salts, calcium carbonate and colloids, nutrient redistribution, gleying and the deposition and loss of materials by erosion, and are very important in soil development and defining soil properties.

6.4 How to Improve the Soil Organic Carbon Content and Ecosystem Services of Agricultural Soils in Mediterranean Regions

Soil organic carbon (SOC) sequestration via agricultural soils has the potential to contribute significantly to climate change mitigation, provided that specific measures are implemented. Sustainable cropland management can play a positive role in reducing greenhouse gases (GHGs) emissions, and in particular carbon dioxide (CO_2), either by decreasing soil organic carbon (SOC) losses, or by increasing the external organic matter input, or with a combination of both.

The adoption of reduced or no-tillage systems, characterized by a lower soil disturbance in comparison with conventional tillage, has proved to impact positively SOC and other physical and chemical processes and functions, e.g. erosion, compaction, ion retention and exchange, buffering capacity, water retention and aggregate stability in the top soils (Novara et al. 2011; Lieskovský and Kenderessy 2014; Marques et al. 2015; Brunori et al. 2016). Moreover, a reduced tillage intensity usually improves soil biological and biochemical processes (Caravaca et al. 2002; Riffaldi et al. 2002; García-Orenes et al. 2010; Marzaioli et al. 2010; Lagomarsino et al. 2011; Laudicina et al. 2011; Novara et al. 2012; Balota et al. 2014; Bevivino et al. 2014; Laudicina et al. 2015).

Table 6.2 Main categories considered in Aguilera et al. (2013). Management practices were grouped in levels, according to their focus on organic inputs, e.g. crop residues, organic amendments, slurry, wastes, or tillage management including conventional and conservation tillage, such as no tillage, reduced or minimum tillage

Category	Organic input	Tillage type	Remarks
Conventional management	None or CR	Conventional	Used as control
Organic farming	All except municipal solid waste/sewage sludge	All except no tillage with herbicides use	No synthetic compounds are used
Organic amendments	Compost, manure, agro-industrial wastes	Conventional	External inputs <10 Mg C ha^{-1} yr^{-1}
Land treatment	Compost, manure, agro-industrial wastes, municipal solid wastes, sewage sludge	Conventional	Organic amendments >10 Mg C ha^{-1} yr^{-1} or municipal solid waste/sewage sludge at any rate
Cover crops	Cover crops	Conventional, RT, NTM	CC replace bare fallow
Slurry	Slurry	Conventional	Raw or digested liquid manures
No tillage	None or CR	No tillage	Organic C inputs may be different from conventional
Reduced tillage	None or CR	Reduced, minimum, subsoil	Organic C inputs may be different from conventional
Combined management practices	Organic amendments and CC/CR	Conventional, RT, NT, NTM	Organic amendments combined with CC, CR, RT or NT
Unfertilized	As in conventional	As in conventional	No synthetic fertilizer

CC: cover crops; CR: crop residues; NT: no tillage; NTM: no tillage by mowing; RT: reduced tillage.

Tillage management together with other Recommended Management Practices (RMPs) in Mediterranean cropping systems were widely studied in a meta-analysis by Aguilera et al. (2013) with the categories described in Table 6.2.

On average, SOC content increased by 98.2% and C sequestration rate by 5.29 Mg C ha^{-1} yr^{-1} in land treatment with the addition of organic amendments at rates >10 Mg C ha^{-1} yr^{-1}, or wastes or sludges at any rate. The application of organic amendments at agronomic rates <10 Mg C ha^{-1} yr^{-1} increased SOC by 23.5% and C sequestration rate by 1.31 Mg C ha^{-1} yr^{-1}. In cover crop category, where the organic matter input is always produced within the system, the average increases in SOC were reduced to 10% and C sequestration averaged 0.27 Mg C ha^{-1} yr^{-1}. In slurry category the differences with conventional management were not significant.

Both the studied conservation tillage categories enhanced C sequestration. No tillage showed an average increase of 11.4% in SOC resulting in a C sequestration rate of 0.44 Mg C ha^{-1} yr^{-1}. Under reduced tillage, SOC content increased by 15% and C sequestration rate by 0.32 Mg C ha^{-1} yr^{-1}. Changes in SOC under no tillage reflect the localization of C derived from crop residues, as most C is gained in the vicinity of the soil surface and might be lost at lower depths. Conversely, under conventional tillage crop residues are distributed throughout the ploughed zone, and can show C levels similar or even higher than no tillage for a given depth (Kay and VandenBygaart 2002; López-Fando and Pardo 2011).

Combined management practices is a mixed category, where organic matter inputs and conservation tillage practices are simultaneously applied. These practices promoted an increase of 49.2% in SOC, and enhanced C sequestration rate by 1.11 Mg C ha^{-1} yr^{-1}.

SOC concentration increased by 19.2% and carbon sequestration rate by 0.97 Mg C ha^{-1} yr^{-1} in organic farming, and carbon increase in organic systems over conventional management was 25% under irrigation, and 13% under rainfed conditions. When analyzed by crop type, the data shows that the best performing organic group is horticulture where SOC is increased by 48%. The type of organic input employed in the organic farming also influences the differences observed among systems. Compost, either applied alone or in combination with cover crops, is the input associated with the highest increases in SOC, equal to 48% for compost alone and 26.2% for mixtures with cover crops. The corresponding carbon sequestration rates were 1.32 and 0.97 Mg C ha^{-1} yr^{-1} respectively. With manure application, the increase in carbon sequestration rate over the conventional fertilization is not significant when the amendment is applied alone. When manure is combined with cover crops, the carbon increase and the sequestration rate over the conventional fertilization are significant, and equal to 35.8% and 0.62 Mg C ha^{-1} yr^{-1} respectively.

The use of cover crops during the winter season, used as green manure or the use of a roller crimper to terminate cover crops, is a successful strategy to increase SOC in horticultural crops and woody crops and for weed control (Smith et al. 2011; Canali et al. 2013; Pardo et al. 2017).

Water management can reduce greenhouse gas emissions in Mediterranean rice paddy fields (Meijide et al. 2017). In Mediterranean areas, strategies to decrease water consumption are recommended as good agricultural practices; thus midseason drainage of the otherwise flooded field can reduce the CH_4 emissions and water consumption, while N_2O emissions are slightly increased and yield reduced. Regulated deficit irrigation in summer crops is able to decrease CO_2 emissions (Zornoza et al. 2016), without any negative effect on crop yields such as tomato (Patanè and Cosentino 2010; Leogrande et al. 2012).

Conservation agriculture (CA) approaches, proposed to reduce the risk of soil erosion and degradation, refer to several practices of soil management in agricultural systems, that minimize the effect on composition, structure and biodiversity. Conservation agriculture includes direct sowing/no-tillage, reduced tillage/ minimum tillage, incorporation of crop residues and establishment of cover crops in both annual and perennial crops.

In the Mediterranean basin, conservation agriculture systems have been widely studied in rainfed conditions (Hernanz et al. 2009; López-Bellido et al. 2010; Farina et al. 2011; Barbera et al. 2012; Mazzoncini et al. 2016). The residues left on the ground help to protect the soil from the wind and the rain (López et al. 1998), improve soil aggregation and fertility (Blanco-Moure et al. 2013) and increase water infiltration in the soil and water availability for the crop (Cantero-Martínez et al. 2007), although a minimum mulch layer is required (Lampurlanés and Cantero-Martínez 2005; Stagnari et al. 2014). In most cases yield is unaffected, although it may increase especially in dry years (Ben-Hammouda et al. 2006; De Vita et al. 2007; Troccoli et al. 2015). In contrast to the extensive research work in rainfed conservation agriculture systems, limited research work is available under irrigation (Lithourgidis et al. 2005; Boulal et al. 2008; Casa and Lo Cascio 2008).

The conventional management of olive groves has been associated with soil erosion (Castro et al. 2008), river and water body pollution (Colombo et al. 2005), degradation of landscape (Parra-López et al. 2009), and climate change (Rodríguez-Entrena et al. 2012). In addition, conventional management reduces soil fertility and olive production, and increases production costs (Calatrava-Leyva et al. 2007), and these effects are more evident in Mediterranean climatic conditions (Gómez et al. 2009).

Organic farming of olive groves at Cordoba (Southern Spain) increased SOC stocks in comparison with a conventional management (Parras-Alcántara and Lozano-García 2014). The highest differences were found in Cambisols, with 74.7 versus 43.8 Mg C ha^{-1}, and Luvisols with 95.4 versus 57.3 Mg C ha^{-1}. In the same area, a cover crop of barley sown in the olive groves had soil losses by erosion of 0.8 t ha^{-1} yr^{-1} and an average annual runoff coefficient of 1.2%; conversely, the conventional management had higher soil losses (2.9 t ha^{-1} yr^{-1}) and the average runoff coefficient was 3.1% (Gómez et al. 2009). Other studies support the use of cover crops or spontaneous natural vegetation to reduce runoff and soil loss in olive groves and vineyards (Bruggeman et al. 2005; Gómez and Giráldez 2007; Novara et al. 2011; Vicente-Vicente et al. 2016).

Temporary ditches have been proposed in the European cross-compliance scheme to reduce soil erosion in sloping soils. During a field monitoring, erosion rates were studied with a UAV-GIS methodology (Bazzoffi 2015) in two farms of central Italy. The results of soil erosion monitoring during two years of observations have shown that temporary ditches were effective in decreasing the cumulated soil erosion in comparison to the field with no ditches, on average by 42.5% (Bazzoffi et al. 2015).

We can conclude that the adoption of sustainable measures for cropland management can decrease soil organic carbon (SOC) losses, increase the external organic matter input, or efficiently combine both. Further measures include sustainable crop management by irrigation, conservation agriculture by no-tillage, residues retention, and cover crops, and soil protection against erosion by water that may increase SOC losses.

References

Acquadro A, Portis E, Scaglione D, Mauro RP, Campion B, Falavigna A, Zaccardelli R, Ronga D, Mauromicale G, Lanteri S (2013) CYNERGIA project: exploitability of *Cynara cardunculus* L. as energy crop. Acta Hort 983:109–150. https://doi.org/10.17660/ActaHortic.2013.983.13

Aguilera E, Lassaletta L, Gattinger A, Gimeno BS (2013) Managing soil carbon for climate change mitigation and adaptation in Mediterranean cropping systems: a meta-analysis. Agric Ecosyst Environ 168:25–36. https://doi.org/10.1016/j.agee.2013.02.003

Aguilera E, Guzmán G, Alonso A (2015) Greenhouse gas emissions from conventional and organic cropping systems in Spain. I. Herbaceous crops. Agron Sustainable Dev 35(2):713–724. https://doi.org/10.1007/s13593-014-0267-9

Bakker MM, Govers G, Jones RA, Rounsevell MDA (2007) The effect of soil erosion on Europe's crop yields. Ecosystems 10:1209–1219. https://doi.org/10.1007/s10021-007-9090-3

Balota EL, Machineski O, Honda C, Yada IFU, Barbosa GMC, Nakatani AS, Coyne MS (2014) Response of arbuscular mycorrhizal fungi in different soil tillage systems to long-term swine slurry application. Land Degrad Dev 27:1141–1150. https://doi.org/10.1002/ldr.2304

Barbera V, Poma I, Gristina L, Novara A, Egli M (2012) Long-term cropping systems and tillage management effects on soil organic carbon stock and steady state level of C sequestration rates in a semiarid environment. Land Degrad Dev 23:82–91. https://doi.org/10.1002/ldr.1055

Baveye PC, Baveye J, Gowdy J (2016) Soil "ecosystem" services and natural capital: critical appraisal of research on uncertain ground. Front Environ Sci 4:41. https://doi.org/10.3389/fenvs.2016.00041

Bazzoffi P (2015) Measurement of rill erosion through a new UAV-GIS methodology. Ital J Agron 10 (Suppl.1). https://doi.org/10.4081/ija.2015.10.s1.708

Bazzoffi P, Francaviglia R, Neri U, Napoli R, Marchetti A, Falcucci M, Pennelli B, Simonetti G, Barchetti A, Migliore M, Fedrizzi M, Guerrieri M, Pagano M, Puri D, Sperandio G, Ventrella D (2015) Environmental effectiveness of GAEC cross-compliance Standard 1.1a (temporary ditches) and 1.2 g (permanent grass cover of set-aside) in reducing soil erosion and economic evaluation of the competitiveness gap for farmers. Ital J Agron 10 (s1). https://doi.org/10.4081/ija.2015.10.s1.710

Beer J, Blodau C (2007) Transport and thermodynamics constrain belowground carbon turnover in a northern peatland. Geochim Cosmochim Ac 71:2989–3002. https://doi.org/10.1016/j.gca.2007.03.010

Ben-Hammouda M, M'Hedhb, K, Abidi L, Rajeh A, Chourabi H, El-Faleh J, Dichiara C (2006) Conservation agriculture based on direct sowing. In: The future of drylands. International scientific conference on desertification and drylands research tunis, Tunisia, pp 647–657

Bennett LT, Mele PM, Annett S, Kasel S (2010) Examining links between soil management, soil health, and public benefits in agricultural landscapes: an Australian perspective. Agric Ecosyst Environ 139(1):1–12. https://doi.org/10.1016/j.agee.2010.06.017

Bernal B, McKinley DC, Hungate BA, White PM, Mozdzer TJ, Megonigal JP (2016) Limits to soil carbon stability; deep, ancient soil carbon decomposition stimulated by new labile organic inputs. Soil Biol Biochem 98:85–94. https://doi.org/10.1016/j.soilbio.2016.04.007

Bernoux M, Branca G, Carro A, Lipper L, Smith G, Bockel L (2010) Ex-ante greenhouse gas balance of agriculture and forestry development programs. Sci Agric 67(1):31–40. https://doi.org/10.1590/S0103-90162010000100005

Bevivino A, Paganin P, Bacci G, Florio A, Pellicer MS, Papaleo MC, Mengoni A, Ledda L, Fani R, Benedetti A, Dalmastri C (2014) Soil bacterial community response to differences in agricultural management along with seasonal changes in a Mediterranean region. PLoS ONE 9 (8):e105515. https://doi.org/10.1371/journal.pone.0105515

Blanco-Moure N, Gracia R, Bielsa AC, Lopez MV (2013) Long-term no-tillage effects on particulate and mineral-associated soil organic matter under rainfed Mediterranean conditions. Soil Use Manage 29:250–259. https://doi.org/10.1111/sum.12039

Boulal H, Gómez-Macpherson H, Gómez JA (2008) Water infiltration and soil losses in a permanent bed irrigated system in Southern Spain. Ital J Agron 3:45–46

Boulal H, Gómez-Macpherson H (2010) Dynamics of soil organic carbon in an innovative irrigated permanent bed system on sloping land in Southern Spain. Agric Ecosyst Environ 139:284–292. https://doi.org/10.1016/j.agee.2010.08.015

Boulal H, Gómez-Macpherson H, Gómez JA, Mateos L (2011) Effect of soil management and traffic on soil erosion in irrigated annual crops. Soil Till Res 115–116:62–70. https://doi.org/10.1016/j.still.2011.07.003

Bruggeman A, Masri Z, Turkelboom F, Zöbisch M, El-Naheb H (2005) Strategies to sustain productivity of olive groves on steep slopes in the northwest of the Syrian Arab Republic. In: Benites J, Pisante M, Stagnari F (eds) Integrated soil and water management for orchard development. Role and importance. FAO Land and Water Bulletin, vol 10. FAO, Rome, Italy, pp. 75–87

Brunori E, Farina R, Biasi R (2016) Sustainable viticulture: the carbon-sink function of the vineyard agro-ecosystem. Agr Ecosyst Environ 223:10–21. https://doi.org/10.1016/j.agee.2016.02.012

Burke IC, Yonker CM, Parton WJ, Cole CV, Flach K, Schimel DS (1989) Texture, climate, and cultivation effects on soil organic matter content in US grassland soils. Soil Sci Soc Am J 53:800–805. https://doi.org/10.2136/sssaj1989.03615995005300030029x

Calatrava-Leyva J, Franco-Martínez JA, González-Roa MC (2007) Analysis of the adoption of soil conservation practices in olive groves: the case of mountainous areas in Southern Spain. Span J Agr Res 5:249–258. https://doi.org/10.5424/sjar/2007053-246

Canali S, Di Bartolomeo E, Tittarelli F, Montemurro F, Verrastro V, Ferri D (2011) Comparison of different laboratory incubation procedures to evaluate nitrogen mineralization in soil amended with aerobic and anaerobic stabilized organic materials. J Food Agric Environ 9:540–546

Canali S, Campanelli G, Ciaccia C, Leteo F, Testani E, Montemurro F (2013) Conservation tillage strategy based on the roller crimper technology for weed control in Mediterranean vegetable organic cropping systems. Eur J Agron 50:11–18. https://doi.org/10.1016/j.eja.2013.05.001

Cantero-Martínez C, Angás P, Lampurlanés J (2007) Long-term yield and water use efficiency under various tillage systems in Mediterranean rainfed conditions. Ann Appl Biol 150:293–305. https://doi.org/10.1111/j.1744-7348.2007.00142.x

Caravaca F, Masciandaro G, Ceccanti B (2002) Land use in relation to soil chemical and biochemical properties in a semiarid Mediterranean environment. Soil Tillage Res 68:23–30. https://doi.org/10.1016/S0167-1987(02)00080-6

Casa R, Lo Cascio B (2008) Soil conservation tillage effects on yield and water use efficiency on irrigated crops in Central Italy. J Agron Crop Sci 194(4):310–319. https://doi.org/10.1111/j.1439-037X.2008.00316.x

Castro J, Fernández-Ondoño E, Rodríguez C, Lallena AM, Sierra M, Aguila J (2008) Effects of different olive-grove management systems on the organic carbon and nitrogen content of the soil in Jaén (Spain). Soil Till Res 98:56–67. https://doi.org/10.1016/j.still.2007.10.002

Cayuela ML, Aguilera E, Sanz-Cobena A, Adams DC, Abalos D, Barton L, Ryals R, Silver WL, Alfaro MA, Pappa VA, Smith P, Garnier J, Billen G, Bouwman L, Bondeau A, Lassaletta L (2017) Direct nitrous oxide emissions in Mediterranean climate cropping systems: emission factors based on a meta-analysis of available measurement data. Agric Ecosyst Environ 238:25–35. https://doi.org/10.1016/j.agee.2016.10.006

Cerdá A, Giménez Morera A, García Orenes F, Morugán A, González Pelayo O, Pereira P, Novara A, Brevik EC (2014) The impact of abandonment of traditional flood irrigated citrus orchards on soil infiltration and organic matter. In: Arnáez J, González-Sampériz P, Lasanta T, Valero-Garcés BL (eds) Geoecología, Cambio Ambiental Y Paisaje: Homenaje Al Profesor José María García Ruiz. Instituto Pirenaico de Ecología, Zaragoza, pp 267–276

Colombo S, Hanley N, Calatrava J (2005) Designing policy for reducing the off farm effects of soil erosion using choice experiments. J Agr Econ 56:81–95. https://doi.org/10.1111/j.1477-9552.2005.tb00123.x

Conant RT, Ryan MG, Ågren GI, Birge HE, Davidson EA, Eliasson PE, Evans SE, Frey SD, Giardina CP, Hopkins FM, Hyvönen R, Kirschbaum MUF, Lavallee JM, Leifeld J, Parton WJ, Megan Steinweg J, Wallenstein MD, Wetterstedt JÅM, Bradford MA (2011) Temperature and soil organic matter decomposition rates synthesis of current knowledge and a way forward. Glob Change Biol 17:3392–3404. https://doi.org/10.1111/j.1365-2486.2011.02496.x

Costantini EAC, Lorenzetti R (2013) Soil degradation processes in the Italian agricultural and forest ecosystems. Ital J Agron 8:233–243. https://doi.org/10.4081/ija.2013.e28

Cowling RM, Ojeda F, Lamont B, Rundel PW, Lechmere-Oertel R (2005) Rainfall reliability, a neglected factor in explaining convergence and divergence of plant traits in fire-prone Mediterranean-climate ecosystems. Global Ecol Biogeogr 14:509–519. https://doi.org/10.1111/j.1466-822X.2005.00166.x

Dabney SM, Delgado JA, Reeves DW (2001) Using winter cover crops to improve soil and water quality. Commun Soil Sci Plant 32(7–8):1221–1250. https://doi.org/10.1081/CSS-100104110

De Vita P, Di Paolo E, Fecondo G, Di Fonzo N, Pisante M (2007) No-tillage and conventional tillage effects on durum wheat yield, grain quality and soil moisture content in Southern Italy. Soil Till Res 92:69–78. https://doi.org/10.1016/j.still.2006.01.012

Di Bene C, Marchetti A, Francaviglia R, Farina R (2016) Soil organic carbon dynamics in typical durum wheat-based crop rotations of Southern Italy. Ital J Agron 11(4):209–216. https://doi.org/10.4081/ija.2016.763

Dominati E, Patterson M, Mackay A (2010) A framework for classifying and quantifying the natural capital and ecosystem services of soils. Ecol Econ 69(9):1858–1868. https://doi.org/10.1016/j.ecolecon.2010.05.002

Falloon P, Jones CD, Ades M, Paul K (2011) Direct soil moisture controls of future global soil carbon changes: an important source of uncertainty. Global Biogeochem Cy 25:GB3010. https://doi.org/10.1029/2010gb003938

Farina R, Seddaiu G, Orsini R, Steglich E, Roggero PP, Francaviglia R (2011) Soil carbon dynamics and crop productivity as influenced by climate change in a rainfed cereal system under contrasting tillage using EPIC. Soil Till Res 112:36–46. https://doi.org/10.1016/j.still.2010.11.002

Farina R, Marchetti A, Francaviglia R, Napoli R, Di Bene C (2017) Modeling regional soil C stocks and CO_2 emissions under Mediterranean cropping systems and soil types. Agric Ecosyst Environ 238:128–141. https://doi.org/10.1016/j.agee.2016.08.015

Fernández-Romero ML, Lozano-García B, Parras-Alcántara L (2014) Topography and land use change effects on the soil organic carbon stock of forest soils in Mediterranean natural areas. Agric Ecosyst Environ 195:1–9. https://doi.org/10.1016/j.agee.2014.05.015

Fisher B, Turner K, Zylstra M, Brouwer R, de Groot R, Farber S, Ferraro P, Green R, Hadley D, Harlow J, Jefferiss P, Kirkby C, Morling P, Mowatt S, Naidoo R, Paavola J, Strassburg B, Yu D, Balmford A (2008) Ecosystem services and economic theory: integration for policy-relevant research. Ecol Appl 18:2050–2067. https://doi.org/10.1890/07-1537.1

Fissore C, Dalzell BJ, Berhe AA, Voegtle M, Evans M, Wu A (2017) Influence of topography on soil organic carbon dynamics in a Southern California grassland. Catena 149 (Part 1), pp 140–149. https://doi.org/10.1016/j.catena.2016.09.016

Francaviglia R, Benedetti A, Doro L, Madrau S, Ledda L (2014) Influence of land use on soil quality and stratification ratios under agro-silvo-pastoral Mediterranean management systems. Agric Ecosyst Environ 183:86–92. https://doi.org/10.1016/j.agee.2013.10.026

Francaviglia R, Bruno A, Falcucci M, Farina R, Renzi G, Russo DE, Sepe L, Neri U (2016) Yields and quality of Cynara cardunculus L. wild and cultivated cardoon genotypes. A case study from a marginal land in Central Italy. Eur J Agron 72:10–19. https://doi.org/10.1016/j.eja.2015.09.014

Francaviglia R, Renzi G, Doro L, Parras-Alcántara L, Lozano-García B, Ledda L (2017) Soil sampling approaches in Mediterranean agro-ecosystems. Influence on soil organic carbon stocks. CATENA 158:113–120. https://doi.org/10.1016/j.catena.2017.06.014

Franzluebbers AJ, Hons FM, Zuberer DA (1998) In situ and potential CO_2 evolution from a Fluventic Ustochrept in south-central Texas as affected by tillage and cropping intensity. Soil Till Res 47:303–308. https://doi.org/10.1016/S0167-1987(98)00118-4

García-Orenes F, Guerrero C, Roldán A, Mataix-Solera J, Cerdà A, Campoy M, Zornoza R, Bárcenas G, Caravaca F (2010) Soil microbial biomass and activity under different agricultural management systems in a semiarid Mediterranean agroecosystem. Soil Tillage Res 109:110–115. https://doi.org/10.1016/j.still.2010.05.005

García-Ruiz JM, Nadal-Romero E, Lana-Renault N, Beguería S (2013) Erosion in Mediterranean landscapes: changes and future challenges. Geomorphology 198:20–36. https://doi.org/10.1016/j.geomorph.2013.05.023

Gómez JA, Giráldez JV (2007) Soil and water conservation. A European approach through ProTerra projects. In: Proceedings of the European congress on agriculture and the environment, seville, 26–28th, 2007

Gómez JA, Sobrinho TA, Giráldez JV, Fereres E (2009) Soil management effects on runoff, erosion and soil properties in an olive grove of Southern Spain. Soil Till Res 102:5–13. https://doi.org/10.1016/j.still.2008.05.005

González-Sánchez EJ, Ordóñez-Fernández R, Carbonell-Bojollo R, Veroz-González O, Gil-Ribes JA (2012) Meta-analysis on atmospheric carbon capture in Spain through the use of conservation agriculture. Soil Till Res 122:52–60. https://doi.org/10.1016/j.still.2012.03.001

Grammelis P, Malliopoulou A, Basinas P, Danalatos NG (2008) Cultivation and characterization of *Cynara cardunculus* for solid biofuels production in the Mediterranean region. Int J Mol Sci 9:1241–1258. https://doi.org/10.3390/ijms9071241

Hernanz JL, Sánchez-Girón V, Navarrete L (2009) Soil carbon sequestration and stratification in a cereal/leguminous crop rotation with three tillage systems in semiarid conditions. Agr Ecosyst Environ 133:114–122. https://doi.org/10.1016/j.agee.2009.05.009

Hillier J, Walter C, Malin D, Garcia-Suarez T, Mila-i-Canals L, Smith P (2011) A farm-focused calculator for emissions from crop and livestock production. Environ Modell Softw 26:1070–1078. https://doi.org/10.1016/j.envsoft.2011.03.014

Infante-Amate J, González de Molina M (2013) The socio-ecological transition on a crop scale: the case of olive orchards in Southern Spain (1750–2000). Hum Ecol 41:961–969. https://doi.org/10.1007/s10745-013-9618-4

Jenny H (1980) The Soil Resource: Origin and Behavior. Ecological Studies, Vol 37, Springer, New York, 377 pp

Kay BD, VandenBygaart AJ (2002) Conservation tillage and depth stratification of porosity and soil organic matter. Soil Till Res 66(2):107–118. https://doi.org/10.1016/S0167-1987(02)00019-3

Kroodsma DA, Field CB (2006) Carbon sequestration in California agriculture, 1980–2000. Ecol Appl 16:1975–1985. https://doi.org/10.1890/1051-0761(2006)016[1975:CSICA]2.0.CO;2
Lag-Brotons A, Gómez I, Navarro-Pedreño J, Bartual-Martos J (2014) Effects of sewage sludge compost on *Cynara cardunculus* L. cultivation in a Mediterranean soil. Compost Sci Util 22:33–39. https://doi.org/10.1080/1065657X.2013.870945
Lagomarsino A, Benedetti A, Marinari S, Pompili L, Moscatelli MC, Roggero PP, Lai R, Ledda L, Grego S (2011) Soil organic C variability and microbial functions in a Mediterranean agro-forest ecosystem. Biol Fertil Soils 47:283–291. https://doi.org/10.1007/s00374-010-0530-4
Lal R (2005) Forest soils and carbon sequestration. For Ecol Manage 220:242–258. https://doi.org/10.1016/j.foreco.2005.08.015
Lal R, Delgado JA, Groffman PM, Millar N, Dell C, Rotz A (2011) Management to mitigate and adapt to climate change. J Soil Water Conserv 66:276–285. https://doi.org/10.2489/jswc.66.4.276
Lampurlanés J, Cantero-Martínez C (2005) Hydraulic conductivity, residue cover and soil surface roughness under different tillage systems in semiarid conditions. Soil Till Res 85:13–26. https://doi.org/10.1016/j.still.2004.11.006
Laudicina VA, Badalucco L, Palazzolo E (2011) Effects of compost input and tillage intensity on soil microbial biomass and activity under Mediterranean conditions. Biol Fertil Soils 47:63–70. https://doi.org/10.1007/s00374-010-0502-8
Laudicina VA, Novara A, Barbera V, Egli M, Badalucco L (2015) Long-term tillage and cropping system effects on chemical and biochemical characteristics of soil organic matter in a Mediterranean semiarid environment. Land Degrad Dev 26:45–53. https://doi.org/10.1002/ldr.2293
Ledda L, Deligios PA, Farci R, Sulas L (2013) Biomass supply for energetic purposes from some *Cardueae* species grown in Mediterranean farming systems. Ind Crop Prod 47:218–226. https://doi.org/10.1016/j.indcrop.2013.03.013
Leogrande R, Lopedota O, Montemurro F, Vitti C, Ventrella D (2012) Effects of irrigation regime and salinity on soil characteristics and yield of tomato. Ital J Agron 7:50–57. https://doi.org/10.4081/ija.2012.e8
Lieskovský J, Kenderessy P (2014) Modelling the effect of vegetation cover and different tillage practices on soil erosion in vineyards: a case study in Vráble (Slovakia) using WATEM/SEDEM. Land Degrad Dev 25:288–296. https://doi.org/10.1002/ldr.2162
Lithourgidis AS, Tsatsarelis CA, Dhima KV (2005) Tillage effects on corn emergence, silage yield, and labor and fuel inputs in double cropping with wheat. Crop Sci 45:2523–2528. https://doi.org/10.2135/cropsci2005.0141
Llasat MC, Llasat-Botija M, Prat MA, Porcú F, Price C, Mugnai A, Lagouvardos K, Kotroni V, Katsanos D, Michaelides S, Yair Y, Savvidou K, Nicolaides K (2010) High-impact floods and flash floods in Mediterranean countries: the FLASH preliminary database. Adv Geosci 23:47–55. https://doi.org/10.5194/adgeo-23-47-2010
López MV, Sabre M, Gracia R, Arrúe JL, Gomes L (1998) Tillage effects on soil surface conditions and dust emission by wind erosion in semiarid Aragón (NE Spain). Soil Till Res 45:91–105. https://doi.org/10.1016/S0167-1987(97)00066-4
López-Bermúdez F (2008) Desertificación: Preguntas y respuestas a un desafío económico, social y ambiental. Fundación Biodiversidad, Madrid, p 129
López-Bellido RJ, Fontán JM, López-Bellido FJ, López-Bellido LL (2010) Carbon sequestration by tillage, rotation, and nitrogen fertilization in a Mediterranean vertisol. Agron J 102:310–318. https://doi.org/10.2134/agronj2009.0165
López-Fando C, Pardo MT (2011) Soil carbon storage and stratification under different tillage systems in a semi-arid region. Soil Till Res 111(2):224–230. https://doi.org/10.1016/j.still.2010.10.011
Lorenz K, Lal R (2016) Environmental impact of organic agriculture. Adv Agron 139:99–152. https://doi.org/10.1016/bs.agron.2016.05.003

Lozano-García B, Parras-Alcántara L (2014) Variation in soil organic carbon and nitrogen stocks along a toposequence in a traditional Mediterranean olive grove. Land Degrad Dev 25:297–304. https://doi.org/10.1002/ldr.2284

Lugato E, Bampa F, Panagos P, Montanarella L, Jones A (2014) Potential carbon sequestration of European arable soils estimated by modelling a comprehensive set of management practices. Glob Chang Biol 20:3557–3567. https://doi.org/10.1111/gcb.12551

Marques MJ, Bienes R, Cuadrado J, Ruiz-Colmenero M, Barbero-Sierra C, Velasco A (2015) Analysing perceptions attitudes and responses of winegrowers about sustainable land management in Central Spain. Land Degrad Dev 26:458–467. https://doi.org/10.1002/ldr.2355

Marzaioli R, D'Ascoli R, De Pascale RA, Rutigliano FA (2010) Soil quality in a Mediterranean area of Southern Italy as related to different land use types. Appl Soil Ecol 44(3):205–212. https://doi.org/10.1016/j.apsoil.2009.12.007

Mazzoncini M, Antichi D, Di Bene C, Risaliti R, Petri M, Bonari E (2016) Soil carbon and nitrogen changes after 28 years of no-tillage management under Mediterranean conditions. Eur J Agron 77:156–165. https://doi.org/10.1016/j.eja.2016.02.011

MEA (2005) Ecosystems and Human Well-being: Synthesis. Island Press, Washington DC, Millennium Ecosystem Assessment, p 137

Meersmans J, Martin MP, Lacarce E, De Baets S, Jolivet C, Boulonne L, Lehmann S, Saby NPA, Bispo A, Arrouays D (2012) A high resolution map of French soil organic carbon. Agron Sustainable Dev 32:841–851. https://doi.org/10.1007/s13593-012-0086-9

Meijide A, Gruening C, Goded I, Seufert G, Cescatti A (2017) Water management reduces greenhouse gas emissions in a Mediterranean rice paddy field. Agric Ecosyst Environ 238:168–178. https://doi.org/10.1016/j.agee.2016.08.017

Mørch HFC (1999) Mediterranean Agriculture—An Agro-Ecological Strategy. Geogr Tidsskr-Den Special Issue 1:143–156

Mukherjee A, Lal R, Zimmerman AR (2014) Effects of biochar and other amendments on the physical properties and greenhouse gas emissions of an artificially degraded soil. Sci Total Environ 487:26–36. https://doi.org/10.1016/j.scitotenv.2014.03.141

Muñoz-Rojas M, Doro L, Ledda L, Francaviglia R (2015) Application of CarboSOIL model to predict the effects of climate change on soil organic carbon stocks in agro-silvo-pastoral Mediterranean management systems. Agric Ecosyst Environ 202:8–16. https://doi.org/10.1016/j.agee.2014.12.014

Nichols JD (1984) Relation of organic carbon to soil properties and climate in the Southern Great Plains. Soil Sci Soc Am J 48:1382–1384. https://doi.org/10.2136/sssaj1984.03615995004800060037x

Nkoa R (2014) Agricultural benefits and environmental risks of soil fertilization with anaerobic digestates: a review. Agron Sustain Dev 34:473–492. https://doi.org/10.1007/s13593-013-0196-z

Novara A, Gristina L, Saladino SS, Santoro A, Cerdà A (2011) Soil erosion assessment on tillage and alternative soil managements in a Sicilian vineyard. Soil Tillage Res 117:140–147. https://doi.org/10.1016/j.still.2011.09.007

Novara A, La Mantia T, Barbera V, Gristina L (2012) Paired-site approach for studying soil organic carbon dynamics in a Mediterranean semiarid environment. CATENA 89(1):1–7. https://doi.org/10.1016/j.catena.2011.09.008

Pardo G, del Prado A, Martínez-Mena M, Bustamante MA, Rodríguez Martín JA, Álvaro-Fuentes J, Moral R (2017) Orchard and horticulture systems in Spanish Mediterranean coastal areas: is there a real possibility to contribute to C sequestration? Agric Ecosyst Environ 238:153–167. https://doi.org/10.1016/j.agee.2016.09.034

Parra-López C, Groot JCJ, Carmona-Torres C, Rossing WAH (2009) An integrated approach for ex-ante evaluation of public policies for sustainable agriculture at landscape level. Land Use Policy 26:1020–1030. https://doi.org/10.1016/j.landusepol.2008.12.006

Parras-Alcántara L, Lozano-García B (2014) Conventional tillage versus organic farming in relation to soil organic carbon stock in olive groves in Mediterranean rangelands (Southern Spain). Solid Earth 5(1):299–311. https://doi.org/10.5194/se-5-299-2014

Parras-Alcántara L, Lozano-García B, Galán-Espejo A (2015a) Soil organic carbon along an altitudinal gradient in the Despeñaperros Natural Park, Southern Spain. Solid Earth 6:125–134. https://doi.org/10.5194/se-6-125-2015

Parras-Alcántara L, Lozano-García B, Brevik EC, Cerdá A (2015b) Soil organic carbon stocks assessment in Mediterranean natural areas: a comparison of entire soil profiles and soil control sections. J Environ Manage 155:219–228. https://doi.org/10.1016/j.jenvman.2015.03.039

Parton WJ, Schimel DS, Cole CV, Ojima DS (1987) Analysis of factors controlling soil organic matter levels in Great Plains grasslands. Soil Sci Soc Am J 51:1173–1179. https://doi.org/10.2136/sssaj1987.03615995005100050015x

Patanè C, Cosentino SL (2010) Effects of soil water deficit on yield and quality of processing tomato under a Mediterranean climate. Agric Water Manag 97:131–138. https://doi.org/10.1016/j.agwat.2009.08.021

Plieninger T, Bieling C, Fagerholm N, Byg A, Hartel T, Hurley P, López-Santiago CA, Nagabhatla N, Oteros-Rozas E, Raymond CM, van der Horst D, Huntsinger L (2015) The role of cultural ecosystem services in landscape management and planning. Curr Opin Environ Sustain 14:28–33. https://doi.org/10.1016/j.cosust.2015.02.006

Porter J, Costanza R, Sandhu HS, Sigsgaard L, Wratten SD (2009) The value of producing food, energy and ES within an agro-ecosystem. Ambio 38:186–193. https://doi.org/10.1579/0044-7447-38.4.186

Powlson DS, Whitmore AP, Goulding KWT (2011) Soil carbon sequestration to mitigate climate change: a critical re-examination to identify the true and the false. Eur J Soil Sci 62–1:42–55. https://doi.org/10.1111/j.1365-2389.2010.01342.x

Ramos ME, Altieri MA, Garcia PA, Robles AB (2011) Oat and oat-vetch as rainfed fodder-cover crops in semiarid environments: effects of fertilization and harvest time on forage yield and quality. J Sustain Agric 35:726–744. https://doi.org/10.1080/10440046.2011.606490

Rasmussen PE, Collins HP (1991) Long-term impacts of tillage, fertilizer, and crop residue on soil organic matter in temperate semiarid regions. Adv Agron 45:93–134. https://doi.org/10.1016/S0065-2113(08)60039-5

Reynaldo V, Banwart S, Black H, Ingram J, Joosten H, Milne E, Noellemeyer E (2012) The benefits of soil carbon. UNEP Yearbook 2012:19–33

Riffaldi R, Saviozzi A, Levi-Minzi R, Cardelli R (2002) Biochemical properties of a Mediterranean soil as affected by long-term crop management systems. Soil Tillage Res 67:109–114. https://doi.org/10.1016/S0167-1987(02)00044-2

Rodeghiero M, Rubio A, Díaz-Pinés E, Romanyà J, Marañón-Jiménez S, Levy GJ, Fernandez-Getino AP, Sebastià MT, Karyotis T, Chiti T, Sirca C, Martins A, Manuel Madeira M, Zhiyanski M, Gristina L, La Mantia T (2011) Soil carbon in Mediterranean ecosystems and related management problems. In: Jandl R, Rodeghiero M, Olsson M (eds), Soil carbon in sensitive european ecosystems: from science to land management. Wiley, pp. 175–218. https://doi.org/10.1002/9781119970255.ch8

Rodríguez-Entrena M, Barreiro-Hurlé J, Gómez-Limón JA, Espinosa-Goded M, Castro-Rodríguez J (2012) Evaluating the demand for carbon sequestration in olive grove soils as a strategy toward mitigating climate change. J Environ Manage 112:368–376. https://doi.org/10.1016/j.jenvman.2012.08.004

Romanyà J, Rovira P (2011) An appraisal of soil organic C content in Mediterranean agricultural soils. Soil Use Manage 27:321–332. https://doi.org/10.1111/j.1475-2743.2011.00346.x

Sandhu HS, Wratten SD, Cullen R, Case B (2008) The future of farming: the value of ecosystem services in conventional and organic arable land: an experimental approach. Ecol Econ 64:835–848. https://doi.org/10.1016/j.ecolecon.2007.05.007

Schnabel RR, Franzluebbers AJ, Stout WL, Sanderson MA, Stuedemann JA (2001) The effects of pasture management practices. In: Follett RF, Kimble JM, Lal R (eds) The Potential of US Grazing Lands to Sequester Carbon and Mitigate the Greenhouse Effect. Lewis Publishers, Boca Raton, FL, pp 291–322

Sims REH, Hastings A, Schlamadinger B, Taylor G, Smith P (2006) Energy crops: current status and future prospects. Glob Change Biol 12:2054–2076. https://doi.org/10.1111/j.1365-2486.2006.01163.x

Smith AN, Reberg-Horton SC, Place GT, Meijer AD, Arellano C, Mueller JP (2011) Rolled rye mulch for weed suppression in organic no-tillage soybeans. Weed Sci 59(2):224–231. https://doi.org/10.1614/WS-D-10-00112.1

Stagnari F, Galieni A, Speca S, Cafiero G, Pisante M (2014) Effects of straw mulch on growth and yield of durum wheat during transition to conservation agriculture in Mediterranean environment. Field Crops Res 167:51–63. https://doi.org/10.1016/j.fcr.2014.07.008

Swinton SM, Lupi F, Robertson GP, Hamilton SK (2007) Ecosystem services and agriculture: cultivating agricultural ecosystems for diverse benefits. Ecol Econ 64:245–252. https://doi.org/10.1016/j.ecolecon.2007.09.020

Troccoli A, Maddaluno C, Mucci M, Russo M, Rinaldi M (2015) Is it appropriate to support the farmers for adopting conservation agriculture? Economic and environmental impact assessment. Ital J Agron 10:169–177. https://doi.org/10.4081/ija.2015.661

Tuomi M, Vanhalaa P, Karhu K, Fritze H, Liski J (2008) Heterotrophic soil respiration. Comparison of different models describing its temperature dependence. Ecol Model 21:182–190. https://doi.org/10.1016/j.ecolmodel.2007.09.003

Vanmaercke M, Poesen J, Verstraeten G, De Vewnte J, Ocakoglu F (2011) Sediment yield in Europe: spatial patterns and scale dependency. Geomorphology 130:142–161. https://doi.org/10.1016/j.geomorph.2011.03.010

Vicente-Vicente J, García-Ruiz R, Francaviglia R, Aguilera E, Smith P (2016) Soil carbon sequestration rates under Mediterranean woody crops using recommended management practices: a meta-analysis. Agric Ecosyst Environ 235:204–214. https://doi.org/10.1016/j.agee.2016.10.024

Von Lützow M, Kögel-Knaber I, Ekschmitte K, Matzner E, Guggenberger G, Marschner B, Flessa H (2006) Stabilization of organic matter in temperate soils: mechanisms and their relevance under different soil conditions—a review. Eur J Soil Sci 57:426–445. https://doi.org/10.1111/j.1365-2389.2006.00809.x

Weber JL (2007) Accounting for soil in the SEEA. European Environment Agency, Rome

Zdruli P, Jones RJA, Montanarella L (2004) Organic Matter in the Soils of Southern Europe. European Soil Bureau Technical Report, EUR 21083 EN, Office for Official Publications of the European Communities, Luxembourg, 16 pp

Zhang W, Ricketts TH, Kremen C, Carney K, Swinton SM (2007) Ecosystem services and dis-services to agriculture. Ecol Econ 64:253–260. https://doi.org/10.1016/j.ecolecon.2007.02.024

Zornoza R, Rosales RM, Acosta JA, de la Rosa JM, Arcenegui V, Faz Á, Pérez-Pastor A (2016) Efficient irrigation management can contribute to reduce soil CO_2 emissions in agriculture. Geoderma 263:70–77. https://doi.org/10.1016/j.geoderma.2015.09.003

Chapter 7
Long-Term Effects of Fertilization on Soil Organism Diversity

Tancredo Augusto Feitosa de Souza and Helena Freitas

Abstract Fertilization applied in long-term farming systems exerts a crucial influence on soil organism diversity and soil properties. This chapter reviews the use of fertilizers for conventional and alternative farming systems in field experiments in order to improve our understanding of the temporal changes on soil organic carbon, soil total nitrogen, arbuscular mycorrhizal fungi diversity and soil macroarthropods during their long-term utilization. We introduce what are the main effects of long-term fertilization systems on several agricultural farming systems around the world. We also present our experimental data about long-term utilization of mineral and organic fertilization from wheat and rapeseed field experiments. Published field studies show that the continuous use of mineral fertilizers might affect negatively soil organic carbon and total nitrogen, which in turn modifies the community composition of macroarthropods, and arbuscular mycorrhizal fungi, whereas organic fertilizers might affect positively these soil properties and soil organism diversity. Our review shows that inputs of organic matter sources can change positively soil properties and annual crop development and yield. Our review also highlights the importance of considering the long-term effect of organic fertilization combined with agricultural management practices, such as stubble retention, fertilization with micronutrient, and inoculation with arbuscular mycorrhizal fungi and N-fixing bacteria on the maintenance of soil fertility and to improve the diversity of soil organisms.

Keywords Fertilization systems · Mineral (NPK) fertilization · Organic fertilization · Sustainable agriculture

---

T. A. F. de Souza (✉)
Agrarian Science Center, Department of Soils and Rural Engineering,
Federal University of Paraiba, Areia, Paraiba 58397-000, Brazil
e-mail: tancredo_agro@hotmail.com

H. Freitas
Centre for Functional Ecology, Department of Life Sciences, University of Coimbra,
3000-456 Coimbra, Portugal
e-mail: hfreitas@uc.pt

S. Gaba et al. (eds.), *Sustainable Agriculture Reviews 28*, Ecology for Agriculture 28,
https://doi.org/10.1007/978-3-319-90309-5_7

7.1 Introduction

During the last decades, the traditional production of food, fiber and energy was based on conventional farming systems (Robertson and Vitousek 2009). This system includes the use of mineral fertilizers, pesticides, herbicides, genetically modified organisms, heavy irrigation, intensive tillage and/or concentrated monoculture production (Souza and Freitas 2017). Accordingly, Tian et al. (2017) mineral fertilization is a kind of soil improvement which uses any material of synthetic origin that is applied to soils to supply one or more plant nutrients, such as nitrogen, phosphorus, potassium, calcium, magnesium, and sulfur. Also, fertilization have been a historical contribution to the impressive crop yield increases realized since the 1950s (Geisseler and Scow 2014). Nowadays, fertilization is recognized as one of the most important activities that influences negatively or positively soil chemical, as well as crop yield and soil organism community structure (LeBauer and Treseder 2008; Allison and Martiny 2008; Mikanová et al. 2013; Souza et al. 2015a, b, c; 2016a, b).

There are evidences that fertilization can affect diversity and function of soil organisms (Carneiro et al. 2015). Soil organisms are classified as any organism that inhabits the soil during part or all of its life. It ranges in size from microorganisms, such as bacteria and fungi to macroarthropods that live primarily on soil surface (Moreira et al. 2010).

While crop production is generally nitrogen or phosphorus limited, soil organisms, such as macroarthropods, soil bacteria, and soil fungi may be carbon and nitrogen limited (Wardle 1992; Allison and Martiny 2008; Liu and Greaver 2010; Lu et al. 2011). According Belay and co-workers (2015) mineral fertilization can affect aboveground community, which in turn affects belowground community structure and their function (Bossio et al. 2005). Treseder (2008) also reports that increasing N inputs by mineral fertilization can suppress soil microorganisms. Conversely, Mikanová and co-workers (2013) reported that the long-term fertilization management, such as practices with the use of farmyard manure can improve soil biological activity and fertility, especially by constant input of organic matter.

Thus, the response of soil organisms and annual crops depends from the kind of fertilizer, its concentration, and its frequency of utilization (Lu et al. 2011; Abdullahi et al. 2013). Considering mineral fertilization as a widespread agricultural practice, we can find several ways to use mineral fertilizers, and in some cases the long-term utilization of this practice could be very dangerous for soil organisms, for environment, and for our survival (Table 7.1) (Gao et al. 2015; Robertson et al. 2015). For example, why the continuous use of mineral fertilizers become less benefic to soil biology, and their interaction with plant yield and soil fertility than the use of organic fertilizers in the same conditions? (See the works done by Zhong et al. 2010, Sharma et al. 2011, Carneiro et al. 2015, and Souza et al. 2015a, b, c to more details). Conversely to mineral fertilization, organic fertilization is a kind of soil improvement which uses any recycled plant- or animal-derived matter that is applied to soils.

The conventional fertilization significantly affects the populations of soil organisms (Mäder et al. 2000). Over time, mineral fertilization cause soil fertility and annual crop yield declines (Muchere-Muna et al. 2007; Gabriel et al. 2010; Drakopoulos et al. 2015). These declines may occur through leaching, soil erosion, crop harvesting, and low input carbon systems (Robertson and Vitousek 2009; Panwar et al. 2010; Conyers et al. 2012; Robertson et al. 2015). Conversely, the continuous input of organic matter source contributes to increase plant yield and soil fertility in several agricultural areas around the world (Muchere-Muna et al. 2007; Hossain et al. 2010; Mikanová et al. 2013). This increase may occur because organic fertilizer nutrients are not readily available, its availability occurs slowly with cumulative effects. So, organic fertilizer is able to provide restoration of nutrients, reconstruction of soil organic matter, soil nutrient levels increasing continuously, low rate of soil erosion and high activity of soil biological component (Bayu et al. 2006; Muchere-Muna et al. 2007; Belay et al. 2015; Gao et al. 2015; Carneiro et al. 2015).

In this chapter, we will illustrate some concepts related to the effects of fertilization systems and agricultural management practices on soil properties, such as soil organic carbon and total nitrogen that in turn affect soil organism community structure, diversity and functioning. By the end of this chapter, we hope to have described some inherent attributes of mineral and organic fertilization that alters soil organic carbon and total nitrogen, which in turn change soil organism diversity, community structure, soil organism activity, plant development and plant yield. In this chapter, we will not consider other soil chemical properties, such as available P, exchangeable K, and soil pH or any soil physical properties, such as texture, moisture and soil density. For additional information about these properties which we will not consider in this chapter, see Souza et al. (2015a, b, c, 2016a, b, c).

7.2 Soil Organisms

Soil organisms can be classified as actinomycetes, algae, arthropods, bacteria, earthworms, fungi, nematodes, and protozoa (Moreira et al. 2010). Each of these groups has characteristics that define them and their functions in soil (Table 7.2). They play a vital role in determining soil fertility, plant growth, and soil structure (Muchere-Muna et al. 2007; Robertson and Vitousek 2009; Gabriel et al. 2010; Panwar et al. 2010; Conyers et al. 2012; Drakopoulos et al. 2015; Robertson et al. 2015).

All soil organism groups can be classified in one or more functional groups (Table 7.2). Each functional group is defined by morphological, physiological, behavioral, and biochemical characters or even by environmental changes and taxonomical characters (Setäla et al. 1998). So, we can define twelve functional groups of soil organism accordingly Brussaard (1998), Swift et al. (2010):

Table 7.1 Summary of main services of ecosystem affected by long-term conventional fertilizations systems

Services of ecosystem	Effect of long-term mineral fertilization on	As a result of long-term mineral fertilization, we found:	Soil organisms affected
Biomass, energy, food and fiber production	Nutrient cycling; Soil organic matter dynamics; Soil structure	Less biomass, food, fiber and energy production; Increase soil organic matter decomposition; Decrease soil functionality, soil organic carbon, and nutrient cycling	Ecosystem engineers (Macroarthropods from Order Hymenoptera and Isoptera, and earthworms); Decomposers or saprotrophs (fungi, bacteria); Soil pathogens (fungi, bacteria); Microfauna (nematodes); Symbionts (N-fixing bacteria and arbuscular mycorrhizal fungi)
Water supply and quality	Soil structure; Nutrient cycling	Increase erosion and soil organic matter decomposition; Increase nitrate leaching on the groundwater	Ecosystem engineers (Macroarthropods from Order Hymenoptera and Isoptera, and earthworms); Decomposers or saprotrophs (fungi, bacteria)
Global climate and atmospheric composition	Soil organic carbon dynamic; CO_2 concentration into the atmosphere	Increase global climate changes; Increase nitrous oxides and CO_2 emissions into the atmosphere	Decomposers or saprotrophs (fungi, bacteria)
Erosion control	Soil structure	Increase soil erosion	Ecosystem engineers (Earthworms, and Macroarthropods species from Order Coleoptera)
Soil pollutants and heavy metals	Nutrient cycling; Decomposition process	Increase soil pollutants and heavy metals by inputs of toxic and dangerous compounds from the fertilizers, herbicides or insecticides	Decomposers or saprotrophs (fungi, bacteria); Soil pathogens (fungi, bacteria); Microfauna (nematodes); Symbionts (N-fixing bacteria and arbuscular mycorrhizal fungi)

(continued)

Table 7.1 (continued)

Services of ecosystem	Effect of long-term mineral fertilization on	As a result of long-term mineral fertilization, we found:	Soil organisms affected
Disease and pest control	Protection against soil pathogens and insect pests	Decrease the benefic soil organism community diversity, functionality and activity by changes in soil pH, soil organic matter, and essential mineral nutrients	Macroarthropods from Order Araneae and Mantodea; Arbuscular mycorrhizal fungi; N-fixing bacteria; Nematodes; Algae; Archaea
Biodiversity conservation	Provision of habitat diversity; Ecosystem maintenance	Changes on soil structure; Increases the use of monoculture farming system	Soil bioindicators[a]

Accordingly, the work done by Swift et al. (2010), Mielniczuk (2008) and Souza et al. (2015a, b, c); [a] accordingly the classification proposed by Kibblewhite et al. (2008) and Oehl et al. (2011)

(1) *Decomposers*—The soil organisms which are capable to produce enzymes that breakdown complex substrates (Organic or inorganic) into their simpler forms to get their energy are included here;
(2) *Ecosystem engineers*—The soil organisms with strong influence on the physical properties of soil are included here;
(3) *Herbivores*—The soil organisms which consume and digest living plant tissues are included here;
(4) *Litter transformers*—The soil organisms which feed and grind organic matter debris making it more accessible to decomposers are included here;
(5) *Micro regulators*—The soil organisms which are capable to regulate biochemical cycles through herbivory are included here. We can also find micro regulators associated with decomposers;
(6) *Pathogens*—The soil organisms which can produce disease in other organisms and obtains their nutrients from living organisms are included here;
(7) *Predators*—The soil organisms which regulate the size of the population of other soil organisms by predation are included here;
(8) *Primary producers*—The soil organisms with photoautotrophic metabolism which assimilate carbon dioxide from the atmosphere are included here;
(9) *Prokaryotic transformers*—The soil organisms which perform specific transformations on carbon, nitrogen, phosphorus and sulfur cycles are included here;
(10) *Regulators*—The soil organisms which control the population of herbivores, pathogens, predators, and other soil organisms by biological control process are included here;

Table 7.2 Resume of soil organism groups and their functions (Adapted from Moreira et al. 2010, Souza et al. 2016a, b, c)

Domain	Kingdom	Phylum	Genus	Functional Group
Archaea	Archaea	Not described	*Not described*	Prokaryotic transformers;
Bacteria	Prokaryote	Actinobacteria	*Actinomyces* *Arthrobacter* *Brevibacterium* *Corynebacterium* *Frankia* *Micrococcus* *Micromonospora* *Nocardia* *Nocardiopsis* *Rhodococcus* *Sphaerisporangium* *Streptomyces* *Streptoverticillium* *Virgisporangium*	Decomposers; Micro regulators; Pathogens; Prokaryotic transformers; Regulators Saprotrophs; Symbionts;
Bacteria	Prokaryote	Bacteroidetes	*Flavobacterium* *Pedobacter*	Pathogens;
Bacteria	Prokaryote	Cyanobacteria	*Not described*	Primary producers; Symbionts;
Bacteria	Prokaryote	Firmicutes	*Anaerobacter* *Bacillus* *Clostridium* *Desulfitobacterium* *Desulfosporosinus* *Desulfosporomusa* *Desulfotomaculum* *Exiguobacterium* *Paenibacillus* *Pasteuria* *Sarcina* *Staphylococcus* *Sporosarcina* *Thermolithobacter*	Decomposers; Micro regulators; Pathogens; Prokaryotic transformers; Regulators Saprotrophs;
Bacteria	Prokaryote	Gemmatimonadetes	*Gemmatimonas*	Prokaryotic transformers;
Bacteria	Prokaryote	Proteobacteria	*Agrobacterium* *Azotobacter* *Desulfovibrio* *Geobacter* *Pseudomonas* *Nitrosomonas* *Nitrobacter* *Rhizobium* *Wolbachia*	Decomposers; Pathogens; Primary producers; Prokaryotic transformers; Symbionts;

(continued)

Table 7.2 (continued)

Domain	Kingdom	Phylum	Genus	Functional Group
Eukarya	Arthropoda	Arachnida Chilopoda Diplopoda Entognatha Insecta Malacostraca	*Not described*	Ecosystem engineers; Herbivores; Litter transformers; Micro regulators; Predators; Regulators;
Eukarya	Animalia	Annelida	*Not described*	Ecosystem engineers; Litter transformers;
Eukarya	Animalia	Nematoda	*Not described*	Herbivores; Predators; Micro regulators; Symbionts;
Eukarya	Fungi	Ascomycota Basidiomycota Blastocladiomycota Chytridiomycota Glomeromycota Microsporidia Neocallimastigomycota	*Acaulospora* *Aspergillus* *Fusarium* *Gigaspora* *Glomus* *Mucor* *Penicillium* *Rhizoctonia* *Rhizoglomus* *Rhizophagus* *Trichoderma*	Decomposers; Pathogens; Predators; Micro regulators; Regulators; Saprotrophs; Symbionts;
Eukarya	Plantae	Chlorophyta Streptophyta	*Not described*	Primary producers;
Eukarya	Protista	Not described because the classification of Protista has been and remains a problematic area of taxonomy	*Not described*	Decomposers; Micro regulators; Pathogens; Primary producers;

(11) *Saprotrophs*—The soil organisms which live on dead and decaying organic material in the soil are included here;

(12) *Symbionts*—The soil organisms which are very closely associated with other organisms (also called *host*) in a positive (mutualistic) relationship are included here;

The classification of functional groups is based on soil organism function and their services to the ecosystem (Brussaard 1998). Basically, there are four main services to the ecosystem:

(a) Decomposition of organic matter;
(b) Nutrient cycling;
(c) Bioturbation;
(d) Control of pests and diseases.

The relationship between these four services determines the balance among the quantity of soil carbon and atmospheric carbon dioxide concentration. So, we can conclude that soil organism has an important role regulating atmospheric composition while at the same time improving soil services, soil fertility and agricultural productivity (Lal and Follett 2009; Robertson et al. 2015). We can find soil organisms exhibiting different community composition in several habitats, such as temperate forests, grasslands, tropical forests, shrublands and anthropogenized habitats (Conyers et al. 2012; Drakopoulos et al. 2015 and Robertson et al. 2015). Their community composition is significantly affected by soil organic carbon and total nitrogen contents (Allison and Martiny 2008), but before to explain how soil organisms are dependents to carbon and nitrogen (Geisseler and Scow 2014), we will introduce you in the next section how fertilization (mineral or organic) affects these soil properties that in turn affects directly or indirectly soil organism community composition.

7.3 Changes in Soil Organic Carbon and Total Nitrogen Content by Fertilization

Increasing soil fertility in annual crop systems by using continuous input carbon practices is a widely suggested way to achieve sustainable grain production by improving soil functioning to increase agricultural productivity (Robertson et al. 2015). It is well established that conventional tillage with mineral fertilization reduced carbon input and accelerated carbon loss, as well as total nitrogen and in some case available P through erosion (Bravo-Garza and Bryan 2005; Chen et al. 2011; Guo et al. 2012). Thus, management of organic matter that slowly increases soil fertility may result in erosion reduction (Lal 2004; Yang et al. 2015).

Although inorganic fertilization could provide essential nutrients for annual crops, it could lead to decrease soil functioning and improve soil acidification and soil hardening. In organic systems, many studies have reported greater levels of soil organic carbon and total nitrogen under organic fertilization that under NPK fertilization (Chan et al. 2011; Conyers et al. 2012; Gao et al. 2015; Souza et al. 2015a). Figure 7.1 illustrate the long-term effects of mineral and organic fertilization on soil organic carbon (Fig. 7.1a), and total nitrogen (Fig. 7.1b) (two important soil properties that modulate soil organism diversity and functionality) from a wheat field cultivated on a Ferralsol.

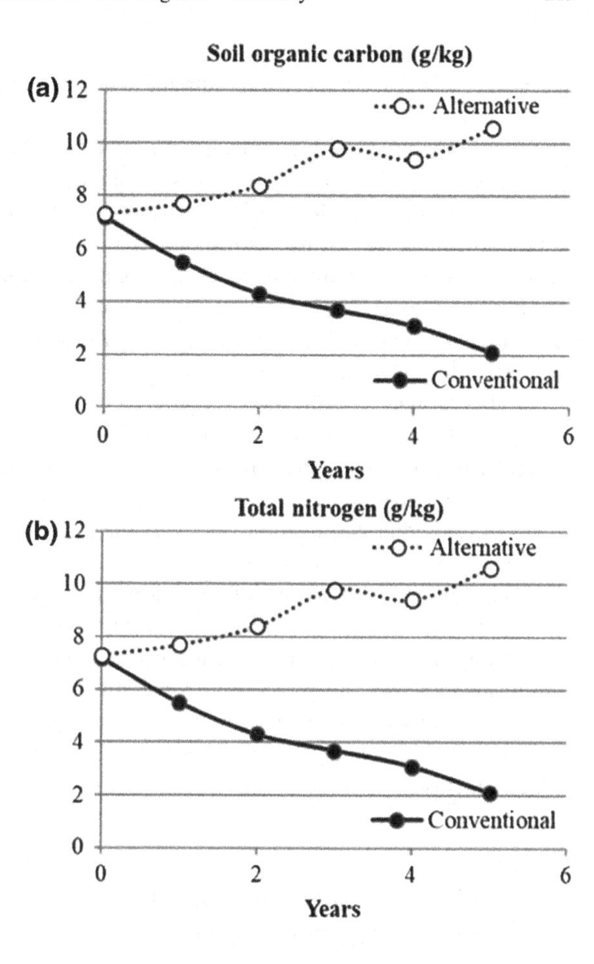

Fig. 7.1 Long-term effects of conventional (black line) and alternative (dotted line) fertilization on soil organic carbon (g/kg, Fig. 7.1a) and total nitrogen (g/kg, Fig. 7.1b) from a wheat field cultivated on a Ferralsol for 5 years (Adapted from Souza et al. 2015a). Conventional fertilization was based accordingly EMBRAPA's recommendation for *Triticum aestivum* cv. BRS-Guamirim, while alternative fertilization was based accordingly regional familiar agriculture sustainable system

These results showed positive effects of organic fertilizers with their long-term utilization on soil organic carbon and total nitrogen (Kimetu et al. 2008; Major et al. 2010; Hossain et al. 2010) and could have been attributed to (i) the lower rate of decomposition and mineralization of organic matter as described by Muchere-Muna et al. (2007), and (ii) these variables are improved as a result of residual effects from continuous use of the alternative fertilization treatment. Saha et al. (2008) reported an increase in soil organic carbon after 3 years of the continuous use of manure application where about 40% carbon was retained in the soil carbon pool. Several previous studies have reported similar positive effects (30 to 54% more soil organic carbon) of cumulative inputs of organic matter by manure and stubble retention on soil organic carbon (Thomas et al. 2007; Kong et al. 2009; Chan et al. 2011; Zhang et al. 2012; Gao et al. 2015). Thomas et al. (2007) reported 21% less total nitrogen in conventional tillage with utilization of mineral fertilizers than no-tillage with stubble retention. Manure and mineral fertilizers have the advantage of supplying essential plant nutrients either directly or indirectly by alleviating some soil stresses like acidic, toxicity or increasing nutrient availability (Lazcano et al. 2012).

The effect of various fertilization systems on soil organic carbon and total nitrogen has been studied in many countries around the world (Treseder 2008; Liu and Greaver 2010; Lu et al. 2011). Despite the considerable research that has been undertaken to measure these effects of fertilization systems on SOC and TN, predicting how these soil properties respond to mineral and organic fertilization remain difficult because the slowness of change and the apparent site specificity of the effects (Zhang et al. 2012; Geisseler and Scow 2014; Gao et al. 2015). Below, we summarize the main effects of different fertilization systems on soil organic carbon and total nitrogen content (Table 7.3).

So, the organic fertilization examined in this section (e.g. Straw bedding impregnated with liquid and solid horse manure, turkey manure, wine-grape solid residue, vermicompost, rabbit manure, farmyard manure, cattle slurry, and biochar) resulted in a fairly wide range of soil organic carbon and total nitrogen, but the effects of individual fertilization were variable, increasing these both soil properties in all studied long-term field experiments (Table 7.3). While, mineral fertilization increased soil organic carbon and total nitrogen in some cases but not in others. We can presume that mineral fertilizations (N, NP, NK, PK, or NPK fertilization) may not reliably increase soil organic carbon and total nitrogen on their own, but that significant increases are possible in some situations through the long-term use of multiple practices combined with the use of mineral fertilizers such as organic fertilizers, stubble management, alternative tillage, legume N input, crop rotation, and elimination of fallow.

7.4 Soil Organisms, Soil Organic Carbon and Soil Total Nitrogen

Several studies reported that soil organisms have been found to be strongly related to soil organic carbon and total nitrogen contents (e.g. Cleveland and Liptzin 2007; Fierer et al. 2009; Kallenbach and Grandy 2011; Geisseler and Scow 2014; Souza et al. 2015a, b, c, Souza et al. 2016a, b, c). This close relationship suggests that changes in soil organic carbon and total nitrogen may alter (positively or not) the dynamic and diversity of soil organisms.

Soil organisms, such as soil macroarthropods, bacteria, fungi and archaea, contain most of the total nitrogen and up to half of the carbon stored in living organisms (Bengtsson et al. 2005; Allison and Martiny 2008). Indeed, this soil biological component carries out the bulk of decomposition and catalyzes important transformation in the nutrient cycles (LeBauer and Treseder 2008; Lu et al. 2011). We also can confer on them a principal role in providing ecosystem services, such as soil fertility, soil quality and water purification (Reich et al. 2007). Despite their importance to the functioning of ecosystems, we must consider that changes in the

Table 7.3 Summary of main effects of different processes and fertilization systems on soil organic carbon and total nitrogen content

Authors (Year)	Kind of fertilization	Frequency of utilization	Main findings
Ai et al. (2013)	Mineral fertilization (Urea—300 kg N ha^{-1}; superphosphate—150 kg P_2O_5 ha^{-1}; and potassium chloride—150 kg K_2O ha^{-1}) and organic fertilization (Straw bedding impregnated with liquid and solid horse manure—37.5 T ha^{-1})	31 years	The soil organic carbon (SOC) and total nitrogen (TN) tended to be greater in the rhizosphere than in bulk soil, and increased in response to long-term organic fertilization in a wheat-maize rotation field cultivated on a fluvo-aquic soil, Hebei Province, China
Barrios-Masias et al. (2011)	Organic fertilization (Turkey manure and wine-grape solid residue—17 T ha^{-1})	12 years	The organic fertilization increased SOC as suggested by high soil microbial biomass and low CO_2 emissions and reduced N leaching potential during winter in a tomato field cultivated on a Paleosol, Yolo County, California
Gao et al. (2015)	Mineral fertilization (Urea—285 and 210 kg N ha^{-1}; Calcium superphosphate—142.5 kg P_2O_5 ha^{-1}; and KCl—71.3 kg K_2O ha^{-1}) and organic fertilization (Manure—11 T ha^{-1})	33 years	The long-term application of fertilizer (organic and inorganic) resulted in a significant increase in the SOC and TN contents in a double wheat-summer maize rotation system cultivated on a fluvo-aquic soil, Tianjin, Northern China
Lazcano et al. (2012)	Organic (vermicompost—*Standard dose* 4.2 and *high dose* 6.3 T ha^{-1}; and rabbit manure—*Standard dose* 5.4 and *high dose* 8.2 T ha^{-1}) and inorganic fertilizers (commercial NPK fertilizer—*Standard dose* 80-24-20 kg ha^{-1} and *high dose* 120-36-30 kg ha^{-1})	3 months	They found significant increases in SOC with the *high dose* of vermicompost in a sweet corn field cultivated on a Humic cambisol, Pontevedra, Northwestern Spain. There were no differences in TN of the soil samples at harvest
Mbuthia et al. (2015)	Inorganic N fertilization (Ammonium nitrate—0, 34, 67 and 101 kg N ha^{-1}), tillage (till and no-till) and cover crops (Hairy vetch	31 years	The no-till treatments under cover crop (vetch or wheat cover) having significantly greater SOC and TN to no cover under the lower N-rates (0, 34 and 67 N).

(continued)

Table 7.3 (continued)

Authors (Year)	Kind of fertilization	Frequency of utilization	Main findings
	and winter wheat, and a no cover control)		At the highest N-rate (101 N kg/ha), there were no significant differences in SOC and TN across tillage or cover crop treatments in a continuous cotton field cultivated on a Ultic Argisol, Jackson, West Tennessee, USA
Mikanová et al. (2013)	Mineral (NPK commercial fertilizer—63 kg N ha^{-1}) and organic fertilization (farmyard manure—35 T ha^{-1} and cattle slurry)	More than 50 years	Organic fertilization (e.g. farmyard manure addition) increased SOC and TN in a crop rotation field (45% cereals, 33% root crops, and 22% fodder crops) cultivated on a Orthic Luvisol, Prague-Ruzyne, Czech Republic
Muchere-Muna et al. (2007)	Organic (manure, *Tithonia diversifolia, Calliandra calothyrsus,* and *Leucaena leucocephala*) and mineral fertilization (30 and 60 kg N ha^{-1})	2 years	After 2 years of trial implementation, SOC and TN contents were improved with the application of organic residues, and manure in an maize field cultivated on Humic Nitosols, Meru South District, Kenya
Olmo et al. (2015)	Biochar addition (0, 0.5, 1, and 2.5% w/w on a dry weight basis)	2 months (Greenhouse conditions)	Biochar addition reduced TN in durum wheat plants cultivated on plastic pots with Haplic Luvisol, Córdoba, Spain. There is no report about SOC
Riley (2016)	Mineral (NPK fertilizer—100-25-120 kg N, P and K ha^{-1} and organic fertilization (Cattle manure 20–60 T ha^{-1})	30 years	Both manure and mineral fertilizer had increased SOC, by 11.3 and 3.4 T ha^{-1} respectively. The author also reports that no residual response of mineral fertilizer was found, but previous manure use gave large effects in a crop rotation field cultivated on Endostagnic Cambisol, Norway

(continued)

Table 7.3 (continued)

Authors (Year)	Kind of fertilization	Frequency of utilization	Main findings
Robertson et al. (2015)	They used three rotation treatments (fallow/wheat, pasture/fallow/wheat, and pasture/ Wheat) combined with two fallow management treatments (traditionally tilled fallow, with crop stubble incorporated by sloughing after harvest, and traditional drilling of the following crop) and a zero-tilled fallow	Three long-term (12, 28 and 94 years) field experiments	They report that the management practices examined in the present study may not reliably increase SOC on their own, but that significant increases in SOC are possible under some circumstances through the long-term use of multiple practices, such as stubble retention + zero tillage + legume N input + elimination of fallow on a Calcarosol and on a Vertosol, Victoria, Australia
Rondon et al. (2007)	Biochar addition (0, 30, 60, and 90 g kg^{-1} soil) and	40 days (Greenhouse conditions)	The SOC and TN significantly increased with the high biochar applications (90 g kg^{-1} soil) by 208.27 and 37.04% respectively in the soils with N-fixing plants cultivated on plastic pots with Latosol, Cali, Colombia

dynamic and diversity of soil organisms can affect terrestrial ecosystem processes and influences ecosystem responses to disturbances such as CO_2 and N addition (Spehn et al. 2005; Reich et al. 2007; Souza et al. 2015a, b, c).

Below, we illustrate the potential impacts of changes in soil organic carbon and total nitrogen contents on soil organism community composition and/or ecosystem processes (Fig. 7.2). Consider fertilization (mineral or organic) or any disturbance, such as biological invasion, fire or drought applied to an annual crop field and the soil organism communities within it. Soil organism composition might be unaltered or resistant to the disturbance, and not change (Fig. 7.2a and b). Alternatively, if the community is sensitive and does change, it could be resilient and quickly recover to its initial composition (Fig. 7.2c). Finally, a community whose composition is sensitive and not resilient might produce process rates similar to the original community if the members of the community are functionally redundant (Fig. 7.2d).

To assess whether soil organism community composition is resistant, resilient, functional redundancy or not change when exposed to fertilization or disturbance as illustrated in the Fig. 7.2, we must consider some points:

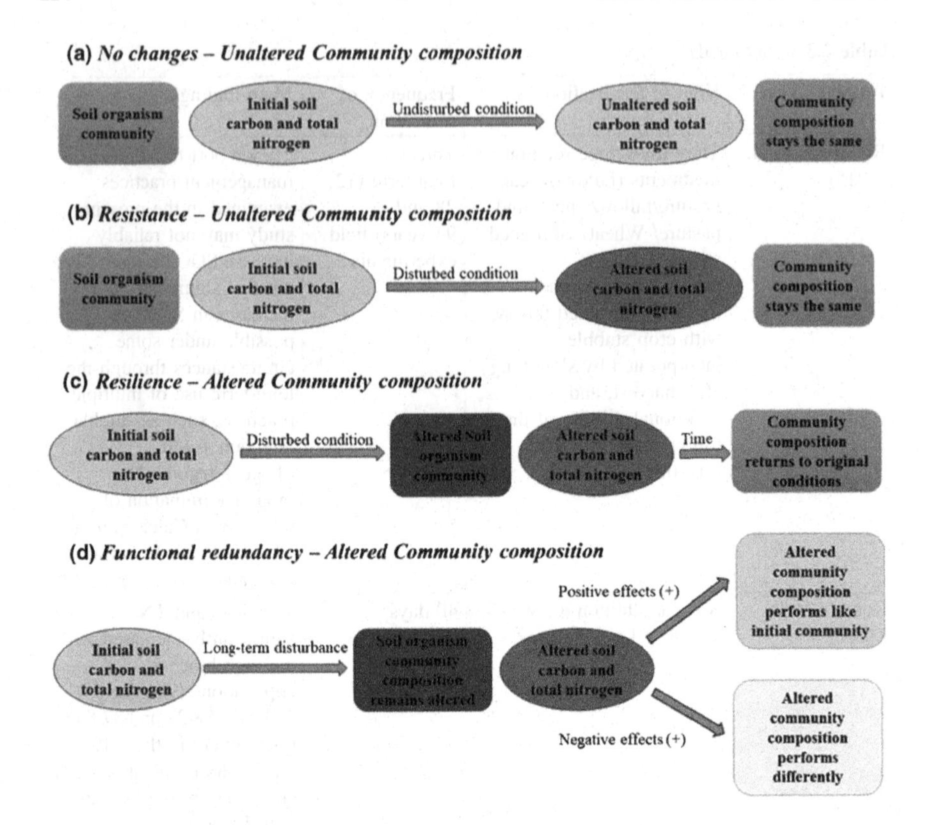

Fig. 7.2 A schematic of how changes in soil organic carbon and total nitrogen content can change soil organism community composition and thereby affect ecosystem processes versus when changes in these soil properties would not have this effect. Here, soil organism community composition are classified as resistant, resilient, or functionally redundant) (Adapted from Allison and Martiny 2008; Geisseler and Scow 2014; Gosling et al. 2016)

(1) The soil organism composition is limited to temperature, fertilization (NPK fertilization or organic fertilization), CO_2 enrichment, and enrichment with C substrates (Allison and Martiny 2008; Armstrong et al. 2015);
(2) The concentration and the kind of the used fertilizer and how often it is applied will have an effect (positive or negative) on the soil organism community composition (Mäder et al. 2000; Hole et al. 2005; Zhong et al. 2010; Carneiro et al. 2015);
(3) Application of organic amendments, increases soil organic carbon and stimulate microbial activity which provides N and P to soil (Abdullahi et al. 2013);
(4) The soil organisms are often limited by energy in soils, litter, and root exudates such as organic acids, sugars and amino acids may stimulate the abundance of soil macroarthropods and growth of microbial populations capable of influencing biogeochemical cycling of C, N, P, and S (Ai et al. 2013);

(5) Generally, organic amendments have positive effects on species richness and abundance of all soil organism groups (macroarthropods, N-fixing bacteria, archaea species, and arbuscular mycorrhizal fungi), except non-predatory insects and soil pathogens (Bengtsson et al. 2005);
(6) Carbon and nitrogen are the limiting nutrients for soil organisms in many terrestrial ecosystems and increased C and N input often leads to higher abundance of soil organisms (Wardle 1992; LeBauer and Treseder 2008; Treseder 2008; Liu and Greaver 2010; Lu et al. 2011);
(7) And finally, maintenance of biodiversity in agricultural landscapes will depend on the preservation, restoration and management of non-cropped areas, such as field margins, edge zones, habitat islands, hedgerows, natural pastures, wetlands, ditches, ponds and other small habitats, are important refuges and source areas for many soil organisms, such as soil macroarthropods (Stopes et al. 1995; Baudry et al. 2000; Tscharntke et al. 2002).

After this section, we asked the following questions? (i) Is soil organism C and N dependents? (ii) Does organic fertilization generally affect positively the diversity within soil organism groups? (iii) Does organic fertilization generally affect positively the abundance of the soil organism community? (iv) Do the effects of organic fertilization differ between soil organism groups? For example, do pest organisms increase more than non-pest groups in organic fertilization?

7.5 Effects of Long-Term Inputs of Nutrients to the Soil on Soil Organism Community Diversity

Soil organism communities and their diversity are also affected by inputs of nutrients provided by fertilization (Geisseler and Scow 2014). According Allison and Martiny (2008) 84% percent of the studies about fertilization and its effects on soil biology components report that belowground communities are sensitive to N, P, and K fertilization. Usually, in conventional systems with mineral NPK fertilization, fertilizer N inputs exceed rates of atmospheric decomposition and fertilizer N is often added in one large application during sowing (Gao et al. 2015). Fertilizer P and K inputs are also added in one large application during sowing, to provide sufficient P and K to plants and to prevent nutrient losses by immobilization, leaching or reaction with clay, and iron and aluminum oxides (Bressan 2001; Sandim et al. 2015).

7.5.1 Soil Macroarthropods

For aboveground communities, Pfiffner and Luka (2000), Gabriel et al. (2010), concluded in their studies that the abundance and diversity of soil macroarthropods depend on farming practices, such as organic versus conventional systems. Organic farming usually increases macroarthropod richness (average 30% higher species richness and 50% higher abundance than conventional farming systems). Usually non-predatory insects and pests respond negatively to organic farming, while predatory insects respond positively (Bengtsson et al. 2005). Macroarthropods contribute to services (e.g., soil fertility) impacting on plant yield in organic farming systems (Pearce and Venier 2006; Gabriel et al. 2010; Mikanová et al. 2013). Macroarthropods actively affect chemical, physical, and biological processes (Lavelle et al. 2006) and believed that they play an important role in nutrient cycling and in the maintenance of good soil quality (Brussaard et al. 1997; Sackett et al. 2010).

Our data indicate that the organic fertilization system changed positively the macroarthropod frequency of occurrence (Fig. 7.3), especially the frequency of predatory insects in the wheat field cultivated on a Ferralsol during 5 years of its utilization. The use of farmyard manure, as an organic fertilizer promoted positive effects whereas the use of mineral fertilizer promoted negative on Order Araneae, Blatodea, Homoptera, Hymenoptera, Mantodea and Orthoptera. So, our findings suggest that inputs of organic matter promoted by organic farming had positive effects on macroarthropod community composition. Plant residues are important to this group of organisms and act as food resource and refuge site to macroarthropods (Costa et al. 2009; Pearce and Venier 2006).

Macroarthropods, especially orders with greater abundance, are widely used to assess the conservation status of ecosystems (Luz et al. 2013). Among the orders that we observed in our study, the most frequent orders were Hymenoptera and Isoptera for all studied treatments (Fig. 7.3b). The first one, especially in the organic fertilization, and the second one is more frequent in the mineral fertilization. Among the Hymenoptera, the family Formicidae were predominant in the mineral fertilization. For the organic fertilization, we found three different families of the order Hymenoptera: Apidae, Formicidae, and Multilidae, but with Formicidae as a dominant group. Our results agree with the works done by Wink et al. (2005) that found Formicidae as a dominant group in different ecosystems and habitats and Luz et al. (2013) that reported higher diversity of ants in habitats with high organic matter contents than disturbed habitats. Among the order Coleoptera, the most frequent families were Carabidae and Scarabaeidae, but the second one only was found in the organic fertilization. Our results agree with the work done by Luz et al. (2013) that reported Scarabaeidae in preserved areas. Beetles of this family are very sensitive to changes in habitat, especially soil organic carbon (Costa et al. 2009; Azevedo et al. 2011). For the mineral fertilization treatments, orders Araneae and Mantodea only occur in the first year of our study, and the release of beneficial macroarthropods was probably more significant, since after its continuous use, there was an increase in number of individuals from order Hymenoptera, family Formicidae, and a decrease in the number of individuals from order Araneae, Mantodea, and Hymenoptera (Souza et al. 2015c). Generally, predators are related to more diverse habitats, with a depth layer of litter that provides hunting and foraging niches and for protection from desiccation (Pearce and Venier 2006).

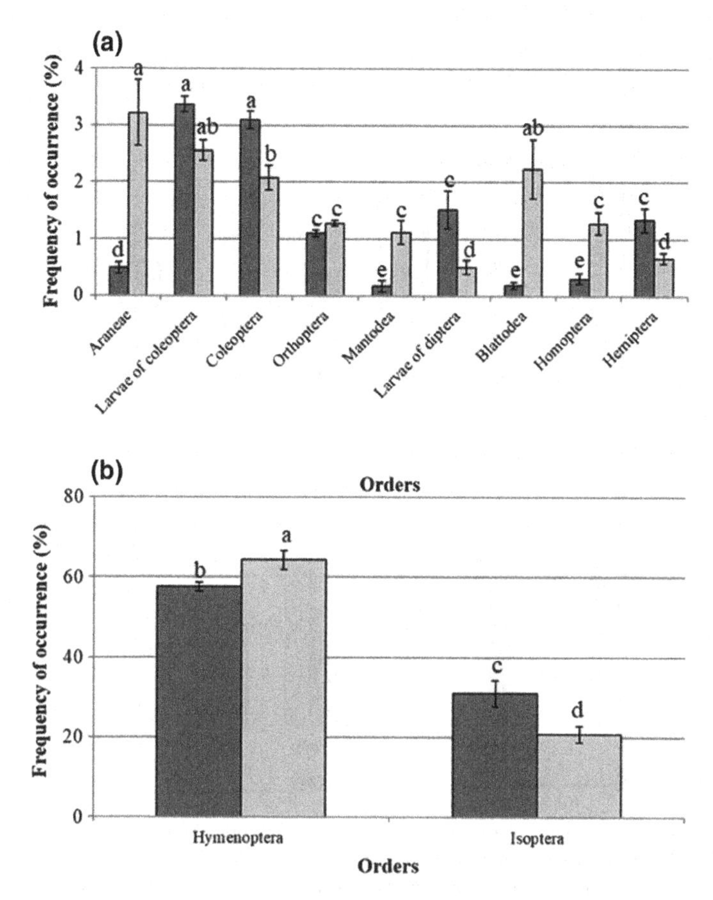

Fig. 7.3 Long-term effects mineral (pale brown) and organic (gray bar) fertilization on macroarthropod frequency of occurrence (%) (Means ± Standard deviation, $N = 90$; Adapted from Souza et al. 2015c) from less frequent Orders of macroarthropod (Fig. 7.3a); and the two most dominant Orders (Fig. 7.3b); Bars of each parameter labeled by different letters indicate significant differences assessed by the Bonferroni test after performing three-way ANOVA ($P < 0.05$)

7.5.2 *Arbuscular Mycorrhizal Fungi*

But high application rates of these macronutrients lead to temporally very high osmotic potentials, potentially toxic concentration, changes in soil pH, and other soil nutrient concentrations by synergic and antagonist interactions (Clark et al. 2007; Cleland and Harpole 2010; Souza et al. 2015a). For example, the supply of mineral nutrients has a strong influence on the AM fungi functionality (Eltrop and Marschner 1996). Accordingly, Smith and Read (2008) and Hodge and Storer (2014), the AM fungi infectivity potential is considerable reduced by high levels of

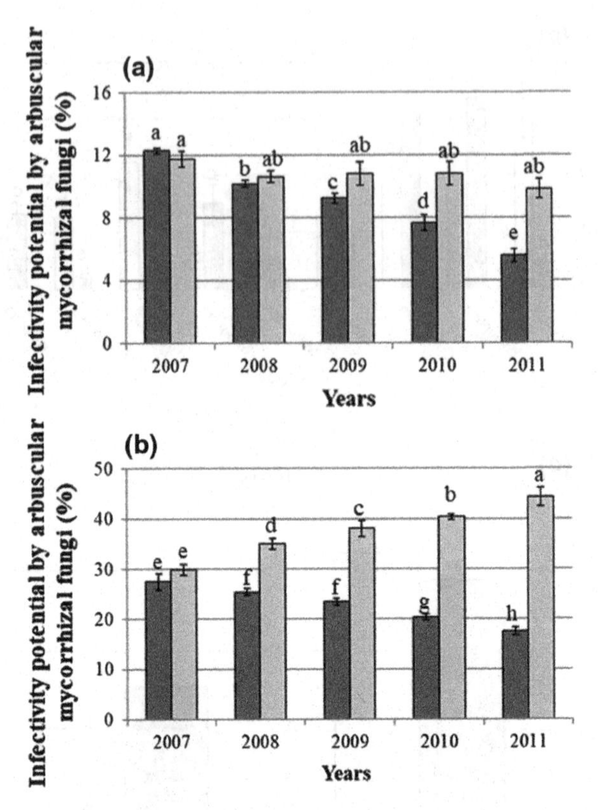

Fig. 7.4 Long-term effects of mineral (pale brown) and organic (gray bar) fertilization on infectivity potential by arbuscular mycorrhizal fungi (%) from soil inoculum from non-mycorrhizal (**a**) and mycorrhizal annual crop root zone (**b**) (Means ± standard deviation, $N = 90$; Adapted from Souza et al. 2016a, b, c). The non-mycorrhizal annual crop considered in this study was rapeseed, while wheat was considered as mycorrhizal annual crop. Mineral fertilization was based accordingly EMBRAPA's recommendation for *Brassica napus* cv. H401, and *Triticum aestivum* cv. BRS-Guamirim, respectively, while organic fertilization was based accordingly regional familiar agriculture sustainable system; Bars of each parameter labeled by different letters indicate significant differences assessed by the Bonferroni test after performing three-way ANOVA ($P < 0.05$)

nitrogen and phosphorous provided by mineral fertilization, as described in the Fig. 7.4.

The Fig. 7.4 describes the potential for subsequent annual crop (maize in our study) to form mycorrhizas provided by soil inoculum from rapeseed (a non-mycorrhizal plant species) and wheat (a mycorrhizal plant species) field. We observed that the infectivity potential by arbuscular mycorrhizal fungi was significantly higher in soil inoculum from wheat field than soil inoculum from rapeseed field (Fig. 7.4a and b). According Lankau et al. (2011) and Warwick (2011) plant species from Family Brassicaceae produce antifungal compounds with negative phytochemical effects on infectivity potential by arbuscular mycorrhizal fungi in soil inoculum from rapeseed field. Among the fertilization treatments, we also found cumulative negative effect of mineral fertilization on infectivity potential by arbuscular mycorrhizal fungi. Bressan (2001) and Siqueira et al. (1982) reported that the metabolites synthesis during the asymbiotic phase of AM fungi from genus *Claroideoglomus* and *Gigaspora* was negatively affect in substrate with high level of nitrogen, and Siqueira et al. (1985) reported that high level of phosphorous reduced significantly energy supply and root colonization during the asymbiotic, pre-symbiotic and symbiotic of arbuscular mycorrhizal fungi from genus *Claroideoglomus, Funneliformis, Gigaspora*, and *Glomus.* For the organic fertilization the positive effects on infectivity potential could be attributed to improvement on mycelium growth, protein synthesis and sporulation during the pre-symbiotic and symbiotic phases of arbuscular mycorrhizal fungi from genus *Claroideoglomus, Gigaspora, Glomus,* and *Scutellospora* as described by Siqueira and Hubbell (1986), Vilariño and Sainz (1997), Fracchia et al. (2001) and Silva et al. (2005).

7.5.3 Soil Bacteria

For soil bacteria, Ai et al. (2013) concluded that mineral fertilization, such as N application increased soil bacteria community, while organic fertilization (manure application) increased archaea community. This study was conducted on a fluvo-aquic soil during two years of a long-term field experiment with wheat and maize. They also report that soil bacteria, such as species from genus *Nitrosospira* was more sensitive to change its structure than archaea community (e.g. species from Phylum Crenarchaeota) after long-term N fertilization and they reported that organic fertilizers can stimulate soil bacteria and archaea activities by providing organic acids, sugars and amino acids that are essential for the growth of their populations (Fontaine and Barot 2005). These two soil organism groups played an important role in nitrogen cycle (soil bacteria more important than soil archaea) (Verhamme et al. 2011) and they are strongly affected by fertilization (Wang et al. 2009).

Soil bacteria and archaea species also can be affected by other soil conditions which are changed by fertilization (mineral or organic) such as pH, total nitrogen content, total organic carbon content and root exudates that determine the services of these soil organisms in agricultural ecosystems (Jia and Conrad 2009). Some studies report that soil bacteria present high activity in near-neutral or alkaline soils, whereas soil archaea presents high activity on acidic soils (Shen et al. 2008; Glaser et al. 2010; Yao et al. 2011). Other studies report that soil bacteria can assimilate more carbon (e.g. CO_2, soil organic matter, soil organic carbon) than soil archaea (Kowalchuk et al. 2000; Ai et al. 2013). So, we can presume three important questions about soil bacteria and soil archaea:

(1) They act in the same services of ecosystem (e.g. nitrogen cycle) (Ai et al. 2013);
(2) They present different metabolic pathways (Verhamme et al. 2011);
(3) They differ in their cell physiology (Walker et al. 2010).

7.6 Changing Soil Organic Carbon, Total Nitrogen and Crop Yields

Current interest in the fertilization management is curious considering that there are many differing interpretations (Gao et al. 2015; Robertson et al. 2015). Numerous field studies have reported greater level of soil organic carbon (SOC) under agricultural systems with continuous input of organic matter source than under traditional tillage with input of mineral fertilizers (Muchere-Muna et al. 2007; Hossain et al. 2010; Mikanová et al. 2013). This has given rise to the widespread view that substantial SOC decreasing may be resulted by the continuous use of mineral fertilization in traditional tillage areas around the world (Saha et al. 2008; Panwar et al. 2010; Conyers et al. 2012; Wezel et al. 2014). However, recent studies suggest that the increasing SOC under organic tillage practices may be less than initially estimated (Luo et al. 2010; Ladha et al. 2011; Körschens et al. 2013; Geisseler and Scow 2014). Other long-term experiments have been reported to increase SOC in some situations under mineral fertilizations (Kallenbach and Grandy 2011; Conyers et al. 2012) and to have no significant effect in other studies (Dalal et al. 2011) compared with mineral fertilization in conventional tillage.

Nitrogen fertilization is one of the most important agricultural N management practices to increase crop yield and to minimize environmental impact of agro-ecosystem (Robertson and Vitousek 2009) by increasing carbon dioxide fixation and root biomass production by annual crops (Rothamsted Research 2008) leading to more crop root residue return to soil (LeBauer and Treseder 2008). But, many studies have reported environmental problems related to the continuous use of mineral N fertilizer on the groundwater and atmosphere (Omar and Ismail 1999; Gao et al. 2015). Therefore, nitrogen management is a key factor in controlling the deterioration of soil quality and the continuous use of chemical fertilizer instead of organic fertilizer have decreased significantly soil organic carbon, total nitrogen and consequently crop yield (Zhang et al. 2008; Wessén et al. 2010; Schroder et al. 2011).

7.6.1 Long-Term Fertilization Systems in a Wheat Field (2007–2011)

The fertilization systems represented by this case are a non-fertilization systems (control), conventional fertilization (mineral NPK fertilization), and alternative fertilization (organic fertilization with farmyard manure) (See more details about fertilizers, doses, and application mode in Table 7.4). The soil of the experimental site is generally quite acid, and low in nutrient status (Table 7.5).

The main objective of this study was to determine whether the continuous use of mineral and organic fertilizers influence plant-soil interaction in wheat plants in field conditions. We used the wheat variety, *Triticum aestivum* var. BRS-Guamirim,

Table 7.4 Experimental setup, fertilizers, doses, and fertilization application mode during the five years of the study

Activities[a]	Control	Conventional	Alternative
Soil prepare (traction)	Yes (Animal)	Yes (Mechanical)	Yes (Animal)
Liming[b]	No	Yes—1.2 T ha^{-1}	Yes—1.2 T ha^{-1}
Mode of application	–	Limestone was incorporated 4 months before planting	
Fertilization	No	Yes—Mineral	Yes—Organic
Fertilizer (doses)	–	Ammonium sulfate (30 kg N ha^{-1}) Triple superphosphate (70 kg P_2O_5 ha^{-1}) Potassium chloride (60 kg K_2O ha^{-1})	Farmyard manure (20 T ha^{-1})
Mode of application	–	Incorporated during planting	Incorporated 2 months before planting
Seeds density	300 seeds/m^2		
Distance between crop lines	17 cm		
Top dressing	No	Yes—N fertilization	No
Fertilizer (doses)	–	Urea (30 kg N ha^{-1})	–
Mode of application	–	Incorporated besides crop lines 30 days after planting	–
After care	Yes–manual control of invasive herbs	Yes—Chemical control of invasive herbs (Glyphosate 2L ha^{-1})	Yes–manual control of invasive herbs

[a]These activities were performed during the five years of the study;
[b]Liming was used two times, during the first year (2007), and in the last year (2011)

Table 7.5 Soil properties of Ferralsol from the experimental area, Areia, PB, Brazil

pH (1:2.5 Soil:H_2O)[a]	4.28
Total organic carbon (g kg^{-1})[b]	7.30
Total nitrogen (g kg^{-1})[a]	0.19
Available P (mg dm^{-3})[c]	4.29

[a]Soil pH and total nitrogen were determined in a suspension of soil and distilled water and using the Kjeldahl method, respectively (Black 1965);
[b]Soil organic carbon was estimated according to the methodology described by Okalebo et al. (1993); and
[c]Available phosphorus (Olsen's P) was determined colorimetrically on spectrophotometer at 882 nm by extraction with sodium bicarbonate for 30 min (Olsen et al. 1954)

which is a highly cultivated wheat variety, particularly in the Southeastern Brazil. We investigated whether the influence of fertilization systems (alternative and conventional) on plant performance, and soil properties in a wheat field cultivated on a Ferralsols after 5 years of continuous use of mineral and organic fertilizers.

The experimental field was under grasses for about 10 years, where signalgrass (*Brachiaria decumbens* Stapf.) was the dominant grass species before to start the experiment. Mineral fertilization significantly increased total nitrogen and wheat yield, but for this last variable the mineral fertilization only promotes positive effects until the second year of its utilization (Fig. 7.5a and c). The wheat yield of the conventional fertilization system was not stable, the yield decreased after the second year of the continuous use of mineral fertilization, and the conventional agriculture system based on the continuous input of mineral NPK fertilization could not be considered sustainable in our study case. Mineral fertilization also significantly decreased soil organic carbon from 5.5 ± 0.3 g/kg in the first year of its utilization to 2.1 ± 0.2 g/kg in the end of the experiment (61.8% less soil organic carbon, Fig. 7.5b). So, based on our results we concluded that the long-term effect of mineral fertilization on wheat yield was soil organic carbon dependent (Fig. 7.5). We also concluded that overtime the mineral fertilization tended to reduce soil organic carbon, while increased total nitrogen in a Ferralsol. The continuous utilization of mineral fertilizers also affected positively soil pH and available phosphorous (Souza et al. 2015a).

In contrast, after 5 years of continuous utilization of the alternative fertilization system (based on the continuous input of an organic matter source) the values of soil organic carbon, total nitrogen and wheat yield were significantly increased (Fig. 7.6). the wheat yield was improved from 840.93 ± 68.01 kg/ha to 1379.85 ± 54.57 kg/ha (Fig. 7.6a), soil organic carbon was improved from values of 7.7 ± 0.4 g/kg in the first studied year to 10.6 ± 0.5 g/kg (37.66% more soil organic carbon) in the end of the experiment (Fig. 7.6b), and total nitrogen was improved from 0.32 ± 0.02 g/kg to 1.14 ± 0.03 g/kg (256.25% more total nitrogen; Fig. 7.6c). This study shows that the continuous use of fertilization systems

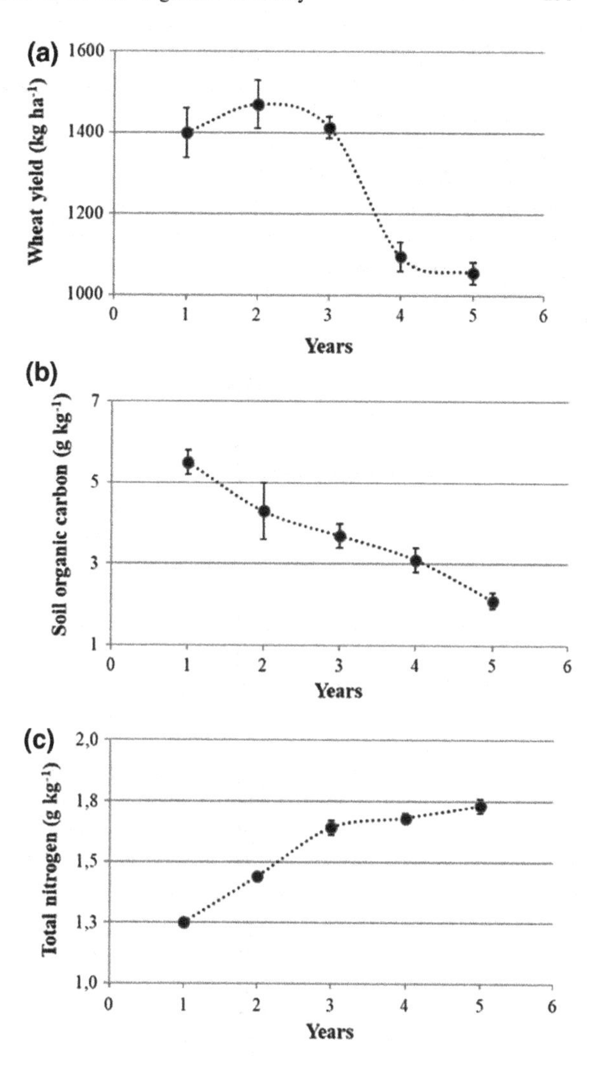

Fig. 7.5 Harvest yield ($kg\ ha^{-1}$) of wheat plants **a**, soil organic carbon ($g\ kg^{-1}$) **b** and total nitrogen ($g\ kg^{-1}$) **c** under conventional fertilization during the 5 years of its utilization (Means ± Standard deviation, $N = 90$; Adapted from Souza et al. 2015c)

changes the plant yield, and soil properties (e.g. soil organic carbon and total nitrogen), while total nitrogen experienced positive effects of both fertilization systems, conventional and alternative, soil organic carbon and wheat yield only experienced positive effects of alternative fertilization system. For the conventional fertilization, we found that after the third year of its use, occurs cumulative negative effects on wheat yield. In conclusion, the alternative fertilization system changed positively the plant performance, and soil properties in the wheat field cultivated on a Ferralsols during 5 years of its utilization. The use of farmyard manure promoted positive effects whereas the use of mineral fertilizers promoted negative on all

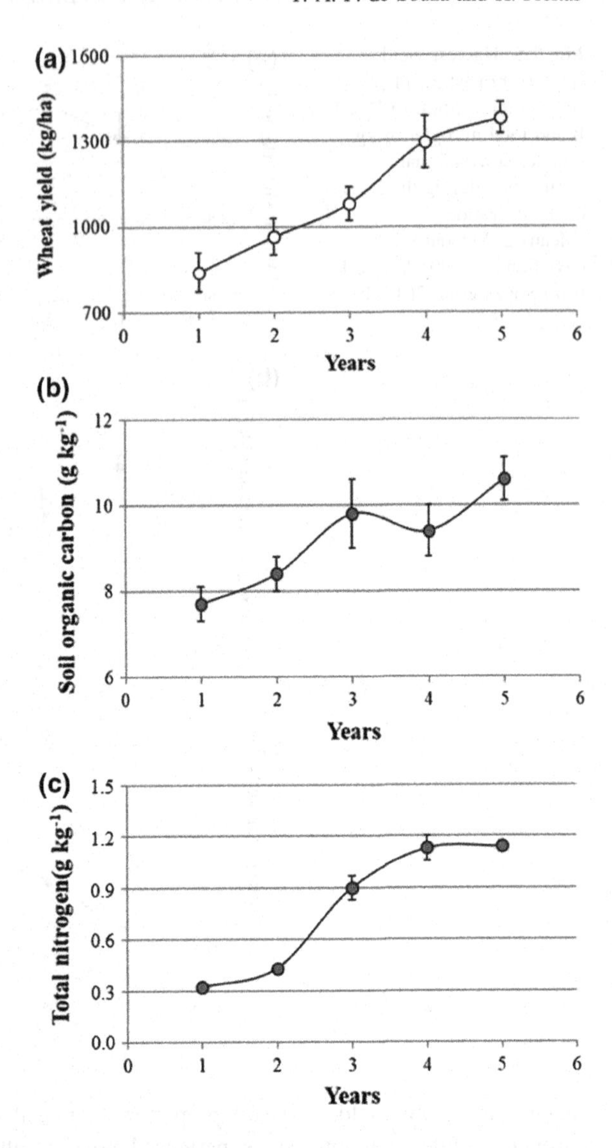

Fig. 7.6 Harvest yield ($kg\ ha^{-1}$) of wheat plants **a** soil organic carbon ($g\ kg^{-1}$) **b** and total nitrogen ($g\ kg^{-1}$) **c** under alternative fertilization during the 5 years of its utilization (Means ± standard deviation, $N = 90$; Adapted from Souza et al. 2015c)

studied variables in our study. The alternative fertilization system enhanced wheat yield after a long-term experiment. The results of our study highlight the importance of considering the long-term effect of fertilizers (mineral or organic) on the growth and mineral status of annual crops, such as *T. aestivum*. Thus, the long-term utilization of an alternative fertilization system with continuous input of organic matter may exploit positive situations of jointly beneficial biotic and abiotic conditions.

7.6.2 Long-Term Fertilization Systems and Stubble Management in a Rapeseed Field (2007–2011)

We performed a long-term study in this area for five years (2007–2011). Thus, we used an area of 28 × 24 m which was under grasses for about 10 years, where signalgrass (*Brachiaria decumbens* Stapf.) was the dominant grass species before to start the experiment. Six treatments were allocated in a factorial design, that consisted of three fertilization systems: (1) Control—non-fertilization; (2) Mineral fertilization—NPK fertilization according EMBRAPA's recommendation for *Brassica napus* L. tillage; and (3) Organic fertilization—Organic fertilization according regional familiar agriculture sustainable system; and two stubble managements: (1) Stubble removed and (2) stubble retention (left standing). Each treatment plot (5.0 × 1.4 m) was replicated in four blocks, and to our analysis we used the central portion (4.0 × 0.75 m) of each plot. The treatments are summarized in the Table 7.6. The objective of the present study was to compare soil chemical properties and rapeseed yield under stubble management and fertilization systems in a long-term field experiment in a Ferralsols of Northeastern Brazil. We used the rapeseed variety, *Brassica napus* L. var. H401, which is a highly cultivated rapeseed variety, particularly in the Southern Brazil. Our hypotheses were that soil organic carbon, total nitrogen, and rapeseed yield are increased by: (1) stubble retained (left standing) rather than stubble removal; and (2) organic fertilization, rather than mineral fertilization.

For the non-fertilization treatment (control), we did not use any fertilizer, herbicide or pesticide to control soil fertility, weed and insect occurrence during the long-term experiment. Rapeseed in the mineral fertilization treatment was sown with 300 kgNPK ha^{-1} as conventional compound field fertilizer (06-24-12). For the organic fertilization treatment, organic fertilizer was farmyard manure. The carbon, nitrogen, and phosphorus contents from the used organic fertilizer were 298.66, 21.19 and 7.29 g kg^{-1} respectively (See more details about fertilizers, doses, and application mode in Table 7.7).

The soil examined was classified as a Ferralsols (WRB 2006). Soils were collected at the beginning of the experiment (March 2007) during the dry period and when the plants were in bud formation stage. The soil of the experimental site is generally quite acid, and low in nutrient status (Table 7.8).

Table 7.6 Summary of experimental treatments during the five years (2007–2011) of the study

Treatment code	Fertilization systems	Stubble management
C-LS	Non-fertilization (Control)	Retained, left standing
C-R	Non-fertilization (Control)	Removed
M-LS	Mineral fertilization	Retained, left standing
M-R	Mineral fertilization	Removed
O-LS	Organic fertilization	Retained, left standing
O-R	Organic fertilization	Removed

Table 7.7 Fertilizers, doses, and fertilization application mode during the five years of the study

Activities[a]	Control	Mineral	Organic
Soil prepare (traction)	Yes (Animal)	Yes (Mechanical)	Yes (Animal)
Liming[b]	No	Yes—1.2 T ha^{-1}	Yes—1.2 T ha^{-1}
Mode of application	–	Limestone was incorporated 4 months before planting	
Fertilization	No	Yes—Mineral	Yes—Organic
Fertilizer (doses)	–	Conventional compound field fertilizer, 06-24-12 (300 kg ha^{-1})	Farmyard manure (10 T ha^{-1})
Mode of application	–	Incorporated during planting	Incorporated 2 months before planting
Top dressing	No	Yes—N fertilization	No
Fertilizer (doses)	–	Urea (120 kg N ha^{-1})	–
Mode of application	–	Incorporated besides crop lines 30 days after planting	–
After care	Yes–manual control of invasive herbs	Yes—Chemical control of invasive herbs (Glyphosate 2L ha^{-1}) Foliar fertilization with micronutrients, when the plants starting flower bud development (Albatroz 300 mL ha^{-1}) Chemical control of *Diabrotica speciosa* Germar and *Myzus persicae* Sulzer, during flowering (Decis-25 160 mL ha^{-1}; Monocrotophos 300 mL ha^{-1})	Yes–manual control of invasive herbs

[a]These activities were performed during the five years of the study;
[b]Liming was used two times, during the first year (2007), and in the last year (2011)

After the five years of our study, the utilization of C-LS, C-R, M-LS, and M-R treatments decrease the soil organic carbon from 3.82 ± 0.18, 3.20 ± 0.25, 4.48 ± 0.18, and 5.04 ± 0.24 g kg^{-1} to 2.84 ± 0.08, 1.32 ± 0.10, 3.46 ± 0.16, and 2.46 ± 0.10 g kg^{-1}, respectively. Conversely, for the continuous use of organic fertilization (O-LS and O-R) we found cumulative positive effects of these treatments on soil organic carbon. The O-LS and O-R treatments improved it from 7.02 ± 0.38 and 5.00 ± 0.42 to 11.06 ± 0.20 and 10.06 ± 0.20 g kg^{-1}, respectively (Fig. 7.7).

After the second year of study, the utilization of mineral fertilization had decreased the total nitrogen from values of 1.73 ± 0.09 to 1.32 ± 0.10 g kg^{-1}. For the organic fertilization, we observed a significant positive effect on total nitrogen

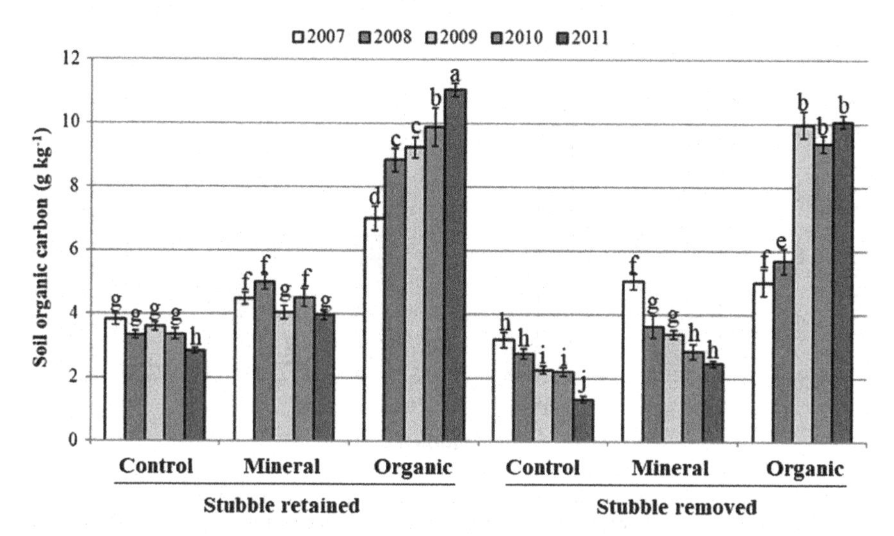

Fig. 7.7 Effects of different fertilization systems (Control, mineral and organic), stubble management (retained and removal), and years of their utilization (2007–2011) on soil organic carbon (g kg^{-1}, means ± Standard deviation, $N = 90$; Bars of each parameter labeled by different letters indicate significant differences assessed by the Bonferroni test after performing three-way ANOVA ($P < 0.05$) Adapted from Souza et al. 2016a)

after the first year of its utilization, and in the end of the experiment we did not find difference between the results of these soil properties from mineral and organic fertilization treatments (Fig. 7.8).

In the mineral, organic and control treatments combined with stubble retained, their continuous use had positive effects on rapeseed yield. Conversely, for control and mineral treatments, both combined with stubble removed treatment, there was a significant negative effect of its continuous use on this variable (Fig. 7.9). These confirm our findings that the organic fertilization combined with stubble retained treatment changed positively the plant yield, and soil properties in the rapeseed field cultivated on a Ferralsols during 5 years of their utilization. The use of farmyard manure promoted positive effects whereas the use of mineral fertilizer promoted negative on soil organic carbon, total nitrogen and available phosphorous in our study. The organic fertilization combined with stubble retained treatment enhanced rapeseed yield after a long-term experiment. The results of our study highlight the importance of considering the cumulative effect of stubble retention and organic fertilizers on the yield of annual crops, like rapeseed. Thus, the long-term utilization of practices with continuous input of organic matter might enhance soil chemical properties and annual crop (rapeseed in our study) development.

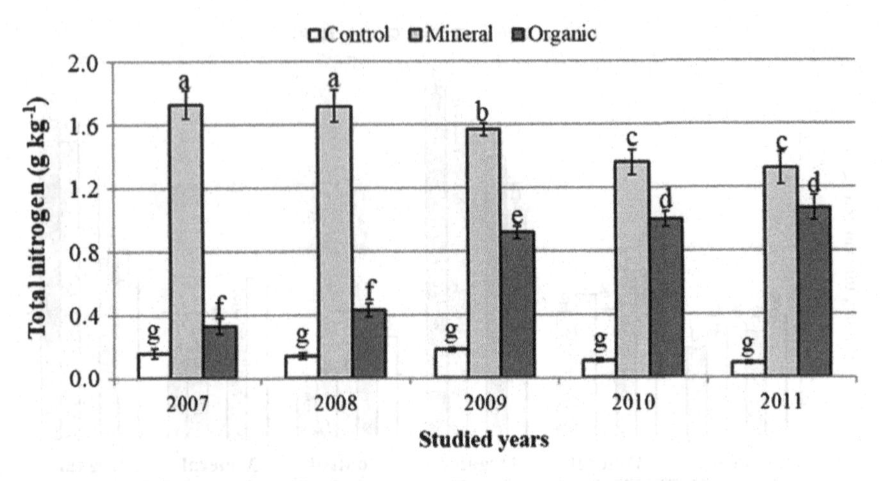

Fig. 7.8 Effects of different fertilization systems (Control, mineral and organic), and years of their utilization (2007–2011) on total nitrogen (g kg^{-1}, means ± Standard deviation, $N = 90$; Bars of each parameter labeled by different letters indicate significant differences assessed by the Bonferroni test after performing three-way ANOVA ($P < 0.05$) Adapted from Souza et al. 2016a)

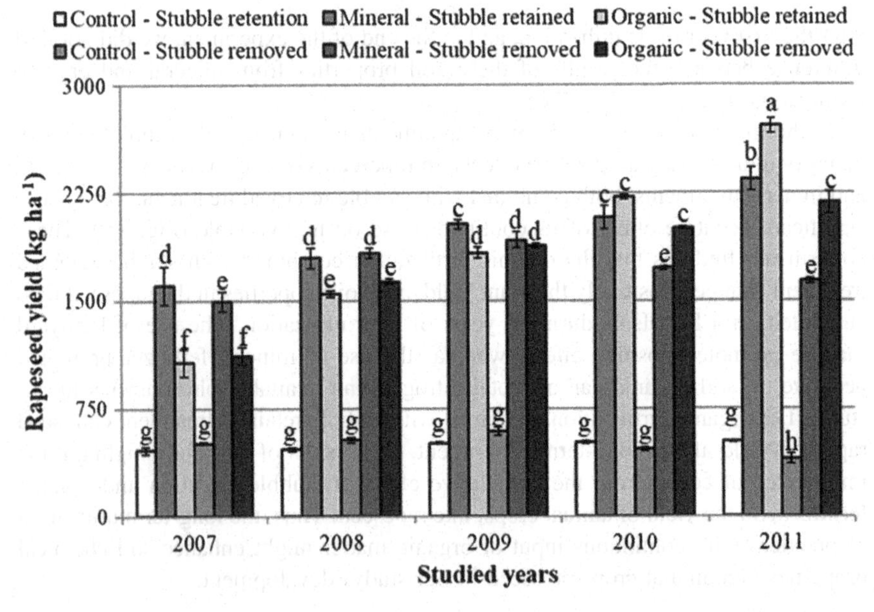

Fig. 7.9 Effects of different fertilization systems (Control, mineral and organic), stubble management (retained and removal), and years of their utilization (2007–2011) on rapeseed yield (kg ha^{-1}, means ± Standard deviation, $N = 90$; Bars of each parameter labeled by different letters indicate significant differences assessed by the Bonferroni test after performing three-way ANOVA ($P < 0.05$) Adapted from Souza et al. 2016a)

7.7 Main Fertilizers Used in the Modern Agriculture

The majority of the mineral fertilizers used in the modern agriculture have been refined to extract nutrients and bind them in specific concentrations with another mineral nutrient or compound (Jones 2012). These fertilizers may be made from petroleum, rocks or even organic sources, and they are refined by massive industries to their pure state with a high consume and demand of energy for their production. Mineral fertilizers have the advantage to provide immediately to the plants essential nutrients, but they do not promote life or soil health (Fan et al. 2008). In the other hand, the organic fertilizers are minimally processed, and the nutrients remain in their natural forms. Organic fertilizers are usually made from animal or plant waste or powdered minerals (e.g. manure, compost, bone and cottonseed meal) and sometimes they are classified as "soil conditioners". Organic fertilizers have the advantage to improve the structure of the soil and increase its ability to hold water and nutrients, and overtime these fertilizers improve soil and plant health (Dittmar et al. 2009). Below, we present the main mineral and organic fertilizers, their main nutrient concentration, and nutrient form for each fertilizer (Table 7.9).

Table 7.8 Soil chemical characteristics (0–20 cm) of the experimental field, March 2007

pH (1:2.5 Soil:H_2O)[a]	4.38
Total organic carbon (g kg^{-1})[b]	7.20
Total nitrogen (g kg^{-1})[a]	0.19
Available P (mg dm^{-3})[c]	3.64

[a]Soil pH and total nitrogen were determined in a suspension of soil and distilled water and using the Kjeldahl method, respectively (Black 1965);
[b]Soil organic carbon was estimated according to the methodology described by Okalebo et al. (1993) and
[c]Available phosphorus (Olsen's P) was determined colorimetrically on spectrophotometer at 882 nm by extraction with sodium bicarbonate for 30 min (Olsen et al. 1954)

Table 7.9 Main mineral and organic fertilizers used in the modern agriculture

Fertilizer	Nutrient %	Nutrient form
Mineral fertilizers		
Ammonium chloride	25% NH_4^+; 62–66% Cl	NH_4^+; Cl
Ammonium molybdate	54% Mo; 5–7% N	Mo; NH_4^+
Ammonium polyphosphate	10% N; 34% P_2O_5	NH_4^+; P_2O_5
Ammonium nitrate	34% N (17% NH_4^+ and 17%)	NH_4^+; NO_3^-
Ammonium sulfate	21% N; 20–24% S	NH_4^+
Anhydrous ammonia	82% N	NH_4^+
Anhydrous ammonium nitrate	32% N	NH_4^+; NO_3^-

(continued)

Table 7.9 (continued)

Fertilizer	Nutrient %	Nutrient form
Boric acid	17% B	B
Calcium ammonium nitrate	20% N (10% NH_4^+, 10% NO_3^-); 2–8% Ca; 1–5% Mg	NH_4^+; NO_3^-; Ca; Mg
Calcium chloride	24% Ca	Ca
Calcium nitrate	14% N; 18–19% Ca; 1–5% Mg	NO_3; Ca; Mg
Calcium sulfate	16% Ca; 13% S	Ca; S
Copper nitrate	22% Cu; 9% N	Cu; NO_3^-
Copper sulfate	13% Cu; 18% S	Cu; S
Diammonium phosphate	18% N; 46% P_2O_5	NH_4^+; P_2O_5
Iron(II) chloride	23% Fe; 30% Cl	Fe; Cl
Iron(II) sulfate	19% Fe; 10–11% S	Fe; S
Iron(III) chloride	15% Fe; 30% Cl	Fe; Cl
Iron(III) nitrate	11% Fe; 8% N	Fe; NO_3^-
Iron(III) sulfate	23% Fe; 18–20% S	Fe; S
Magnesium sulfate	9% Mg; 12–14% S	Mg; S
Manganese oxide	41% Mn	Mn
Manganese(II) chloride	35% Mn; 45% Cl	Mn; Cl
Manganese(II) nitrate	16% Mn; 8% N	Mn; NO_3^-
Manganese(II) sulfate	26% Mn; 14–15% S	Mn; S
Monoammonium phosphate	11% N; 52% P_2O_5	NH_4^+; P_2O_5
Monocalcium phosphate	18% P_2O_5; 18–20% Ca; 10–12% S	P_2O_5; Ca; S
Muriate of potash	60–62% K_2O	K_2O
Nitrogen solutions	28–32% N	NH_4^+; NO_3^-; $CO(NH_2)_2$
Nitrophosphate	14% N; 18% P_2O_5; 8–10% Ca	NO_3^-; P_2O_5; Ca
Phosphate rock	24% P_2O_5; 23–27% Ca	P_2O_5; Ca
Potassium chloride	58% K_2O; 45–48% Cl	K_2O; Cl
Potassium hydroxide	70% K_2O	K_2O
Potassium nitrate	44% K_2O; 13% N	K_2O; NO_3^-
Potassium sulfate	50% K_2O; 15–17% S; 1–1.2% Mg	K_2O; S; Mg
Sodium molybdate	39% Mo	Mo
Sodium nitrate	15% N	$NaNO_3$
Sulfate of potash magnesia	22% K_2O; 4.5% Mg; 22–24% S; 1–2.5% Cl	K_2O; Mg; S; Cl
Thomas slag	12% P_2O_5; 20–29% Ca; 0.4–3% Mg	P_2O_5; Ca; Mg
Triple superphosphate	46% P_2O_5	P_2O_5
Urea	46% N	$CO(NH_2)_2$
Zinc chloride	40% Zn; 44% Cl	Zn; Cl
Zinc nitrate	19% Zn; 8% N	Zn; NO_3^-
Zinc sulfate	20% Zn; 16–18% S	Zn; S

(continued)

Table 7.9 (continued)

Fertilizer	Nutrient %	Nutrient form
Organic fertilizers		
Blood meal	70% Organic matter; 10% N	–
Bone meal	6% Organic matter; 1.5% N; 20% P_2O_5	–
Cattle manure	57% Organic matter; 1.7% N; 0.9% P_2O_5; 1.4% K_2O	–
Chicken manure	50% Organic matter; 3.0% N; 3.0% P_2O_5; 2.0% K_2O	–
Compost	31% Organic matter; 1.4% N; 1.4% P_2O_5; 0.8% K_2O	–
Crop residue (sugarcane, rice)	36% Organic matter; 20:1 C:N; 1% N	–
Crop residue (coffee)	46% Organic matter; 20:1 C:N; 1.3% N	–
Crop residue (cotton, peanut, castor bean, soybean)	70% Organic matter; 5% N	–
Fish meal	50% Organic matter; 4% N; 6% P_2O_5	–
Horse manure	46% Organic matter; 1.4% N; 0.5% P_2O_5; 1.7% K_2O	–
Peat	30% Organic matter; 18:1 C:N ratio; 1% N	–
Sheep manure	65% Organic matter; 1.4% N; 1.0% P_2O_5; 2.0% K_2O	–
Swine manure	53% Organic matter; 1.9% N; 0.7% P_2O_5; 0.4% K_2O	–

Accordingly, the Agronomy Guide (2016)

7.8 Conclusion

In our analysis of organic and mineral fertilization in different crop systems, we found evidence that the continuous use of organic fertilizer promoted positive effects on soil organic carbon and total nitrogen, whereas the continuous use of mineral fertilizer promoted negative on these soil properties. For both organic and mineral fertilization, we show that soil organisms (for example, macroarthropods, arbuscular mycorrhizal fungi, and N-fixing bacteria) are carbon and nitrogen dependents while plants are only mineral nutrient dependents. Overall, arbuscular mycorrhizal fungal infectivity potential was very sensitive in long-term mineral fertilization systems than other soil organism group. Additionally, we conclude that the long-term utilization of organic fertilizations systems, based on organic farming without use of pesticides, herbicides, and inorganic fertilizers changed positively the macroarthropod and microorganism community composition.

References

The Agronomy Guide (2016) College of Agricultural Sciences, PennState

Abdullahi R, Sheriff HH, Lihan S (2013) Combine effect of bio-fertilizer and poultry manure on growth, nutrients uptake and microbial population associated with sesame (*Sesamum indicum* L.) in North-eastern Nigeria, IOSR. J Environ Sci Toxicol Food Technol 5:60–65

Ai C, Liang G, Sun J, Wang X, He P, Zhou W (2013) Different roles of rhizosphere effect and long-term fertilization in the activity and community structure of ammonia oxidizers in a calcareous fluvo-aquic soil. Soil Biol Biochem 57:30–42

Allison SD, Martiny JBH (2008) Resistance, resilience and redundancy in microbial communities. Proc Nat Acad Sci USA 105:11512–11519

Armstrong A, Waldron S, Ostle NJ, Richardson H, Whitaker J (2015) Biotic and abiotic factors interact to regulate northern peatland carbon cycling. Ecosystems 18:1395–1409. https://doi.org/10.1007/s10021-015-9907-4

Azevedo FR, Moura MAR, Arrais MSB, Nere DR (2011) Composição da entomofauna da floresta nacional do Araripe em diferentes vegetações e estações do ano. Ceres 58(6):740–748

Barrios-Masias FH, Cantwell MI, Jackson LE (2011) Cultivar mixtures of processing tomato in an organic agroecosystems. Org Agric 1(1):17–30

Baudry J, Burel F, Thenail C, Le Coeur D (2000) A holistic landscape ecology study of the interactions between activities and ecological patterns in Brittany, France. Landscape Urban Plann 50:119–128

Bayu W, Rethman NFG, Hammes PS, Alemu G (2006) Effects of farmyard manure and inorganic fertilizers on sorghum growth, yield, and nitrogen use in a semi-arid area of Ethiopia. J Plant Nutr 29:391–407

Belay Z, Vestberg M, Assefa F (2015) Diversity and abundance of arbuscular mycorrhizal fungi across different land use types in a humid low land area of Ethiopia. Trop Subtrop Agroecosyst 18:47–69

Bengtsson J, Ahnström J, Weibull AC (2005) The effects of organic agriculture on biodiversity and abundance: a metaanalysis. J Appl Ecol 42:261–269

Berner EK, Berner RA (1998) Global environment: water, air, and geochemical cycles. Prentice Hall, Upper Saddle River, New Jersey

Black CA (1965) Methods of soil analysis, part 2. In: Black CA (ed) Agronomy monograph No 9. American Society of Agronomy, Madison, pp 771–1572

Bossio DA, Girvan MS, Verchot L, Bullimore J, Borelli T, Albrecht A, Scow KM, Ball AS, Pretty JN, Osborn AM (2005) Soil microbial community response to land use change in an agricultural landscape of western Kenya. Microb Ecol 49:50–62

Bravo-Garza MR, Bryan RB (2005) Soil properties along cultivation and fallow time sequence on Vertisols in northeastern Mexico. Soil Sci Soc Am J 69:473–481. https://doi.org/10.2136/sssaj2005.0473

Bressan W (2001) Interactive effect of phosphorus and nitrogen on in vitro spore germination of *Glomus etunicatum* Becker & Gerdemann, root growth and mycorrhizal colonization. Braz J Microbiol 32:276–280

Brussaard L (1998) Soil fauna, guilds, functional groups and ecosystem processes. Appl Soil Ecol 9:123–135

Brussaard L, Behan-Pelletier VM, Bignell DE, Brown VK, Didden W, Folgarait P, Fragoso C, Freckman DW, Gupta VVSR, Hattori T, Hawksworth DL, Klopatek C, Lavelle P, Malloch DW, Rusek J, Soderstrom B, Tiedje JM, Virginia RA (1997) Biodiversity and ecosystem functioning in soil. Ambio 26:563–570

Carneiro MAC, Ferreira DA, Souza ED, Paulino HB, Saggin Junior OJ, Siqueira JO (2015) Arbuscular mycorrhizal fungi in soil aggregates from fields of “murundus” converted to agriculture. Pesq agropec bras 50:313–321. https://doi.org/10.1590/s0100-204x2015000400007

Chan KY, Conyers MK, Li GD, Helyar KR, Poile G, Oates A, Barchia IM (2011) Soil carbon dynamics under different cropping and pasture management in temperate Australia: results of three long-term experiments. Soil Res 49:320–328. https://doi.org/10.1071/sr10185

Chen R, Hu J, Dittert K, Wang J, Zhang J, Lin XG (2011) Soil total nitrogen and natural 15Nitrogen in response to longterm fertilizer management of a corn-wheat cropping system in Northern China. Commun Soil Sci Plant 42:323–331

Clark CM, Cleland EE, Collins SL, Fargione JE, Gough L, Gross KL, Pennings SC, Suding KN, Grace JB (2007) Environmental and plant community determinants of species loss following nitrogen enrichment. Ecol Lett 10:596–607

Cleland EE, Harpole WS (2010) Nitrogen enrichment and plant communities. Ann N Y Acad Sci 1195:46–61

Cleveland CC, Liptzin D (2007) C:N: P stoichiometry in soil: is there a "Redfield ratio" for the microbial biomass? Biogeochemistry 85:235–252

Conyers M, Newton P, Condon J, Poile G, Mele P, Ash G (2012) Three long-term trials end with a quasi-equilibrium between soil C, N, and pH: an implication for C sequestration. Soil Res 50:527–535. https://doi.org/10.1071/sr12185

Costa CMQ, Silva FAB, Farias AI, Moura RC (2009) Diversidade de scarabaeidae (coleoptera, scarabaeidae) coletados com armadilha de interceptação de vôo no refúgio ecológico Charles Darwin, igarassu-PE-brasil. Rev Bras de Entomol 53(1):88–94

Dalal RC, Allen DE, Wang WJ, Reeves S, Gibson I (2011) Organic carbon and total nitrogen stocks in a Vertisol following 40 years of no-tillage, crop residue retention and nitrogen fertilisation. Soil Tillage Res 112:133–139. https://doi.org/10.1016/j.still.2010.12

Dittmar H, Drach M, Vosskamp R, Trenkel ME, Gutser R, Steffens G (2009) Fertilizers, 2: types in ullmann's encyclopedia of Ind. Chemistry. Wiley, Weinheim. https://doi.org/10.1002/14356007.n10_n01-324

Drakopoulos D, Scholberg JMS, Lantinga EA, Tittonell PA (2015) Influence of reduced tillage and fertilization regime on crop performance and nitrogen utilization of organic potato. Org Agric. https://doi.org/10.1007/s13165-015-0110-x

Eltrop L, Marschner H (1996) Growth and mineral nutrition of non-mycorrhizal and mycorrhizal Norway Spruce (*Picea doies*) seedlings grow in semi-hydroponic sand culture II: carbon partitioning in plants supplied with ammonium or nitrate. New Phytologist 133:474–486

Fan MS, Zhao FJ, Fairweather-Tait SJ, Poulton PR, Dunham SJ, McGrath SP (2008) Evidence of decreasing mineral density in wheat grain over the last 160 years. J Trace Elem Med Biol 22:315

Fierer N, Carney KM, Horner-Devine MC, Megonigal JP (2009) The biogeography of ammonia-oxidizing bacterial communities in soil. Microb Ecol 58:435–445

Fontaine S, Barot S (2005) Size and functional diversity of microbe populations control plant persistence and long-term soil carbon accumulation. Ecol Lett 8:1075–1087

Fracchia S, Menendez A, Godeas A, Ocampo JA (2001) A method to obtain monosporic cultures of arbuscular mycorrhizal fungi. Soil Biol Biochem 33:1283–1285

Gabriel D, Sait SM, Hodgson JA, Schmutz U, Kunin WE, Benton TG (2010) Scale matters: the impact of organic farming on biodiversity at different spatial scales. Ecol Lett 13:858–869

Gao W, Yang J, Ren S, Hailong L (2015) The trend of soil organic carbon, total nitrogen, and wheat and maize productivity under different long-term fertilizations in the upland fluvo-aquic soil of North China. Nutr Cycl Agroecosyst. https://doi.org/10.1007/s10705-015-9720-7

Geisseler D, Scow KM (2014) Long-term effects of mineral fertilizers on soil microorganisms—A review. Soil Biol Biochem 75:54–63

Glaser K, Hackl E, Inselsbacher E, Strauss J, Wanek W, Zechmeister-Boltenstern S, Sessitsch A (2010) Dynamics of ammonia-oxidizing communities in barley-planted bulk soil and rhizosphere following nitrate and ammonium fertilizer amendment. FEMS Microbiol Ecol 74:575–591

Gosling P, Jones J, Bending GD (2016) Evidence for functional redundancy in arbuscular mycorrhizal fungi and implications for agroecosystems management. Mycorrhiza 26:77–83. https://doi.org/10.1007/s00572-015-0651-6

Guo SL, Wu JS, Coleman K, Zhu HH, Li Y, Liu WZ (2012) Soil organic carbon dynamics in a dryland cereal cropping system of the Loess Plateau under long-term nitrogen fertilizer applications. Plant Soil 353:321–332
Hodge A, Storer K (2014) Arbuscular mycorrhizal and nitrogen: implications for individual plants through to ecosystems. Plant Soil 386:1–19
Hole DG, Perkins AJ, Wilson JD, Alexander IH, Grice PV, Evans AD (2005) Does organic farming benefit biodiversity? Biol Conserv 122:113–130
Hossain MK, Strezov V, Chan KY, Nelson PF (2010) Agronomic properties of wastewater sludge biochar and bioavailability of metalsin production of cherry tomato (*Lycopersicon esculentum*). Chemosphere 78:1167–1171
Jia Z, Conrad R (2009) Bacteria rather than Archaea dominate microbial ammonia oxidation in an agricultural soil. Environ Microbiol 11:1658–1671
Jones Jr B (2012) Inorganic chemical fertilizers and their properties. In: Plant nutrition and soil fertility manual, 2nd edn. CRC Press. ISBN 978-1-4398-1609-7. eBook ISBN 978-1-4398-1610-3
Kallenbach C, Grandy AS (2011) Controls over soil microbial biomass responses to carbon amendments in agricultural systems: a meta-analysis. Agr Ecosyst Environ 144:241–252
Kibblewhite MG, Ritz K, Swift MJ (2008) Soil health in agricultural systems. Philos Trans R Soc Series B 363:685–701
Kimetu J, Lehmann J, Ngoze SO, Mugendi DN, Kinyangi JM, Riha S, Verchot L, Recha JW, Pell AN (2008) Reversibility of soil productivity decline with organic matter of differing quality along a degradation gradient. Ecosystems 11:726–739
Kong X, Dao TH, Qin J (2009) Effects of soil texture and land use interactions on organic carbon in soils in North China cities' urban fringe. Geoderma 154:86–92
Körschens M, Albert E, Armbruster M, Barkusky D, Baumecker M, Behle-Schalk L, Bischoff R, Cergan Z, Ellmer F, Herbst F, Hoffmann S, Hofmann B, Kismanyoky T, Kubat J, Kunzova E, Lopez-Fando C, Merbach I, Merbach W, Pardor MT, Rogasik J, Rühlmann J, Spiegel H, Schulz E, Tajnsek A, Toth Z, Wegener H, Zorn W (2013) Effect of mineral and organic fertilization on crop yield, nitrogen uptake, carbon and nitrogen balances, as well as soil organic carbon content and dynamics: results from 20 European long-term field experiments of the twenty-first century. Arch Agron Soil Sci 59:1017–1040
Kowalchuk GA, Stienstra AW, Heilig GHJ, Stephen JR, Woldendorp JW (2000) Molecular analysis of ammonia-oxidising bacteria in soil of successional grasslands of the Drentsche A (The Netherlands). FEMS Microbiol Ecol 31:207–215
Ladha JK, Reddy CK, Padre AT, van Kessel C (2011) Role of nitrogen fertilization in sustaining organic matter in cultivated soils. J Environ Qual 40:1756–1766
Lal R, Follett RF (Eds) (2009) Soil carbon sequestration and the greenhouse effect. Soil Science Society of America Inc., Madison, WI)
Lal R (2004) Soil carbon sequestration to mitigate climate change. Geoderma 123:1–22. https://doi.org/10.1016/j.geoderma.2004.01.032
Lankau RA, Wheeler E, Bennett AE, Strauss SY (2011) Plant-soil feedbacks contribute to an intransitive competitive network that promotes both genetic and species diversity. J Ecol 99:176–185. https://doi.org/10.1111/j.1365-2745.2010.01736.x
Lavelle P, Decaëns T, Aubert M, Barot S, Blouin M, Bureau F, Margerie P, Mora P, Rossi JP (2006) Soil invertebrates and ecosystem services. Eur J Soil Biol 42:3–15
Lazcano C, Gómez-Brandón M, Revilla P, Domínguez J (2012) Short-term effects of organic and inorganic fertilizers on soil microbial community structure and function: A field study with sweet corn. Fert Soils, Biol. https://doi.org/10.1007/s00374-012-0761-7
LeBauer DS, Treseder KK (2008) Nitrogen limitation of net primary productivity in terrestrial ecosystems is globally distributed. Ecology 89:371–379
Liu L, Greaver TL (2010) A global perspective on belowground carbon dynamics under nitrogen enrichment. Ecol Lett 13:819–828
Lu M, Yang Y, Luo Y, Fang C, Zhou X, Chen J, Yang X, Li B (2011) Responses of ecosystem nitrogen cycle to nitrogen addition: a meta-analysis. New Phytol 189:1040–1050

Luo Z, Wang E, Sun OJ (2010) Can no-tillage stimulate carbon sequestration in agricultural soils? A meta-analysis of paired experiments. Agr Ecosyst Environ 139:224–231. https://doi.org/10.1016/j.agee.2010.08.006

Luz RA, Fontes LS, Cardoso SRS, Lima ÉFB (2013) Diversity of the arthropod edaphic fauna in preserved and managed with pasture áreas in Teresina-piauí-brazil. Braz J Biol 73(3):483–489

Mäder P, Edenhofer S, Boller T, Wiemken A, Niggli U (2000) Arbuscular mycorrhizae in a long-term field trial comparing low-input (organic, biological) and high-input (conventional) farming systems in a crop rotation. Biol Fertil Soils 31:150–156

Major J, Rondon M, Molina D, Riha SJ, Lehmann J (2010) Maize yield and nutrition during 4 years after biochar application to a Colombian savanna oxisol. Plant Soil 333:117–128

Mbuthia LW, Acosta-Martínez V, DeBryun J, Schaeffer S, Tyler D, Odoi E, Mpheshea M, Walker F, Eash N (2015) Long term tillage, cover crop, and fertilization effects on microbial community structure, activity: Implications for soil quality. Soil Biol Biochem 89:24–34

Mielniczuk J (2008) Matéria orgânica e a sustentabilidade de sistemas agrícolas. In: Santos GA, Silva LS, Canellas LP, Camargo FAO (eds) Fundamentos da matéria orgânica do solo: Ecossistemas tropicais e subtropicais. Porto Alegre, Fundação Agrisus, pp 1–5

Mikanová O, Šimon T, Kopecký J, Ságová-Marečková M (2013) Soil biological characteristics and microbial community structure in a field experiment. Open Life Sci 10:249–259

Moreira FMS, Huising EJ, Bignell DE (2010) Manual de biologia dos solos tropicais: Amostragem e caracterização da biodiversidade. Lavras, UFLA

Muchere-Muna M, Mugendi D, Kung'u J, Mugwe J, Bationo A (2007) Effects of organic and mineral fertilizer inputs on maize yield and soil chemical properties in a maize cropping system in Meru South District, Kenya. Agroforest Syst 69:189–197

Oehl F, Jansa J, Ineichen K, Mäder P, van der Heijden M (2011) Arbuscular mycorrhizal fungi as bioindicators in Swiss agricultural soils. Recherche Agronomique Suisse 2:304–311

Okalebo JR, GathuaKW,Woomer PL (1993) Laboratory methods of plant and soil analysis: a working manual. Technical Bulletin No. 1 Soil Science Society East Africa

Olmo M, Villar R, Salazar P (2015) Changes in soil nutrient availability explain biochar's impact on wheat root development. Plant Soil 339:333–343

Olsen SR, Cole CV, Watanable FS, Dean LA (1954) Estimation of available phosphorous in soils by extraction with Sodium bicarbonate. US Department of Agriculture, Washington, DC (Circular 939)

Omar SA, Ismail M (1999) Microbial populations, ammonification and nitrification in soil treated with urea and inorganic salts. Folia Microbiol 44:205–212

Panwar NR, Ramesh P, Singh AB, Ramana S (2010) Influence of organic, chemical, and integrated management practices on soil organic carbon and soil nutrient status under semiarid tropical conditions in central India. Commun Soil Sci Plant Anal 41:1073–1083

Pearce JL, Venier LA (2006) The use of ground beetles (Coleoptera: Carabidae) and spiders (Araneae) as bioindicator of sustainable forest management: a review. Ecol Indic 6:780–793

Pfiffner L, Luka H (2000) Overwintering of arthropods in soils of arable fields and adjacent semi-natural habitats. Agric Ecosyst Environ 78(3):215–222

Reich PB, Wright IJ, Lusk CH (2007) Predicting leaf physiology from simple plant and climate attributes: a global GLOPNET analysis. Ecol Appl 17:1982–1988

Riley H (2016) Residual value of inorganic fertilizer and farmyard manure for crop yields and soil fertility after long-term use on a loam soil in Norway. Nutr Cycl Agroecosyst 104(1):25–37

Robertson GP, Vitousek PM (2009) Nitrogen in agriculture: balancing the cost of an essential resource. Annu Rev Environ Resour 34:97–125

Robertson F, Armstrong R, Partington D, Perris R, Oliver I, Aumann C, Cravwíord D, Rees D (2015) Effect of cropping practices on soil organic carbon: evidence from long-term field experiments in Victoria, Australia. Soil Res 53:636–646

Rondon MA, Lehmann J, Ramírez J, Hurtado M (2007) Biological nitrogen fixation by common beans (Phaseolus vulgaris L.) increases with bio-char additions. Biol Fertil Soils 43(6):699–708

Rothamsted Research (2008) Guide to the classical and other long-term experiments, datasets and sample archive. Premier Printers Ltd., Bury St Edmunds, Suffolk, UK

Sackett TE, Classen AT, Sanders NJ (2010) Linking soil food web structure to above- and below-ground ecosystem processes: a meta-analysis. Oikos 119:1984–1992

Saha S, Mina BL, Gopinath KA, Kundu S, Gupta HS (2008) Relative changes in phosphatase activities as influenced by source and application rate of organic composts in field crops. Bioresour Technol 99:1750–1757

Sandim AS, Souza TAS, Ferreira-Eloy NR (2015) Manual de práticas agrícolas. Botucatu, Biblioteca Nacional. eBook ISBN 978-85-920166-2-3

Schroder JL, Zhang H, Girma K, Raun WR, Penn CJ, Payton ME (2011) Soil acidification from long-term use of nitrogen fertilizers on winter wheat. Soil Sci Soc Am J 75:957–964

Setäla H, Laakso J, Mikola J, Huhta V (1998) Functional diversity of decomposer organisms in relation to primary production. Appl Soil Ecol 9:25–31

Sharma MP, Reddy UG, Adholeya A (2011) Response of arbuscular mycorrhizal fungi on wheat (*Triticum aestivum* L.) grown conventionally and on beds in a sandy loam soil. Indian J. Microbiol 51:384–389

Shen J, Zhang L, Zhu Y, Zhang J, He J (2008) Abundance and composition of ammonia-oxidizing bacteria and ammonia-oxidizing archaea communities of an alkaline sandy loam. Environ Microbiol 10:1601–1611

Silva FSB, Yano-Melo AM, Brandão JA, Maia LC (2005) Sporulation of arbuscular mycorrhizal fungi using Tris-HCl buffer in addition to nutrient solutions. Braz J Microbiol 36:327–332

Siqueira JO, Hubbell DH (1986) Effect of organic substrates on germination and germ tube growth of vesicular-arbuscular mycorrhizal fungus spores in vitro. Pesquisa Agropecuária Brasileira 21:523–527

Siqueira JO, Hubbell DH, Schenck NC (1982) Spore germination and germ tube growth of a vesicular-arbuscular mycorrhizal fungus "in vitro". Mycologia 74:952–959

Siqueira JO, Sylvia D, Gibson J, Hubbel D (1985) Spores, germination, and germ tubes of vesicular-arbuscular mycorrhizal fungi. Can J Microbiol 31:965–997

Smith SE, Read DJ (2008) Mycorrhizal symbiosis. Academic, San Diego

Souza TAF, Freitas H (2017) Arbuscular mycorrhizal fungal community assembly in the Brazilian tropical seasonal dry forest. Ecol Process 6(1)

Souza TAF, Rodrigues AF, Marques LF (2015a) Long-term effects of alternative and conventional fertilization. II: Effects on *Triticum aestivum* L: II development and soil properties from a Brazilian Ferralsols. Russ Agric Sci (in press)

Souza TAF, Rodrigues AF, Marques LF (2015b) Long-term effects of alternative and conventional fertilization. I: Effects on arbuscular mycorrhizal fungi community composition. Russ Agric Sci (in press)

Souza TAF, Rodrígues AF, Marques LF (2015c) Long-term effects of alternative and conventional fertilization on macroarthropod community composition: a field study with wheat (*Triticum aestivum* L) cultivated on a Ferralsol. Org Agric. https://doi.org/10.1007/s13165-015-0138-y

Souza TAF, Rodrigues AF, Marques LF (2016a) The trend of soil chemical properties, and rapeseed productivity under different long-term fertilizations and stubble managements in a Ferralsols of Northeastern Brasil. Org Agric (in press)

Souza TAF, Rodrígues AF, Marques LA (2016b) The trend of soil chemical properties, and rapeseed productivity under different long-term fertilizations and stubble management in a Ferralsols of Northeastern Brasil. Org Agr. https://doi.org/10.1007/s13165-016-0164-4

Souza TAF, Rodriguez-Echeverria S, Andrade LA, Freitas H (2016c) Could biological invasion by Cryptostegia madagascariensis alter the composition of the arbuscular 497 mycorrhizal fungal community in semi-arid Brazil? Acta Bot Bras 30 (1). https://doi.org/10.1590/0102-3306201abb0190

Spehn EM et al (2005) Ecosystem effects of biodiversity manipulations in European grasslands. Ecol Monogr 75:37–63

Stopes C, Measures M, Smith C, Foster L (1995) Hedgerow management in organic farming. In: Isart J, Llerena JJ (eds) Biodiversity and land use: the role of organic farming, pp 121–125. Multitext, Barcelona, Spain

Swift MJ, Bignell D, Souza FM, Huising J (2010) O inventário da diversidade biológica do solo: conceitos e orientações gerais. In: Moreira FMS, Huising EJ, Bignell DE (eds) Manual de biologia dos solos tropicais: Amostragem e caracterização da biodiversidade. Editora UFLA, pp 23–41

Thomas GA, Titmarsh GW, Freebairn DM, Radford BJ (2007) No-tillage and conservation farming practices in grain growing areas of queensland: a review of 40 years of development. Aust J Exp Agric 47:887–898

Treseder KK (2008) Nitrogen additions and microbial biomass: a meta-analysis of ecosystem studies. Ecol Lett 11:1111–1120

Tscharntke T, Steffan-Dewenter I, Kruess A, Thies C (2002) Contribution of small habitat fragments to conservation of insect communities of grassland–cropland landscapes. Ecol Appl 12:354–363

Verhamme DT, Prosser JI, Nicol GW (2011) Ammonia concentration determines differential growth of ammonia-oxidising archaea and bacteria in soil microcosms. ISME J 5:1067–1071

Vilariño A, Sainz MJ (1997) Treatment of *Glomus mosseae* propagules with sucrose increases spore germination and inoculum potential. Soil Biol Biochem 29:1571–1573

Walker C, De La Torre J, Klotz M, Urakawa H, Pinel N, Arp D, Brochier-Armanet C, Chain P, Chan P, Gollabgir A (2010) *Nitrosopumilus maritimus* genome reveals unique mechanisms for nitrification and autotrophy in globally distributed marine crenarchaea. Proc Natl Acad Sci 107:8818–8823

Wang Y, Ke X, Wu L, Lu Y (2009) Community composition of ammonia-oxidizing bacteria and archaea in rice field soil as affected by nitrogen fertilization. Syst Appl Microbiol 32:27–36

Wardle DA (1992) Impacts of disturbance on detritus food webs in agro-ecosystems of contrasting tillage and weed management practices. In: Begon M, Fitter AH (eds) Advances in Ecological Research. Academic Press, pp 105–182

Warwick SI (2011) Brassicaceae in agriculture. In: Schmidt R, Bancrof I (eds) Genetics and genomics of the Brassicaceae. Springer, New York, NY, pp 33–50

Wessén E, Nyberg K, Jansson JK, Hallin S (2010) Responses of bacterial and archaeal ammonia oxidizers to soil organic and fertilizer amendments under long-term management. Appl Soil Ecol 45:193–200

Wetzel K, Silva G, Matczinski U, Oehl F, Fester T (2014) Superior differentiation of arbuscular mycorrhizal fungal communities from till and no-till plots by morphological spore identification when compared to T-RFLP. Soil Biol Biochem 72:88–96

Wink C, Guedes JVC, Fagundes CK, Rovedder AP (2005) Insetos edáficos como indicadores da qualidade ambiental soilborne insects as indicators of environmental quality. Revista de Ciências Agroveterinárias 4(1):60–71

Yang J, Gao W, Ren S (2015) Long-term effects of combined application of chemical nitrogen with organic materials on crop yields, soil organic carbon and total nitrogen in fluvoaquic soil. Soil Till Res 151:67–74

Tian J, Lou Y, Gao Y, Fang H, Liu S, Xu M, Blagodatskaya E, Kuzyakov Y (2017) Response of soil organic matter fractions and composition of microbial community to long-term organic and mineral fertilization. Biol Fertil Soils 53:523–532. https://doi.org/10.1007/s00374-017-1189-x

Yao H, Gao Y, Nicol GW, Campbell CD, Prosser JI, Zhang L, Han W, Singh BK (2011) Links between ammonia oxidizer community structure, abundance, and nitrification potential in acidic soils. Appl Environ Microbiol 77:4618–4625

Zhang HM, Wang BR, Xu MG (2008) Effects of inorganic fertilizer inputs on grain yields and soil properties in a long-term wheat-corn cropping system in South China. Commun Soil Sci Plant Anal 39:1583–1599

Zhang JY, Zhang WJ, Xu MG, Huang QH, Luo K (2012) Response of soil organic carbon and its particle-size fractions to different long-term fertilizations in red soil of China. Plant Nutr Fert Sci 18:868–875 (In Chinese)

Zhong W, Gu T, Wang W, Zhang B, Lin X, Huang Q, Shen W (2010) The effects of mineral fertilizer and organic manure on soil microbial community and diversity. Plant Soil 326:511–522

Chapter 8
Agroecological Protection of Mango Orchards in La Réunion

Jean-Philippe Deguine, Maxime Jacquot, Agathe Allibert, Frédéric Chiroleu, Rachel Graindorge, Philippe Laurent, Guy Lambert, Bruno Albon, Marlène Marquier, Caroline Gloanec, Luc Vanhuffel, Didier Vincenot and Jean-Noël Aubertot

Abstract Mango is one of the world's major tropical crops. In Réunion, the crop is plagued by pests, which have, over several decades, led to an over-reliance on agrochemicals. These expensive treatments have limited efficacy, and negative effects on the environment are associated with health risks and ecological imbalances. In addition, these agroecosystems are not ecologically sustainable. That is why in recent years, Agroecological Crop Protection has been applied to mango production in Réunion. The Biophyto project (2012–2014) was co-designed between 2010 and 2011 and brought together a number of agricultural partners in a collective approach to the crisis. It was followed until 2017 with further studies and experiments, as well as an extension phase to transfer knowledge to production areas. The experience broke new ground in Agroecological Crop Protection. A large quantity of data was collected which enabled comparisons between conventional and agroecological orchards. The results of this pioneering experience were very encouraging and the major points are (1) The

J.-P. Deguine (✉) · M. Jacquot · A. Allibert · F. Chiroleu
Cirad, UMR PVBMT, 97410 Saint-Pierre, France
e-mail: jean-philippe.deguine@cirad.fr

R. Graindorge
ARMEFLHOR, 1 chemin de l'IRFA, 97410 Saint-Pierre, France

P. Laurent
University Institute of Technology, University of Réunion, 40 avenue de Soweto, 97455 Saint-Pierre, France

B. Albon · M. Marquier
FDGDON, 23 Rue Jules Thirel - Cour de l'Usine de Savanna, 97460 Saint Paul, France

C. Gloanec · L. Vanhuffel · D. Vincenot
Chambre d'agriculture de La Réunion, B.P. 134, 97463 Saint-Denis Cedex, France

J.-N. Aubertot
INRA, UMR AGIR 1248, 24 chemin de Borderouge–Auzeville, 31320 Castanet-Tolosan, France

G. Lambert
Aix Marseille Université, LESA, 13621 Aix en Provence, France

S. Gaba et al. (eds.), *Sustainable Agriculture Reviews 28*, Ecology for Agriculture 28,
https://doi.org/10.1007/978-3-319-90309-5_8

Agroecological Crop Protection practices, mainly the suppression of pesticides, use of prophylaxis and permanent vegetal cover, which are the bases of conservation biological control, have been widely adopted by farmers. (2) These practices were found to reduce populations of pests and damage caused, e.g. mango bug, mealybugs, and had no negative impact on flowering level. (3) The treatment frequency index (TFI) decreased from 22.4 before the project to 0.3 after the project. Production costs were reduced by 35% without any loss of yield, except in a few specific circumstances. (4) The 124,001 arthropods identified from the 126,753 arthropods collected in orchards belong to 4 classes, 23 orders, 215 families, 451 genera and 797 morphotypes. The parasitoids formed the richest trophic group. (5) Bottom-up and top-down controls of biodiversity within a single community of arthropods were observed, and the role of ants, including invasive species, was quantified. (6) There was a negative effect of parasitoid diversity on the abundance of the Seychelles mealybug; the proportion of mango orchards in the landscape had a positive effect of on the abundance of South African citrus Thrips. (7) This experience also produced a large number of transfer assistance tools, particularly in the field of professional training. For example, a University Certificate of Professional Qualification (UCPQ) entitled "Agroecological Crop Protection", and aimed at growers, technicians and agricultural advisers, has been available since 2013. (8) Other tools have been implemented by agricultural agencies and policy makers in order to facilitate the extension of agroecological practices: demonstration Dephy Ferme plots, a Biophyto Agri-Environment and Climate Measure and an Economic and Environmental Interest Group.

Keywords Agroecology · Crop protection · Co-design · Conservation biological control · Functional biodiversity · Training · Transfer assistance
Research and development

8.1 Introduction

The mango tree *Mangifera indica* L. is of the Anacardiaceae family, native from India. Mango is the world's main tropical fruit crop after the banana, with 33 million tonnes produced annually (Food and Agriculture Organization 2015). The main production area is Asia and the main producing countries are India, China and Thailand. Mango cultivation began to develop in La Réunion from the 1980 s onwards. Today, nearly 400 ha are intensively cultivated: on average, more than 250 trees are planted per hectare.

This monoculture system is very favourable to the development of pests and over the last two decades growers have turned to agrochemical protection. Today, mango growers in Réunion are faced with major phytosanitary problems. Several pests attack mango trees and, without directly threatening the life of the tree, can cause serious damage to the crop (Amouroux and Normand 2013). During the flowering period, the tree is particularly vulnerable to pests (Fig. 8.1). The early winter blooms are the first to be affected: the coolness of the austral winter favours

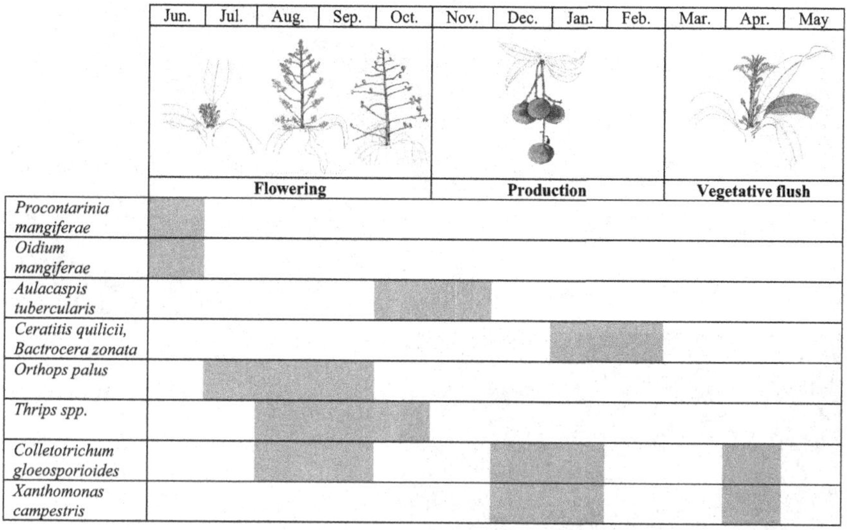

Jun. Jul. Aug. Sep. Oct. Nov. Dec. Jan. Feb. Mar. Apr. May

Flowering Production Vegetative flush

Procontarinia mangiferae
Oidium mangiferae
Aulacaspis tubercularis
Ceratitis quilicii, Bactrocera zonata
Orthops palus
Thrips spp.
Colletotrichum gloeosporioides
Xanthomonas campestris

Fig. 8.1 Risk periods (in pink) of major mango tree pests in Réunion

the development of Oidium (*Oidium mangiferae* Berthet), a fungus capable of destroying flowers. It is also during this period that the blossom mango gall midge, *Procontarinia mangiferae* (*Erosomyia mangiferae* Felt), mango bug (*Orthops palus* Taylor) and various species of thrips proliferate. These early flowerings are very desirable for the growers because they guarantee an entire crop before the hurricane season (December to March) which fetches a good price. Chemical protection started to be used heavily during this period and has now become routine. Furthermore, the fruits (mangoes) are susceptible to attack from different species of scale insects including the Seychelles cochineal *Icerya seychellarum* (Westwood) and mango scale *Aulacaspis tubercularis* Newstead and several species of fruit fly (*Bactrocera zonata* (Saunders), *Ceratitis rosa* (Wiedemann), *Ceratitis quilicii* (de Meyer)). Immediately after the first rains of the hurricane season, bacterial blight (*Xanthomonas campestris* pv. *Mangiferae indica*) and anthracnose (*Colletotrichum gloeosporioides* (Penz)) begin to develop. These diseases can quickly devalue the market price of the crop. Figure 8.2 shows some of the pests that cause economic damage during the flowering and fruiting period of the mango.

To manage populations and the damage caused by these pests, growers have become accustomed to using agrochemicals on their orchards, mainly synthetic insecticides (Normand et al. 2011). However, chemical protection is a far from satisfactory solution. Pests quickly become resistant to pesticides in most cases and very few commercial substances are certified in France for use on mango. Agrochemical protection has also shown socioeconomic limitations (limited efficacy, high costs, labour-intensive) and poses significant risks to human health and the environment (Deguine et al. 2009), especially since mango orchards can be located near urbanized areas. They are also planted in filtering soils sensitive to

Fig. 8.2 Some pests and pest damage. **a** Flower midge *Procontarinia mangiferae*; **b** Seychelles mealybug *Icerya seychellarum*; **c** Damage caused by oidium (*Oidium mangiferae*) on mango inflorescences; **d** Mango bug *Orthops palus*; **e** Natal *Ceratitis rosa* Fly; **f** *Thrips spp*. (Photos: a, b, d and f: A. Franck, c and e: D. Vincenot)

erosion and increase the risks of diffuse pollution in drinking water catchments and coral reefs. Finally, it is also known that agroecosystems using chemicals are not ecologically sustainable (Deguine et al. 2016). Conversely, there is a growing emphasis on the intensification of ecological processes in agroecosystems, and feedback from experience confirms the value of this approach (Tittonnell et al. 2016) which is consistent with the principles of agroecology (Brym and Reeve 2016). This approach is particularly relevant in the field of pest management (Coll and Wajnberg 2017; Deguine et al. 2009, 2016, 2017; Reddy 2017).

At the end of the 2000 s, stakeholders in the Réunion mango industry met and mobilized for an agroecological approach to protect mango orchards against pests as an alternative to unsatisfactory agrochemical protection. In this context, a dozen partners from Réunion in agricultural research, experimentation, training and development, as well as a dozen volunteer mango growers, began to collaborate with the aim of reducing or eliminating use of agrochemicals and testing the principles of Agroecological Crop Protection (Gloanec et al. 2017). Agroecological Crop Protection (ACP) is an innovative and systematic approach, focusing on the application of the principles of agroecology to crop protection, in which ecology is given the utmost importance (Deguine et al. 2009). ACP aims to reconcile the effectiveness of crop protection against pests and diseases with the socio-economic, ecological, environmental and health sustainability of agroecosystems (Deguine et al. 2017). ACP also aims to assist the transition from agrochemical to agroecological farming.

This summary reports the findings of a mango agroecological protection program in Réunion, following 6 years of decline. It presents the observation and data collection systems used in the Biophyto project, the results obtained and the steps taken to transfer knowledge to growers. This unique experience has helped to promote agroecological transition.

8.2 Design, Organization and Collective Dynamics of the Agroecological Experience

8.2.1 Context and Co-design

This experience took place in a specific context: many agroecological techniques have already been adopted in La Réunion after the success of the Gamour project and the success of the experience in reducing the number of Fruit flies attacking Cucurbitaceae (Deguine et al. 2015a). Moreover, this experience also represents a coherent and tangible adaptation of the national agroecology guidelines (the agroecological plan for France) and pesticide reduction (Ecophyto 1 and 2).

A first phase of joint discussion and consultation took place as early as 2009 and continued until 2011 between agricultural organizations and the volunteer mango growers, placing them at the centre of the scheme as the first beneficiaries of the expected results. These discussions focused on the co-design and co-construction of the Biophyto project (Sustainable production of insecticide-free mangoes in Réunion) (Deguine et al. 2017). This project was submitted for funding under the 2012 Casdar Innovation and Partnership call for projects, proposed by the French Ministry of Agriculture, Food and Forestry. Biophyto won this call for projects, and it subsequently took place from 2012 to 2014, testing conservation biological control (Ferron and Deguine 2005) in mango orchards: suppression of chemical protection and use of vegetal cover to promote functional biodiversity (especially crop auxiliaries) in orchards. The Biophyto project focused on the following aspects: the implementation of these innovative techniques in pilot sites, the characterization of functional biodiversity in orchards, economic analysis of the sector and the assessment of the commercial viability of the production. This project marked a break with traditional farming and represented a major step towards the development of organic mango production satisfying Organic Farming specifications. In view of the good results obtained in this project, Biophyto was followed by another project, Biophytomang[2], financed in 2015 under the Ecophyto national plan. This new project aimed to provide additional knowledge to Biophyto through further research and experimentation. The agroecological experiment in Réunion for the protection of mango orchards then continued under the dual supervision of agricultural organizations (responsible for knowledge transfer to growers and the training of professionals) and the public authorities assisting in regulation, financing and the extension of new agroecological knowledge to other sectors.

Finally, years of research, experimentation and agroecological protection of the mango allowed the growers and researchers in Réunion to familiarize themselves with a field that remained poorly understood during the agrochemical period: the role of functional biodiversity in the ecology of agroecosystems in general, and mango orchards in Réunion in particular. This area is the subject of considerable discussion in this article, taking into account new knowledge acquired on functional biodiversity in orchards and the factors influencing it, within a framework of natural pest regulation.

8.2.2 Issues and Objectives of the Biophyto Project

Biophyto took into consideration economic issues related to productivity and gross margins for the growers. It also focused on commercial issues such as ways of strengthening the sector. Technical issues were also taken into account, especially the reduction of the use of conventional chemicals. This project answered an ecological challenge: to promote ecological processes and interactions between plant and animal communities whilst respecting the principles of biological conservation control. In addition, Biophyto has contributed to the preservation of biodiversity in Réunion, one of the world's biodiversity hotspots (Myers et al. 2000). Finally, the project has strengthened the image of the mango, a traditional and emblematic fruit in Réunion. Biophyto aimed to adapt agroecological techniques for orchard protection, while eliminating insecticides in pilot sites, i.e. conventional and organic growers, using the benefits of renewed functional biodiversity in the absence of insecticide treatments. Downstream, Biophyto aims to study commercial production for different markets (including short circuits and export).

The objectives of the Biophyto project were to bring together partners in Agroecological Crop Protection, to evaluate the performance of agroecological protection in a network of pilot farms and to study the transfer of agroecological protection to growers. Biophyto improved technical routes for mango growers, contributing to the development of good quality, healthy mangoes; it added new scientific knowledge on the agroecology of mango orchards and training modules with innovative teaching. Finally, through this project, agroecological protection techniques already used in vegetable crops in Réunion are being introduced into orchards.

8.2.3 Biophyto Project Organization and Partnership

Activities and coordination are divided into 3 clusters: the administrative and financial management of the project is entrusted to the Réunion Chamber of Agriculture (development division), the project leader is CIRAD (a research centre) and technical coordination is provided by AROP-FL.

Eleven technical partners involved in the project have received funding from CASDAR (agricultural and rural development trust account): the Réunion Insectarium Association (INSECTARIUM), the Réunion Association of Fruit Growers (AROP-FL), the Réunion Chamber of Agriculture, the Centre for International Cooperation in Agronomic Research for Development (CIRAD), the Réunion Technical Institute of Fruit, Vegetable and Horticulture (ARMEFLHOR), the Local Public Institution for Agricultural Education and Training (EPLEFPA) in St-Paul, the Departmental Federation of Defense Against Pests (Réunion) (FDGDON), the Forum of Reasoned Agriculture Respecting the Environment (FARRE), the Réunion Island Organic Farming Association (GAB), the Tropical Certifying Body of Réunion Indian Ocean (OCTROI) and the University of Réunion (IUT de St Pierre).

Other technical partners involved in the implementation of the project are (excluding CASDAR funding): the Directorate for Food, Agriculture and Forestry (DAAF), the Technological Mixed Network "Development of Organic Agriculture" (RMT DévAB) and the Réunion Water Office. The steering committee comprises the Directorate for Food, Agriculture and Forestry (DAAF) and the QUALITROPIC Competitiveness Cluster. In addition to the Ministry of Agriculture, Agri-Food and Forestry (through the CASDAR grant), financial partners include the Regional Council of Réunion, the General Council of Réunion, the French State and the European Union. The ECOPHYTO planners have given considerable support to the project, and have actively participated in the financing of various operations and communication projects.

8.2.4 Other Studies and Experiments

Given the encouraging results achieved in 2012 to 2014, Biophyto was followed by another project, Biophytomang[2], financed in 2015 under the Ecophyto national plan. The project had two objectives:

(a) To provide complementary knowledge, by finely analyzing data that had not been analyzed within the time allowed for the Biophyto project. This includes the data collected by FDGDON from 2012 to 2014 on the Biophyto pilot sites relating to certain pests (mealybugs, thrips, bugs, and oidium) and mango tree phenology (flowering). These analyses were carried out by CIRAD. The results are presented in this summary and the methods used are recalled on a case by case basis.
(b) To develop a spray irrigation system between rows of mango trees for the vegetal cover in the Biophyto plots. This irrigation system had to be complementary to the pre-existing drip irrigation system designed for the mango trees. This experiment was carried out by Armeflhor. The experimental approach and the results are presented in this summary.

The Biophyto project also served as a framework for two Ph.D. theses at the University of Réunion, the results of which are also presented in this summary.

One thesis focused on the *Orthops palus*, a mango pest: the main results are presented here (Atiama 2016). The other thesis focused on biodiversity and orchard ecology and, given the originality of the results, they are presented in detail in this summary (Jacquot 2016). Finally, this agroecological experiment and the majority of the results obtained can be integrated harmoniously into new fruit production systems that go beyond the challenges of crop protection (Simon et al. 2016).

8.2.5 The Foundations of the Dephy Ferme Network

The Dephy Ferme network is a network of demonstration farms managed under the Ecophyto National Plan. Launched in 2010 in Réunion, this national scheme aims to reduce the use of pesticides by agricultural professionals, individuals and communities. The Dephy Ferme mango network brings together mango growers aided by a representative from the Réunion Chamber of Agriculture, the official agricultural development agency. Growers in the Dephy Ferme network voluntarily took part in a pesticide reduction programme favouring alternative pest control methods and were thus familiar with Agroecological Crop Protection.

Each farm is a space for information exchange, demonstration and training for professionals and those studying agriculture. Although the Dephy Ferme Mango network does not perfectly match Biophyto's network of pilot sites, it relies on a number of Biophyto projects. For example, mango growers in the Dephy Ferme network were among the first to use agroecological protection in their orchards by halting herbicide and insecticide treatments and planting permanent vegetal cover. They also opened their farms to growers curious about this new technique. Biodiversity, the Treatment Frequency Index (TFI), yields and harvest quality are important indicators for evaluating the effectiveness of agroecological protection and are subject to annual technical assessments.

8.3 Methodology, Types of Data Collected and Types of Results Presented

8.3.1 Selecting Pilot Sites and Agroecological Techniques in the Biophyto Project

The Biophyto project was based on a network of 13 pilot orchards (http://www.agriculture-biodiversite-oi.org/Biophyto): an orchard belonging to the agricultural college of Saint-Paul), 7 orchards adhering to Growers Organisations under AROP-FL and 2 orchards outside professional organizations and 2 organic orchards. Figure 8.3 shows the location of these orchards in the mango production areas of Réunion. The growers responsible for these pilot sites were at the centre of the

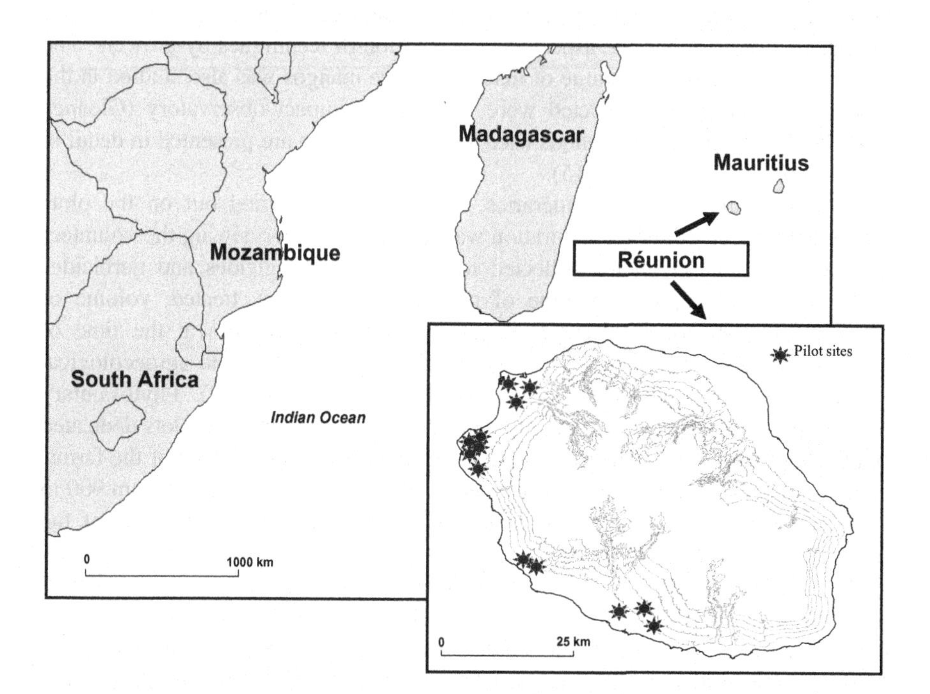

Fig. 8.3 Location of Réunion and location of Biophyto project pilot orchards

scheme and played a major role both in the design of the project and in its implementation and evaluation.

The selection of orchards took into account three main criteria: the motivation and commitment of the volunteer growers and the associated monitoring structures (agricultural college, growers' organisations, Chamber of Agriculture, ABM) to guarantee the smooth running of operations; the representativeness of the network in relation to the mango producing areas; and the possibility of setting up a "control" plot close to the pilot orchard. In each pilot site, a couple of plots were identified and compared: a Biophyto plot using agroecological farming (see characteristics below) and a control plot using conventional techniques. The "pilot" plot is referred to as the "agroecological" plot or the "Biophyto" plot, and the control plot is sometimes referred to as "agrochemical" or "conventional". The Biophyto plot was characterized by the cessation of insecticide treatments and the installation of a spontaneous vegetal cover, which resulted in the cessation of herbicide treatments. It was compared to the control plot, where the farmer continued to cultivate as usual.

8.3.2 Collecting and Managing Data in the Biophyto Project

Data collection in the Biophyto network covered many fields: agronomic and phytosanitary data on farming techniques, plant and animal biodiversity monitoring in orchards, socio-economic aspects and perception of techniques by growers. The commercial and economic value of insecticide-free mangos was also studied in the Biophyto project. Data collected were fed into an impact observatory (Gloanec et al. 2017). The different methods used to collect the data are presented in detail in other documents (Gloanec 2015).

As regards the technical itineraries, each operation carried out on the plots during the project's three-year duration was recorded by interviewing the volunteer growers. Information relates collected to insecticides, fungicides and herbicides (date of product application, type of product used, surface treated, volume of product used, phenological stage observed on the mango trees at the time of application). The information also concerns farm management and agroecological practices (mowing of vegetal cover, use of flower strips, etc.). Phytosanitary monitoring was carried out on the thirteen pilot sites, each with two plots dedicated to the Biophyto project. The plots were, depending on the constraints of the farms, adjacent or separated by a few tens of meters. Their surface area varied from 900 to 2000 m^2. This monitoring of the plots was carried out over the lifetime of the project.

Throughout the Biophyto project, FDGDON regularly monitored mealybug, thrip, bug and oidium populations (Brun-Vitelli 2015). Pests were described using criteria based on the pest type (abundance class, damage and/or presence/absence). Methodologies differed from one pest to another and are described case by case below.

The data concerning mango tree phenology were collected by FDGDON throughout the project. When the trees were not all at the same stage, the majority phenological stage was noted. The ten phenological stages in the growth of the tree are: light green olive-shaped swollen buds with unopened protective scales; Elongations and beginning of bud openings, appearance of inflorescences (bracts); Opening of the buds with spreading scales which fall, more visible bracts of inflorescences; Elongation of inflorescences, more bracts present; End of elongation of inflorescences with clearly visible secondary axes but no or few open flowers; Flowering with open flowers and flower buds on the inflorescence; Late flowering with few open flowers, many dry flowers and the presence of small green fruits; Fruit growth; Fruits; Post-fruiting and flowering stage with vegetative stages of the tree.

Some data were the subject of specific collection and analysis methods that are not presented here but are described in the paragraphs devoted to them (e.g. functional biodiversity).

8.3.3 *A Holistic Approach with Varied and Encouraging Results*

At the end of the Biophyto project, a seminar entitled "Biodiversity and Agroecological Protection of Crops" was organized in St-Pierre (Réunion) from 21 to 24 October 2014 (Deguine et al. 2015b). It was an opportunity to exchange ideas, discuss the Biophyto project and present the results available at that time. It united 169 participants from different backgrounds: scientists, students, professionals in agriculture and environmental management, representatives of local authorities and institutions, etc. The minutes (Deguine et al. 2015b) can be consulted here: http://www.agriculture-biodiversite-oi.org/Biophyto.

Many results were presented at the seminar: growers' adoption of techniques such as vegetal cover, characterization and evolution of functional biodiversity, tools for marketing insecticide-free mangos, production of information, training and teaching tools. These results have already attracted considerable attention, both from the scientific community and professionals such as agricultural technicians and growers.

Knowledge has been acquired on the characterization of functional biodiversity in orchards, the impact of agroecological farming (vegetal cover) and landscape on functional biodiversity and the structure and function of food webs. Other scientific findings include key pests that were not well known at the beginning of the project such as the mango bug. The spatial and temporal evolution of levels of several pests and major diseases, with or without chemical pesticides, were studied in depth. Interactions between plants, focusing on floral species, trap plants and refuge plants were also studied. By extension, the studies also involved organic farming, as most of the results were obtained in situations where organic specifications were met.

The results of the project were presented in scientific publications, posters and communications at national and international congresses. The research provided numerous internships for students (PhD, Master, Agronomy, and University Technology Degree). These results cannot all be presented here; this summary presents the agroecological techniques used, the socio-economic results, the prospects for marketing of insecticide-free mangos, data acquired on pests and functional biodiversity, and the main collaborative tools for the transfer of agroecological knowledge.

8.4 Agroecological Practices in the Field

Agroecological protection techniques applied to mango orchards were based on conservation biological control, which included eliminating, or severely reducing insecticides, prophylaxis, and creation of plant biodiversity in orchards (e.g. permanent vegetal cover, flower strips or trap plants) and discontinuing of herbicide treatments. Figure 8.4 presents the objectives, strategy and examples of Conservation Biological Control (Deguine et al. 2016).

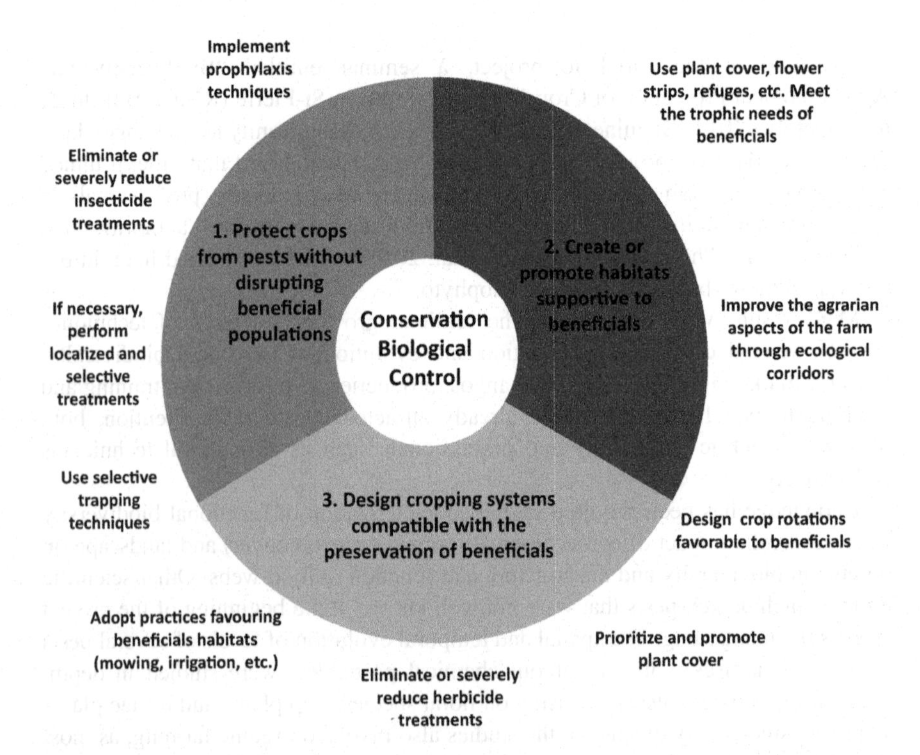

Fig. 8.4 Biological conservation control. The first circle represents strategies. The second one highlights tactics. The third circle shows some examples of practices (Adapted from Deguine et al. 2017)

8.4.1 Installation and Management of Vegetal Cover

Vegetal cover was the preferred method for creating plant biodiversity. Figure 8.5 shows, on the left, a conventional orchard with herbicidal treatments and no vegetal cover and, on the right, an agroecological orchard with vegetal cover and no herbicides. Vegetal cover was used on all pilot farms. Growers have widely adopted this practice, to the extent that some have extended it to some or all of their operations. In itself, this represents a guarantee of adoption by the growers, but it has sometimes hampered comparisons between Biophyto plots and control plots. Technical enhancement of vegetal cover (adapted irrigation systems) and specific aspects related to possible new interactions (e.g. impact of irrigation on water supply and mango phenology) was carried out during the Biophytomang[2] project, using an experimental approach. The results are described below. The spray irrigation system can now be recommended. Further studies are still being carried out to optimize this irrigation, especially as regards the supply of water to mango trees. The satisfaction survey carried out at the end of the project shows that all growers are very satisfied with the implementation of permanent plant cover in orchards (Gloanec 2015). Growers appreciate its usefulness; it promotes functional biodiversity, helps protect the environment and combats erosion, and installation is straightforward.

Other techniques for the creation of plant biodiversity in orchards, and in particular flower strips, have been studied in some parts of the system but are not presented in detail in this section. We give the main characteristics in this summary. The implantation of flower strips in orchards has been the subject of additional tests with some Biophyto growers, in particular organic growers who are particularly interested in this method of habitat management. We have characterized the interactions between selected flowering plants and harmful and useful arthropods present in general and Parasitica (Hymenoptera) in particular. A large proportion of the Parasitica species were parasitoids of other insects, including pests. The plant with the greatest abundance and diversity of Parasitica was *Lobularia maritima*

Fig. 8.5 **a** Conventional orchard with herbicide treatments and no vegetal cover; **b** agroecological orchard, without herbicidal treatment and with vegetal cover. (Photos: a and b: J.-P. Deguine)

(Brassicaceae). These results confirmed the worth of using flowering plants as a conservation biological control to promote parasitoid pests in organic farming (Gloanec et al. 2017). Flower strips are an ideal complement to vegetal cover to promote functional biodiversity. An evaluation carried out in a Biophyto orchard in 2013 using Malaise traps showed that plots using these two techniques had 10 times more parasitoids than the agrochemical plot (Gloanec et al. 2017). Detailed results of flower strip studies are available in other papers (Deguine et al. 2015b; Gloanec et al. 2017).

8.4.2 Elimination of Insecticides and Herbicides

Insecticides were eliminated on all Biophyto plots. This necessary agroecological practice promoted functional biodiversity and restored ecological balances. In total, over all the Biophyto plots and over the three years of the project, only two insecticide treatments were necessary, with the agreement of the growers and the project managers. The elimination of insecticide treatments resulted, in a few cases, in production losses due to attacks by certain pests such as the gall midge or plant bugs, but did not have an impact on yield in most cases according to growers (Gloanec and Guignard 2015). It should be noted that establishing ecological equilibriums, i.e. optimizing the interactions between plant communities such as mango or introduced plants, as well as animals such as harmful arthropods or useful ones, requires a long period of time that exceeds the duration of the project. Finally, herbicides were also eliminated; vegetal cover took the place of herbicide grass management. Certain fungicides targetting oidium, although compatible with organic specifications, were not always used; this contributed, in some situations, to production losses.

8.4.3 Appropriation: Moving from Conventional to Agroecological Farming

Using the agricultural data collected over three years, we described the two plot types (Biophyto and Control) using phytosanitary and farming characteristics (Fig. 8.6). The Biophyto plots required, as expected, fewer phytosanitary products than the control plots. In our graphical representation, we separated the conventional insecticides from the adulticide bait called Syneis-Appat. This adulticide bait is authorized in organic farming, and is used locally as an attractant but in a very different way from conventional insecticides (Deguine et al. 2015a). This bait was applied to both the Biophyto plots and the control plots. Fungicides were more widely used on control plots than on Biophyto plots. Vegetal cover trimming took

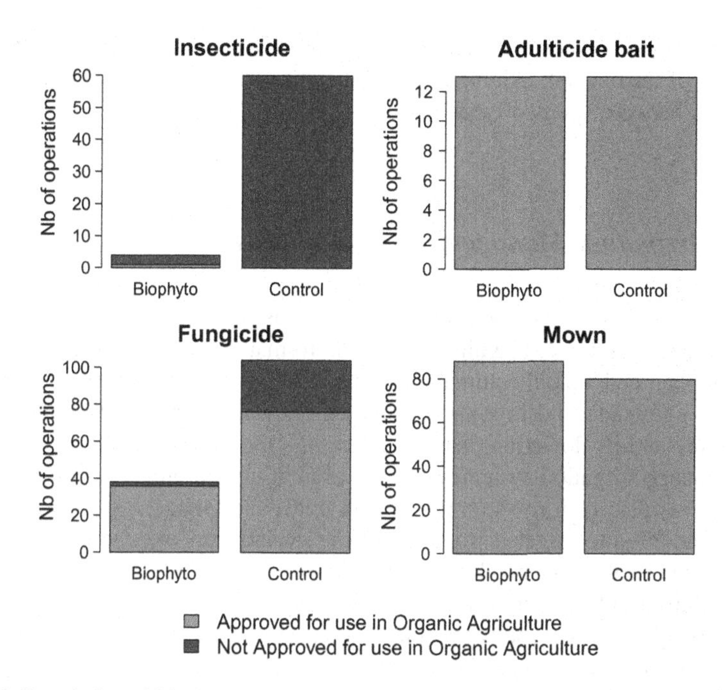

Fig. 8.6 Description of Biophyto plots and control plots according to phytosanitary and cultural characteristics. The numbers represent insecticide, fungicide and adulticide applications, distinguishing what is approved in Organic Agriculture and what is not, and of verge trimming. Total operations on all farms during the project

place on both plot types; this was due to the gradual introduction, during the project, of vegetal covers from 2012 on the majority of plots.

Due to the growers' enthusiasm to set up permanent vegetal cover in the orchards, another study aimed to describe the evolution of the plots during the Biophytomang[2] project. Cultivation techniques were described for each plot and each year by a Factorial Correspondence Analysis (FCA), a descriptive statistical technique. This analysis describes the evolution of plots over time simply and graphically, based on a large number of data. The FCA was carried out on a contingency table and makes it possible to highlight the links between rows and columns of the table (Brun-Vitelli 2015). We used the number of each type of system per year and per plot. This analysis enabled us to highlight changes in farming techniques (Brun-Vitelli 2015):

- for certified organic farms, there were few differences between the Biophyto plot and techniques did not change during the life of the Biophyto project;
- on some farms, there were differences between the two plots, present from the beginning to the end of the project;
- on the majority of farms, there was a change in techniques: during the project, the control plots increasingly began to use Biophyto techniques. This reflected

the satisfaction of growers and the adoption of techniques (vegetal cover, cessation of treatments), although some difficulties were encountered in comparing results between the two types of plot.

8.4.4 Irrigation Management in Orchards

The move towards agroecological practices—the use of irrigation and vegetal cover in particular—contributed significantly to the formation of bio-ecological balances between plant and animal communities. The aim is natural control of pests but this can have unintended positive or negative consequences on the phenology of the mango tree and its flowering phase (see Sect. 5.4).

The mango irrigation systems currently used in mango orchards in Réunion are drip systems placed in rows on the ground under the foliage. Two irrigation systems, in addition to targeted irrigation of the mango tree, were tested to promote plant growth in the orchard: long-range micro-sprinklers positioned at the foot of the trees; dripper lines placed between rows (Fig. 8.7).

The mango tree is a species requiring a warm and humid season (austral summer) and a cool dry season (austral winter). Four important phenological stages are generally defined: vegetative growth, vegetative rest, flowering and harvesting (Vincenot et al. 2009). Summer is favourable to the vegetative growth of the tree while winter induces flowering, then fruiting and fruit growth. To induce flowering, conditions must therefore be unfavourable to vegetative growth. Additionally,

Fig. 8.7 The two vegetal cover irrigation systems tested: **a** long-range micro-sprinklers positioned at the base of the trees; **b** lines of drippers placed between the rows. (Photos: a and b: J. Bouriga)

the decline in temperatures and precipitation beginning in May in Réunion generally limits late vegetative growth. This phase of vegetative rest is also the period during which water stress is introduced to allow the growth units to acquire sufficient maturity to flower (Nuñez-Elisea and Davenport 1994). Growers then reinforce these drought conditions by stopping irrigation, creating genuine water stress. Considering the phenological characteristics of the mango tree in Réunion, maintaining a vigorous vegetal cover in the whole orchard requires irrigation of the vegetal cover without affecting the different phenological stages of the mango tree. The vegetal cover needs irrigation during the dry season, during which water stress may be required for the mango trees in order to induce flowering. The effects of irrigation on a diversified and permanent vegetal cover could thus be either negative: a resumption of vegetative growth of the mango tree and inhibition of flowering, or positive, by a minimum water supply to the trees, limiting the negative effects of severe water stress which are often observed (Vincenot et al. 2009).

The results of this experiment made it possible to validate the effectiveness of the irrigation systems (adapted irrigation of the vegetal cover, limited inconvenience for other farming work and easy maintenance). In addition, adaptation of vegetal cover management operations is necessary, particularly mowing. Technical improvements in mowing methods are planned (use of lawnmowers/strimmers, grazing, etc.).

To summarize, the growers involved in the Biophyto project were trained in agroecological practices such as suppression of pesticides, prophylaxis, permanent vegetal cover, which are the bases of conservation biological control and have generally proved to be effective in protecting against pests. An irrigation system maintaining vegetal cover in orchards was developed, in addition to the traditional mango irrigation system. This irrigation needs to be carried out in conjunction with mango irrigation, in order not to disturb some of its phenological stages, such as flowering.

8.5 Effectiveness of Agroecological Farming on Mango Pests and Phenology

8.5.1 New Knowledge on the Biology of Mango Bugs to Improve Recommendations on Agroecological Management of Bug Populations

The results that follow were acquired within the framework of a thesis (Atiama 2016). In the early 2010s, *Orthops palus*, which was wrongly referred to as *Lygus palus* or *Taylorilygus palus*, and referred to by growers as the mango bug, was considered the number one mango pest in Réunion and very little knowledge was available on the insect. Knowledge about taxonomy, bioecology and genetic diversity were obtained during the course of the thesis, forming much of the knowledge available today in the literature on *O. palus*. This knowledge now makes *O. Palus* one of the most studied mirids in tropical fruit crops.

The main bio-ecological findings were as follows: The miridofauna in the mango orchards was inventoried and of the 13-recorded species of mirids, *O. palus* was the most abundant on the tree inflorescences during flowering. To assist in identifying *O. palus* in the laboratory and recognizing it in the field, three original tools were used (an identification key, Cytochrome c Oxydase I sequences and a field recognition form). An *O. palus* breeding program was developed; it has helped to study its life cycle and to measure the duration of its development stages (Atiama 2016). In addition, the inventory of *O. palus* (15 species in Réunion) showed that it is a polyphagous species and that it is mostly a "flower bug", likely to spend the year moving, depending on the availability of resources, from one flowering plant to another (Fig. 8.8).

Finally, in addition to the precautionary measures to avoid population flows (invasions) between Indian Ocean islands and other sub-regions in light of these findings proposals for the agroecological management of *O. palus* may now be established (Fig. 8.9). Further results on mango bug populations were obtained from observations carried out by FDGDON throughout the Biophyto project (see 5.2).

8.5.2 Noteworthy Results on the Agroecological Management of Certain Pests

A detailed analysis of the pest results (univariate and multivariate) was carried out by CIRAD, based on agricultural data (Biophyto observatory) and field data collected by FDGDON. A factor analysis of correspondence for each pest (Brun-Vitelli 2015) gave the following results.

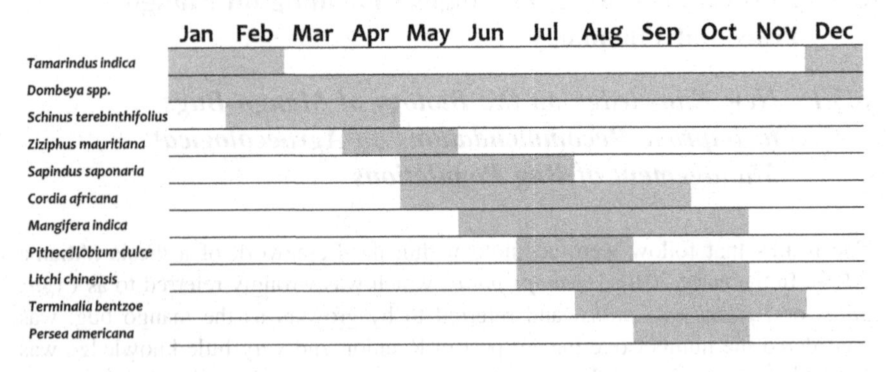

Fig. 8.8 Representation of the flowering periods (green) of *Orthops palus* host plants in Réunion, showing that this species can potentially spend the year moving from one flowering plant to another (Atiama 2016)

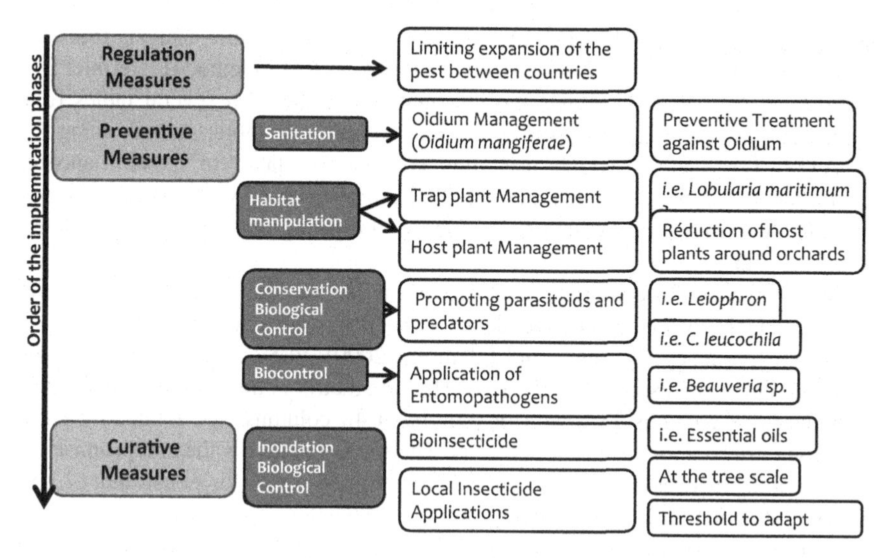

Fig. 8.9 Representation of the agroecological management strategy for *Orthops palus* populations in Réunion (Adapted from Atiama 2016)

Firstly, damage caused by *Orthops palus* and by oidium was simultaneous. In 80% of the cases where there was no oidium (2387 observations over 3 years), there was no *O. palus*. This observation makes it necessary to test the hypothesis that managing oidium would at the same time manage *O. palus* populations (Atiama 2016).

Moreover, with regard to *Thrips* (notably the South African Citrus Thrips, *Scirtothrips aurantii*), there was no difference in infestation between the control and the Biophyto plots, which makes it possible to propose two hypotheses that should be tested in the future: the ineffectiveness of insecticide and the effectiveness of a permanent vegetal cover in orchards.

Finally, during the large outbreak of the mealybug Seychelles *Icerya seychellarum* in 2012, the main results, presented in this summary, show that frequent insecticide treatments cause mealybug populations to increase.

8.5.3 *Predicting Pest Profiles in Different Agricultural Practices*

It is possible to identify agricultural practices using pest profiles. Plots are classified into three groups according to their status and trajectory: Biophyto plots, control plots and transition plots (control for Biophyto). Using observed pests we sought to predict the status of the plots by placing them into one of these three groups. To do this, we used the STATIS-LDA method, which is a multi-table analysis (Sabatier et al. 2013). This approach has two steps: first, the STATIS method (L'Hermier Des Plantes 1976), which integrates the information contained in several tables (one table per year for each pest profile) into one compromise table; second, using the STATIS-LDA method which uses this compromise table to explain another (Brun-Vitelli 2015). Using this method, plots are reclassified with a success rate of 72.2%, a good level of prediction. This shows the strong link between agricultural practices and pests.

With the STATIS compromise table, we can use the PLS2 method (Vivien and Sabatier 2001) to predict the abundance of each pest as a function of insecticide treatment frequency (Varmuza and Filzmoser 2009). PLS2 is an extension of the multiple linear regressions for matrices. The first matrix is the compromise for pest abundance; it contains 18 rows (for plots) and 48 columns (for abundance compromise variables for each pest per year). The second matrix is the compromise for agricultural practices over three years; it contains 18 rows (for plots) and 12 columns (for agricultural practice compromise variables).

Conversely, predicting is also possible. In this case, the PLS2 method is used between the two compromises issued from STATIS and is based on a summary of the data on agricultural practices over three years and a summary of pest abundance data over three years. This method has given excellent results predicting scale insect populations for different agricultural practices, according to the results presented above: the plots least damaged by scale insects are those not treated with insecticides, the most damaged plots are those using most insecticides (Fig. 8.10). Abundance of mango bugs and thrips is also relatively well predicted.

8.5.4 *Implications of Changing Agricultural Practices on Mango Flowering*

Several studies have been conducted on the effects of vegetal cover irrigation on the phenology of the mango tree and its water requirements.

Early experiments carried out during the project showed that agricultural practices could have a significant effect on flowering, causing a late and less intense flowering on plots with vegetal cover. By providing trees with water more regularly and in greater quantity, the vegetative rest phase did not occur. The rate of fruiting was difficult to evaluate because of the numerous pests attacks on the agroecological plots (Normand et al. 2015).

FDGDON concurrently monitored flowering during the Biophyto project on all pilot sites. Descriptive analyses showed that the flowering period in the Biophyto plots was comparable to that in the control plots (Brun-Vitelli 2015) over the 3 years of the project (Fig. 8.11). However, there were differences during the three years. In 2012 and 2013, flowering was identical between both types of plots on all farms, except one where in 2013 flowering in the control plot took place slightly

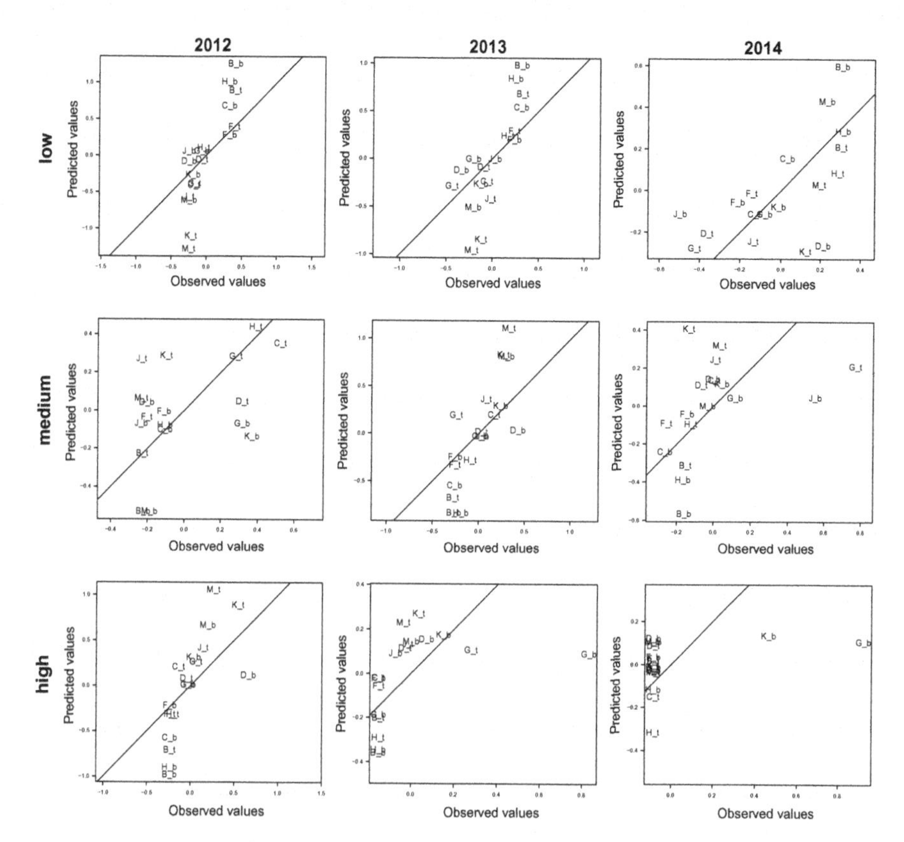

Fig. 8.10 Representation of predicted values versus mealybug abundance values, issued from STATIS analysis on pest abundance, by year and according to the observed level (low, medium or high). The predictions were calculated by PLS2 method (Vivien and Sabatier 2001) and leave-one-out validation from data issued from STATIS analysis on cultural practices. Each point corresponds to a coded plot (for confidentiality) with the following code: 'grower code plot code'. The colors correspond to the groups defined in multivariate analysis (green: Biophyto, violet: transition, red: control). The line corresponds to the first bisector. The closer a plot is to this line, the better the prediction: low and medium mealybug abundance are correctly predicted from cultural practices, unlike high abundance, due to a lack of annual data by plot

earlier. Conversely, in 2014, the flowering period was broader. Identical flowering was observed on both plots on five farms; early flowering on the Biophyto plot was noted on two farms; earlier flowering on the control plot was observed on two farms. This is certainly due to the year and not due to farming practices, as earlier blooming plots were sometimes the control plots, sometimes the Biophyto plots. Further observations showed that the flowering period is broader with the José variety than the Cogshall.

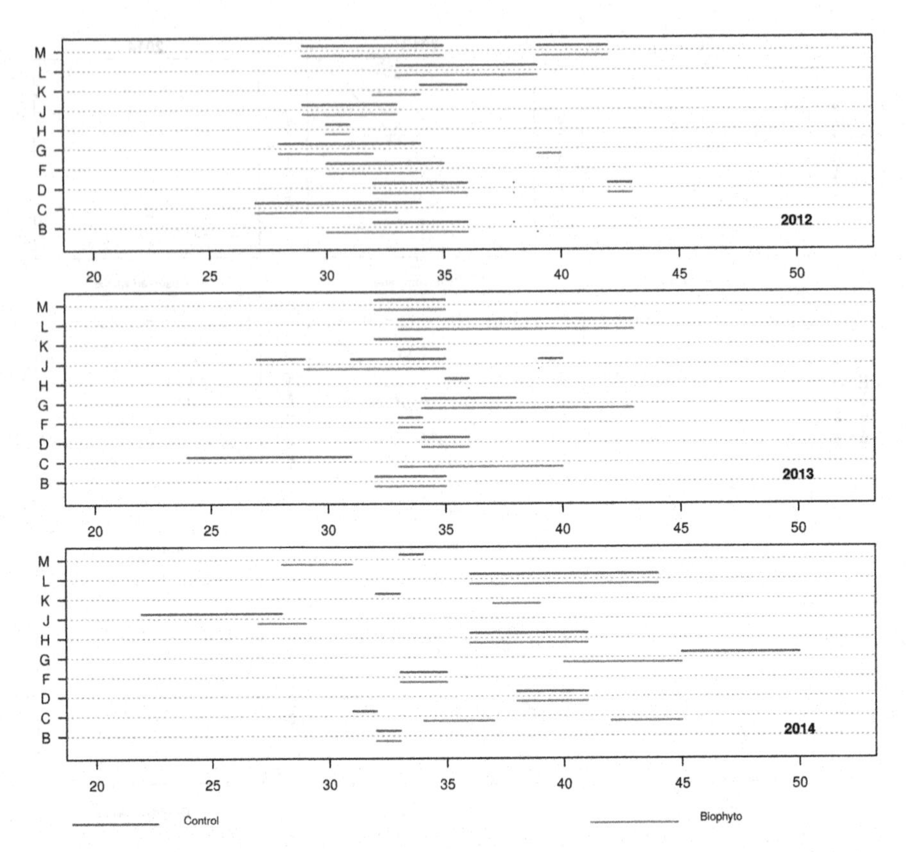

Fig. 8.11 Flowering periods on 10 pilot Biophyto farms from 2012 to 2014, on the control plots (in red) and the Biophyto plots (in green). The x-axis gives the week numbers in the year. The y-axis lists the various pilot orchards, identified by letters

Finally, other experiments were carried out in 2015 within the framework of the Biophytomang[2] project. The three test plots were located in a large mango production area where the low precipitation of the southern winter is not conducive to maintaining a vegetal cover. Measurements focused on irrigation management based on soil moisture status, monitoring of flowering and yield, and monitoring of vegetal cover (Bouriga and Graindorge 2016). Experts agreed to maintain the useful reserve of the soil at 25% and it has been hypothisized that short and regular watering would be more favourable to the vegetal cover than to the tree. In 2015, unusually high precipitation in May and June allowed the vegetal cover to remain on the plots without additional irrigation. However, the decision rules tested for vegetal cover irrigation methods showed its effectiveness during the dry season and the irrigation favoured the vegetal cover during July and August. The condition of the vegetal cover was heterogeneous in terms of flower height and diversity.

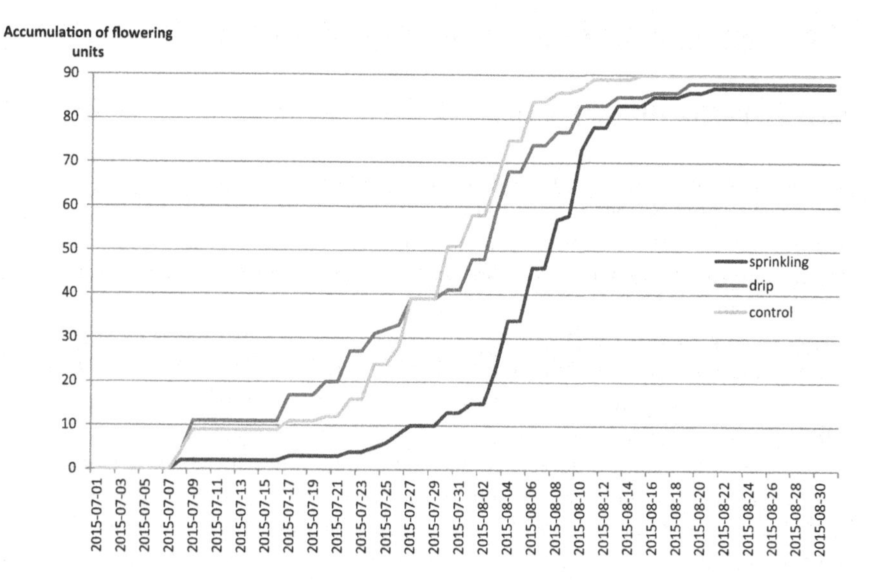

Fig. 8.12 Flowering data by irrigation type (drip: orange, control: green, sprinkling: green). The curves represent the accumulation of labeled units (90 units per modality) reaching the first flowering stage according to the irrigation method

Significant rainfall during the "water "stress"" period caused a slight shift in bud burst in all orchards during the season. The results also showed that vegetal cover irrigation appeared to induce a slight delay in flowering (Fig. 8.12), but flowering was greater on two of the plots. Yield estimates did not reveal any difference, irrespective of vegetal cover irrigation (Bouriga and Graindorge 2016).

To sum up, Agroecological Crop Protection practices were found to have a positive impact on the reduction of pest populations and damage, especially mango bugs and mealybugs. In addition, they have no negative impact on the flowering levels and fruit production.

8.6 Biodiversity, a Key Component of Orchard Ecological Functioning

The previous section dealt with pests of mango orchards. This section focuses on the biodiversity and ecology of mango orchards, with the aim of naturally regulating populations of harmful arthropods and the damage they cause. The summary of the data and information presented below is based on a thesis on this subject (Jacquot 2016) and the resulting publications.

8.6.1 The Role of Biodiversity in Agroecosystems

8.6.1.1 Biodiversity in Communities

At the community level, biodiversity corresponds to the variety and abundance of the species present (Magurran 2004). Specifically, biodiversity has two components: species richness and abundance. Species richness is the total number of species present in the unit under study. Abundance, on the other hand, expresses the relative abundance of species in a community. A community with equal species richness and abundance is one in which each species has the same number of individuals (equitable community) and is considered to be more diverse than a community dominated by one or more abundant species in which other species are rare.

On the other hand, biodiversity and the communities in which it is expressed also have structure and function attributes (Noss 1990). Community structuring is often carried out on the basis of trophic interactions between consumers and resources that occur between species or groups of species. These are known as trophic networks (Morin 2011). Species can be grouped into functional and trophic entities, depending on the resources they use (guilds), the functions they perform (functional groups or trophic groups) or how they acquire energy (trophic levels) (Morin 2011).

Biodiversity has a close relationship with ecosystem function and in particular with ecosystem goods, processes and services (Concepts defined from Christensen et al. 1996). Ecosystem processes are water cycle and storage, biological productivity, biogeochemical cycles and storage, decomposition and maintenance of biological diversity. The other two concepts are anthropocentric. On the one hand, ecosystem goods are to the properties of ecosystems that have a direct market value, such as food, building materials, medicinal products or genes for plant improvement and biological control agents). On the other hand, ecosystem services are the properties of ecosystems from which man derives direct or indirect benefits and which historically have no market value. The values of these services have nevertheless been measured to emphasize their importance in decision making (Costanza et al. 1997). From the late 1980 s onwards, an intensive period of research began to study the effect of biodiversity loss on ecosystem function. This research has been regularly summarized (Cardinale et al. 2009; Duffy et al. 2007; Hooper et al. 2005; Loreau et al. 2001). It is now recognized that biodiversity contributes to ecosystem function, diversity loss being one of the main causes of ecosystem malfunction, comparable to climate change (Hooper et al. 2012). The results of this research also emphasize that biodiversity produces many goods and services for humans (Cardinale et al. 2012; Diaz et al. 2006; Mace et al. 2012). According to Brose and Hillebrand (2016), future studies on the relationship between biodiversity and ecosystem function should take into account this multi-trophic community component and consider these relationships within wider spatial and time scales.

8.6.1.2 Biodiversity in Agroecosystems

Agricultural habitats occupy more than 30% of the land surface, forming one of the largest terrestrial biomes (Foley et al. 2005). Based on data for the year 2000, cropland and pasture represent 12% and 22% of land area, respectively (Ramankutty et al. 2008). The importance of agriculture on a global scale has two implications. First, agricultural habitats represent areas of interest for biodiversity conservation and ecosystem services (Iverson et al. 2014; Pimentel et al. 1992; Tscharntke et al. 2005). Second, agriculture is one of the most important factors in global change and biodiversity loss (Tscharntke et al. 2005). For example, agricultural activities account for 48% of diversity loss due to land use change (Murphy and Romanuk 2014). The impact of agriculture is not confined to land use: agricultural practices also influence the nature of agroecosystems in which they are conducted as well as adjacent ecosystems. Thus, agricultural intensification, through increased use of chemical inputs and homogenization of landscapes, is a major cause of diversity loss (Tscharntke et al. 2005). The impacts of this intensification exceed the negative impact on biodiversity. For example, the use of pesticides has negative effects both on the environment and on human health, to the point where it is necessary to quantify the costs of these externalities in assessing the overall value of pesticide use (Bourguet and Guillemaud 2016).

In the face of environmental and health challenges, and alongside research on the effect of diversity loss on ecosystem function, scientists have promoted sustainable agriculture that conserves biodiversity, while its various services as well as those provided to agroecosystems (Pimentel et al. 1992; Altieri 1999). This concept of a more sustainable and environmentally friendly agriculture has taken many forms: organic farming (Bellon and Penvern 2014; Darnhofer et al. 2010), agroecology (David and Wezel 2012; Deguine et al. 2016; Wezel et al. 2009) and the ecological intensification of agriculture (Doré et al. 2011; Gaba et al. 2014; Geertsema et al. 2016). Management of biodiversity and its services is central to these forms of sustainable agriculture and is the subject of specific techniques such as habitat management (Gurr et al. 2004). In agroecosystems, humans benefit from ecosystem services. Estimates of the value of these services show that they represent a total of $92/ha/year, including food supply ($54/ha/year) and two regulatory services: biological pest control ($24/ha/year) and pollination ($14/ha/year) (Costanza et al. 1997). This estimate underlines the importance of biological pest control in agroecosystems.

8.6.2 Objectives of Biodiversity Studies on Mango Orchards in Réunion

In this section, we are interested in biodiversity—more specifically in the biological regulation of arthropod pests of mango trees in Réunion. In communities, crop-damaging arthropods are subject to two types of trophic control: bottom-up control through plant diversity; top-down control through the diversity of natural enemies. The diversity of natural enemies and plant diversity favour pest control; In addition, agricultural practices and the landscape are known to influence natural enemies, plants and, directly, pests (Rusch et al. 2010). However, the underlying mechanisms are rarely elucidated. Moreover, the relative importance of these different factors is seldom quantified, thus limiting the knowledge necessary for effective pest control.

In order to develop biodiversity-based agriculture, it is necessary to acquire knowledge on agroecosystem function, in particular the factors influencing the conservation of biodiversity and the provision of its services. Community ecology (multi-trophic approaches on a large spatial and temporal scale) and crop protection (mechanistic approaches at multiple scales) provide a relevant framework for understanding both the mechanisms that would allow better pest control and also the relationship between biodiversity and ecosystem function in multi-trophic contexts and at larger spatial scales.

The studies presented in this section address these issues. Through multi-trophic and multi-scale approaches, and using the Biophyto project network, we studied the role of biodiversity on ecological function of mango-based agroecosystems in Réunion, with particular attention paid to insect pest control, taking into account the effect of agricultural practices and landscape. The three objectives were (1) to describe the agroecosystem studied: arthropod communities, plants, farming practices and landscape; (2) to identify and understand the role of biodiversity in multi-trophic interactions in food webs in different farming practices and landscapes; and (3) to identify the services provided by natural enemy diversity and the effect of farming practices and landscape on pest control.

8.6.3 *Acquisition of Biodiversity Data and the Methodology Used*

To meet these objectives, a systemic scientific approach was used, studying objects and factors simultaneously at multiple scales including arthropod and plant communities, agricultural practices and landscape types. We chose to perform global "snapshot" sampling. Once a year (2012–2014), in August when all the plots were in flower, we collected the data over a 2-week period for the different scales of the study. Of the 13 farms in the Biophyto network, ten were involved in these studies. We followed the agroecological and control plots at each site. The average size of the two plots was 1404 m^2.

8.6.3.1 Description of Arthropod Communities

The purpose of sampling arthropod communities was to determine their taxonomic composition and to quantify the abundance and diversity of trophic groups, bringing together species with similar functions (Morin 2011). Sampling was conducted in different strata of mango orchards, the soil surface and the canopy of the mango trees, in order to obtain a more comprehensive view of the communities. A detailed methodology of arthropod sampling is presented elsewhere (Jacquot et al. in prep).

Major work was carried out to identify as many species as possible. The study of arthropods in Réunion agroecosystems was previously mainly concerned with crop pests and their specialist natural enemies. At the beginning of the study, we did not know which taxonomic groups were representative of the trophic groups we wanted to study. First, the samples were sorted to separate the collected arthropods from the debris and to classify the arthropods by order. Identifications were then carried out with the help of the available bibliography and the entomological collection of the UMR PVBMT (Plant Protection Pole) of Saint-Pierre. For the majority of the groups, the experts have checked identifications by the CIRAD team. Different species which were too morphologically similar to be distinguished were grouped into a single species group (genus name followed by "spp.").

Assigning species to trophic groups was done using the literature. We examined 10 trophic groups of arthropods including 6 main trophic groups: (1) herbivorous pests: species known to be harmful to mango, i.e. its feeding or development on mango trees causes economic losses; (2) non-pest herbivores: herbivorous species (phytophagous *s.l.*), not known to be mango pests; (3) detritivores: species of detritivores, microherbivores and mycetophagous; (4) omnivores: species feeding on other arthropods as well as on plants and detritus; (5) parasitoids: species that parasitize or destroy detritivorous or herbivorous arthropods; (6) predators: strictly predatory species feeding almost exclusively on other arthropods.

8.6.3.2 Description of Agricultural Practices and Landscape

We examined agricultural practices using two indicators: plant communities, which reflect weed management (mowing, irrigation, mulching, herbicide treatments) and insecticide treatments.

Each year between 2012 and 2014, vascular plant species were identified in segments where arthropod sampling by aspiration had taken place one week prior. We measured the linear distance covered by each plant species and the abundance and diversity of plants along the transect.

During the Biophyto project, the dates of work done on the plots were analyzed. This information began to be collected in July 2012. Therefore, only the number of insecticide treatments carried out during the 2-month period before the arthropod collection (July and August) was used as an indicator, although all data are available from July 2012 to December 2014.

In addition, the use of ArcGis Geographic Information System (GIS) software, 2012 IGN aerial photographs, and annual in situ checks enabled habitats to be mapped in the landscapes for each sampling period. Landscape was studied within an 800 m diameter circle around each plot. Landscape indicators were measured using two tools. The Fragstats software (version 4.2) allowed the indicators to be quantified on a large scale. These included habitat diversity and border lengths. In addition, the R software quantified the indicators at habitat level including the proportion of semi-natural habitats, proportion of mango orchards and other similar aspects.

8.6.4 *Arthropod Communities*

8.6.4.1 Composition and Importance of Taxonomic Groups

A total of 126,753 arthropods were collected and 124,001 were identified. They belonged to 4 classes, 23 orders, 215 families, 451 genera and 797 morphotypes. The accuracy of the morphotype identification varies, for example: 217 morphotypes for which we were able to identify to species level, 389 morphotypes to genus and 156 morphotypes to family or superfamily level.

The order with the most variety and abundance was Hymenoptera, with two distinct groups: ants (Hymenoptera: Formicidae) and Hymenoptera Parasitica, most of "Hymenoptera non Formicidae") (Fig. 8.13). Ants alone accounted for a large number of arthropods, with 65,487 individuals collected, for only 21 species. Conversely, the other Hymenoptera represented more than 300 species for only 7,500 individuals collected. Thrips (Thysanoptera) and Hemiptera were two groups present in high numbers. This can be explained by the fact that three mango insect pests belong to these two groups: one species of Thrips and two species of Hemiptera. Spiders (Araneae) and Diptera are also important taxonomic groups with more than 4,000 individuals identified. In terms of species richness, four orders were distinguished in addition to Hymenoptera: Diptera, Coleoptera, Hemiptera and Spiders.

Our results are an important contribution to the knowledge of arthropods in mango orchards in Réunion, and may be useful for other horticultural systems on the island and for mango orchards in other areas. This study discovered species new to Réunion and new to Science. The species new to Réunion mainly consist of spiders in 11 new families and at least 32 new species (Jacquot et al. 2016). Added to this are four morphotypes that are potentially species new to Science. Among Hymenoptera parasitoids, there are two new families, 144 new species (Muru et al. 2017) for Réunion. Photographic catalogues of the new species were created for a number of orders, in particular for beetles (Fig. 8.14).

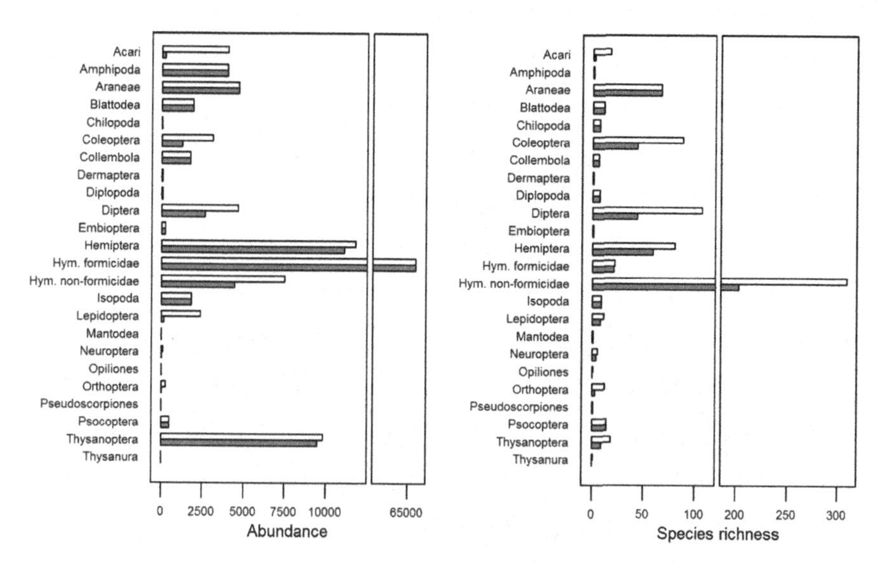

Fig. 8.13 Abundance and number of species by order of arthropods identified: in mango orchards in Réunion. The white bars indicate the number of species or the number of individuals collected, the green bars indicate the number of species or the number of individuals collected and attributed to trophic groups according to the literature

8.6.4.2 Composition and Importance of Trophic Groups

In total, 109,741 individuals from 523 species were assigned to 10 trophic groups, including 109,079 individuals from 504 species of arthropod assigned to the six main trophic groups (Table 8.1). The numbers of individuals or species that may have been attributed to trophic groups based on bibliography or isotopic analyses differ from one order to another. In terms of abundance and species richness, the resolution is low for mites (Acari), Coleoptera, Diptera, Non-formicidae Hymenoptera, Lepidoptera and Orthoptera. The resolution is relatively good for Hemiptera in terms of abundance, but less in terms of species number.

The main trophic groups have different taxonomic structure (Table 8.1). Herbivore pests were represented by four insect species, mostly Hemiptera (the Seychelles mealybug *Icerya seychellarum* and mango bug *Orthops palus*), species of mango midge (*Procontarinia spp.*) and a species of Thrips (*Scirtothrips aurantii*). Non-harmful herbivores were an important trophic group in terms of individuals and species, mostly Hemiptera, Thrips and, to a lesser extent, Coleoptera. The detritivores were mainly represented by individuals belonging to terrestrial crustaceans (Isopods or cloportes and Amphipods), Cockroaches and Collembola. Psoques and Diptera dominated this trophic group in species richness. For their part, omnivores were almost exclusively represented by 19 species of ants. Among them, three invasive exotic species were present in high numbers:

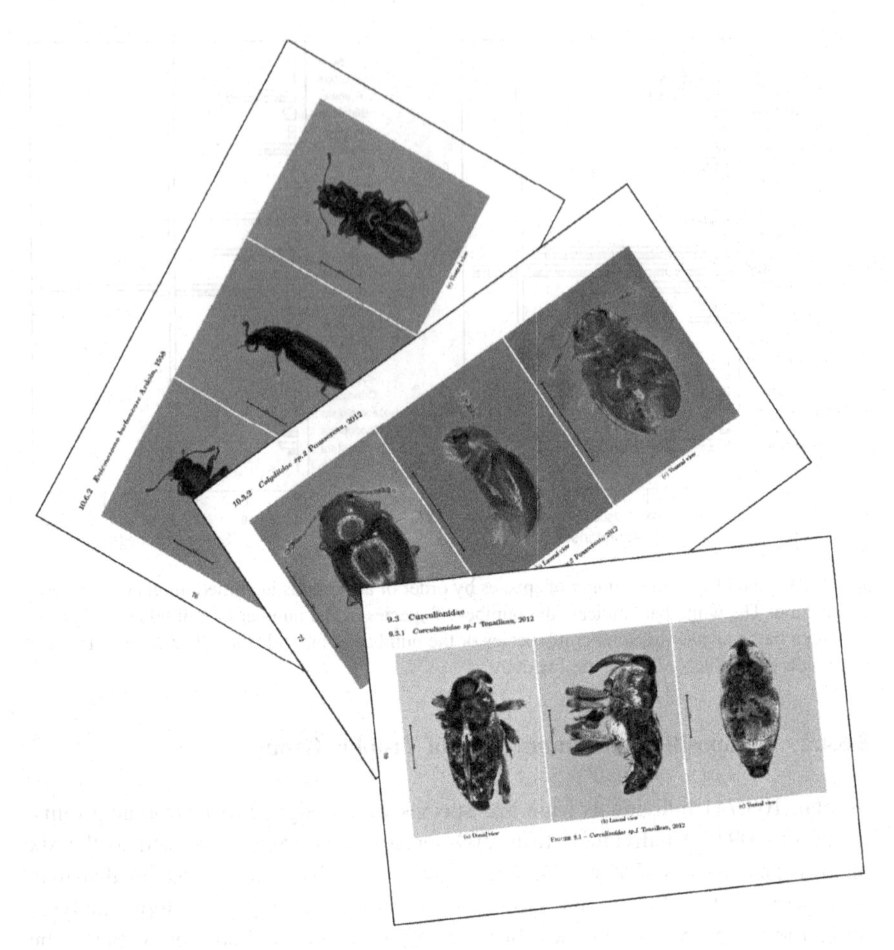

Fig. 8.14 Example of a photographic catalogue: the Coleoptera identified in mango orchards in Réunion. (Photo: M. Jacquot)

Pheidole megacephala (55,608 individuals), *Brachymyrmex cordemoyi* (4,440 individuals) and *Solenopsis geminata* (1,081 individuals). They were an important component of the system, as they influence functional biodiversity and its services. The parasitoids formed the richest trophic group, composed of 182 species of Hymenoptera and 8 species of Diptera. Parasitoid Hymenoptera diversity was very high surprisingly so. It can be explained by the complementarity of the three sampling techniques, whereas most of the studies in this group used malaise traps or the breeding of host arthropods (often of economic importance). Finally, predators were mostly represented by spiders.

Table 8.1 Taxonomic composition of the six main trophic groups of arthropods collected in mango orchards

Class/Order	Herbivorous pests		Herbivorous non-pests		Detrivores		Parasitoïds		Predators		Omnivores		Total abundance	Total species richness
	N	RS	N	RS	N	RS	N	RS	N	RS	N	RS		
Arachnida														
Acari	–	–	–	–	–	–	–	–	25	1	184	1	209	2
Araneae	–	–	–	–	–	–	–	–	4176	65	–	–	4176	65
Pseudoscorpiones	–	–	–	–	–	–	–	–	1	1	–	–	1	1
Insecta														
Blattodea	–	–	–	–	1906	11	–	–	–	–	–	–	1906	11
Coleoptera	835	35	–	–	73	7	–	–	355	1	1	1	1264	44
Collembola	–	–	–	–	1735	6	–	–	–	–	–	–	1735	6
Dermaptera	–	–	–	–	–	–	–	–	–	–	62	1	62	1
Diptera	46	10	974	1	884	24	736	8	1	1	–	–	2641	44
Embioptera	–	–	–	–	228	1	–	–	–	–	–	–	228	1
Hemiptera	7522	56	3596	2	–	–	–	–	–	–	2	1	11120	59
Hym. formicidae	–	–	–	–	–	–	–	–	5	2	65482	19	65487	21
Hym. non-formicidae	84	4	–	–	–	–	4147	182	–	–	–	–	4231	186
Lepidoptera	18	7	–	–	129	1	–	–	–	–	–	–	147	8
Mantodea	–	–	–	–	–	–	–	–	1	1	–	–	1	1
Neuroptera	–	–	–	–	–	–	–	–	24	3	3	1	27	4
Orthoptera	13	3	–	–	–	–	–	–	–	–	–	–	13	3
Psocoptera	–	–	–	–	481	14	–	–	–	–	–	–	481	14
Thysanoptera	8777	6	718	1	–	–	–	–	–	–	12	2	9507	9

(continued)

Table 8.1 (continued)

Class/Order	Herbivorous pests		Herbivorous non-pests		Detrivores		Parasitoïds		Predators		Omnivores		Total abundance	Total species richness
	N	RS	N	RS	N	RS	N	RS	N	RS	N	RS		
Malacostraca														
Amphipoda	–	–	–	–	3970	1	–	–	–	–	–	–	3970	1
Isopoda	–	–	–	–	1819	9	–	–	–	–	–	–	1819	9
Myriapoda														
Chilopoda	–	–	–	–	–	–	–	–	14	7	–	–	14	7
Diplopoda	–	–	–	–	40	7	–	–	–	–	–	–	40	7
Total	17295	121	5288	4	11265	81	4883	190	4602	82	65746	26	109079	504

8.6.5 Ecological Functioning of Mango Orchards

8.6.5.1 Interactions Between Trophic Groups

Understanding the factors that maintain diversity in communities and the effects of this diversity on the ecological functioning of agroecosystems is essential for managing biodiversity and its services. For this, we have used three years' worth of data and studied the six main trophic groups of our system. We studied the relationships between trophic groups and the effects of plant diversity, the frequency of insecticide treatments and landscape complexity (Shannon index of habitat diversity, percentage of semi-natural elements) on these trophic groups. The results presented below are from a work in progress (Jacquot et al. in prep).

These results, derived from a comparison of models of structural equations, show that diversity exerts simultaneous bottom-up and top-down controls within the same arthropods community (Jacquot et al. in prep). These results diverge from the historical opposition between systems dominated either by bottom-up or top-down controls (Hairston et al. 1960; Oksanen et al. 1981). The results of the analyses show that the relationships between trophic groups are different on the soil surface and in the tree canopy. Several mechanisms appear to be involved. First, on the soil surface (a complex stratum containing numerous plants species) the specialization of secondary consumers appears to be a key factor explaining the relationship with primary consumers. Moreover, there is a diversity cascade in this stratum, bottom-up and positive, with plant diversity indirectly supporting parasitoid diversity. Secondly, in the canopy (a mono-specific stratum), the bottom-up control exerted by primary consumer diversity on the diversity of secondary consumers may be due to a "birdfeeder" effect (Eveleigh et al. 2007). Mango inflorescences attract primary consumers which in turn attract secondary consumers.

8.6.5.2 Influence of Invasive Ants on Predation

Invasive ants, including *Pheidole megacephala* and *Solenopsis geminata*, are an important component of arthropod communities in mango orchards in Réunion. They are omnivorous (Holway et al. 2002). In order to identify the species involved in predation and the role of invading ants, we conducted a sentinel prey experiment in two strata of the orchards (soil surface and canopy). The results that follow are from works published elsewhere (Jacquot et al. 2017). Analyses were conducted on data from the 2014 arthropod collections and the sentinel prey predation rate of the same year. We constructed a hierarchical Bayesian model to quantify the effect of the abundance of dominant invasive ant species on the predation rate and on the respective diversity of predators and omnivores, as well as the relationship between the diversity of these trophic groups and predation rate.

The results show that predation of sentinel prey mainly takes place on the soil surface. In this stratum, *Pheidole megacephala* provides predation supplied by omnivore diversity. At the same time, there is direct predation, which reduces omnivore diversity. This results in a negative interaction between omnivore diversity and predation. Conversely, *Pheidole megacephala* does not influence predator diversity, and this has a positive relationship with predation rate. Simultaneously, *Solenopsis geminata* predates directly without affecting the diversity of the two trophic groups of natural enemies (omnivores and predators). The role of invasive ants in predation (Crowder et al. 2010; Hogg and Daane 2011; Snyder et al. 2006) and biodiversity reduction (Hogg and Daane 2011; Kenis et al. 2009; Snyder et al. 2006) is already known. But, to our knowledge, the influence of invading natural enemies on the predation service provided by diverse natural enemy communities has never previously been demonstrated, either for ants or for other taxa.

8.6.5.3 Effects of Natural Enemies, Practices and Landscape on Pests

This study aims to identify the factors influencing the abundance of insect pests, of which four species, or groups of species are likely to cause damage during the mango flowering period. These factors include natural enemies, agricultural practices and landscape.

Mixed linear models were obtained from the 10 pairs of plots, sampled once per year from 2012 to 2014. We estimated the abundance of each pest species in each plot based on data from three types of sampling method presented (Jacquot et al. 2016). We tested the effects of (1) the diversity of three groups of natural enemies (parasitoids, predators and omnivores) with two indicators per group (estimated richness at the soil surface and in the mango canopy); (2) the frequency of insecticide treatments; (3) the proportion of semi-natural elements and mango orchards within a radius of 400 m around each plot.

The results confirm the existence of only two effects (Table 8.2). The first is a negative effect of parasitoid diversity on the abundance of the Seychelles scale. This suggests the importance of parasitoids in pest control in our system. The other is a positive effect of the proportion of mango orchards in the landscape on the abundance of South African citrus fruit Thrips.

Table 8.2 Estimated parameters for abundance models for the four species (or groups of species) of insect pests present during mango flowering. (Semi-natural = proportion of semi-natural habitats in landscape; mango orchards = proportion of mango orchards in landscape)

	Scirtothrips aurantii		*Procontarinia spp.*		*Orthops palus*		*Icerya seychellarum*	
	Estimate ± SE	p-value	Estimate ± SE	p-value	Estimate ± SE	p-value	Estimate ± SE	p-value
Intercept	0.03 ± 0.39	0.94	0.01 ± 0.31	0.97	0.01 ± 0.26	0.96	0.02 ± 0.13	0.89
Parasitoid diversity								
Canopy	0.24 ± 0.19	0.21	0.13 ± 0.16	0.42	0.24 ± 0.21	0.25	−0.02 ± 0.08	0.76
Ground surface	−0.05 ± 0.12	0.65	0.00 ± 0.04	0.94	−0.01 ± 0.06	0.84	−0.42 ± 0.16	0.01
Predator diversity								
Canopy	0.01 ± 0.04	0.89	0.01 ± 0.04	0.90	0.26 ± 0.23	0.26	0.00 ± 0.03	0.98
Ground surface	0.00 ± 0.04	0.95	0.01 ± 0.06	0.83	0.00 ± 0.04	0.95	0.03 ± 0.09	0.73
Omnivore diversity								
Canopy	0.01 ± 0.05	0.86	0.00 ± 0.04	0.95	0.12 ± 0.16	0.44	−0.01 ± 0.04	0.88
Ground surface	0.01 ± 0.05	0.84	0.10 ± 0.15	0.48	0.00 ± 0.03	0.94	−0.19 ± 0.17	0.25
Insecticide spraying	0.01 ± 0.03	0.87	−0.07 ± 0.12	0.53	0.00 ± 0.03	0.91	0.00 ± 0.02	0.95
Landscape								
Mango orchard	0.45 ± 0.18	0.01	−0.01 ± 0.05	0.88	0.03 ± 0.08	0.74	−0.01 ± 0.06	0.82
Semi-nat.	−0.03 ± 0.09	0.75	−0.03 ± 0.09	0.75	0.00 ± 0.04	0.98	0.06 ± 0.13	0.62

8.6.6 Implications for Agroecological Management of Mango Orchards

8.6.6.1 Importance of Diversity of Natural Enemies

Previous studies aimed to understand the role of biodiversity in the ecology of mango orchards and to propose strategies for the natural regulation of insect pests. We studied three groups of natural enemies: parasitoids, predators and omnivores. Our results confirm the importance of using insect pest control provided by the high diversity of these natural enemies. In fact, high predator diversity favours predation on the soil surface (Jacquot et al. 2017). High predator diversity also influences insect pest communities in the canopy: the higher the predator diversity, the higher the diversity of insect pests (Jacquot et al. in prep). Finally, analyses of the abundance of insect pests have shown that higher parasitoid diversity reduces the abundance of the Seychelles scale.

Our results do not allow us to identify practices that would favour predator and omnivore diversity. The results also suggest that understanding the interactions between trophic groups is necessary to identify these practices. For example, the fact that an ascending cascade effect of plant diversity on predators was not highlighted rejects the possibility of using this lever to favour predators. In traditional studies, which do not take into account the multi-trophic structure of networks, plant diversity directly favours predator diversity (Dassou and Tixier 2016; Haddad et al. 2009; Hertzog et al. 2016; Scherber et al. 2010). Understanding the factors that directly affect natural enemies in mango orchards is therefore essential for designing cropping systems that exploit their services.

8.6.6.2 Impact of Invasive Ants

Invasive ants are a major problem worldwide (Lowe et al. 2000; Suarez et al. 2010). The role of these invasive natural enemies needs to be understood to assess their impact on the agroecosystems they invade, as well as to explain the compromises that may be necessary to reduce their numbers.

In our system, *Pheidole megacephala* and *Solenopsis geminata* were predators performing a more important role than predator diversity (Jacquot et al. 2017). However, they also create disservices by reducing omnivore diversity (*Pheidole megacephala*) (Jacquot et al. 2017), inflicting painful bites on growers (*Solenopsis geminata* (Wetterer 2010) and generally, promoting populations of harmful Hemiptera (mealybugs, aphids) (Holway et al. 2002; Offenberg 2015). In addition, these invasive ant species represent a major threat to biodiversity in the areas they invade (Lowe et al. 2000). We therefore do not recommend the use of techniques aiming specifically at using ant species as a predation service in mango orchards.

In order to identify controls for invasive ants and for the conservation of natural enemies, it would be interesting to identify which agricultural practices and landscape characteristics influence ant invasion and the diversity of natural enemies.

8.6.6.3 Importance of Plant Diversity in Insect Pest Control

Of the major results of our work on functional biodiversity, only one technique, plant diversity in orchard grasses, was highlighted as having any effects. However, implications for insect pest management during mango bloom can be identified. The effect of plant diversity in orchard grasses on arthropod communities on the soil surface is fascinating, both by its nature and when it is compared with the result on effect of parasitoid diversity. In fact, the results of the structural equation models show that plant diversity has an ascending diversity cascade effect (Jacquot et al. in prep). Plant diversity favours the diversity of non-harmful herbivores, which in turn promotes a more diversified parasitoid community (Jacquot et al. in prep). However, the greater the parasitoid diversity, the lower the abundance of the Seychelles scale. As a result, managing diversified grasses in orchards would allow a biological control of the Seychelles scale, probably by preserving parasitoid diversity.

8.6.6.4 Diversification of Crops at the Landscape Scale

On a broader scale, the proportion of mango orchards in the landscape promotes the South African citrus fruit Thrip populations. The main mango production area in Réunion is expanding. During the Biophyto project, we observed the planting and replanting of many orchards, which now cover the majority of cultivated areas. In the medium term, landscape homogenization and the preponderance of mango orchards could thus increase the damage caused by Thrips and, most likely, favour the development of other pests by concentrating the same resource in space (Root 1973). Moreover, even if our studies do not show this, landscape simplification could have negative effects on the diversity of natural enemies, which is often observed (Bianchi et al. 2006; Chaplin-Kramer et al. 2011; Veres et al. 2013).

To compensate for this landscape simplification via mango monoculture, two solutions are available. First, crop diversification would result in a higher landscape complexity and a "dilution" of mango orchards in production basins. Crop diversification is a medium-term objective in areas occupied mainly by perennial crops. It can be achieved by the planting of alternative fruit species when mango orchards are renewed (usually every 20–30 years). Diversification could also be envisaged in orchards by mixing several fruit species (Simon et al. 2010). Using the mixed forest model, the cultivation of phylogenetically distant species would ensure resistance by association with specialist and generalist insect pests (Castagneyrol et al. 2014). Second, the creation and maintenance of semi-natural elements (hedges, groves, etc.) in production areas could increase landscape complexity and, consequently, numbers of natural enemies and pollinators. Studies conducted in mango orchards in South Africa validate such concepts. They show that natural vegetation found at greater distances from the orchard result in poorer control of insect pests such as fruit flies and midges (Henri et al. 2015) and a poorer pollination service provided (Carvalheiro et al. 2012). Furthermore, Carvalheiro et al. (2010) have also shown that cultivation of small areas (25 m^2) of native flowers (Asphodelaceae and Acanthaceae) alongside the two types of mango orchards increases the diversity and number of visitors to mango inflorescences and increases mango production.

To conclude, the role of biodiversity in the ecology of mango orchards was poorly understood before the Biophyto project. This biodiversity was the subject of a bibliographic review and detailed studies, which yielded new and valuable results. The 124,001 arthropods identified of the 126,753 arthropods collected in orchards belong to 4 classes, 23 orders, 215 families, 451 genera and 797 morphotypes. Parasitoids form the richest trophic group. Bottom-up and top-down controls of biodiversity within a single community of arthropods were observed, and the role of ants, including invasive species, was described. There was a negative effect of parasitoid diversity on the abundance of the Seychelles mealybug. Inversely, the proportion of mango orchards in the landscape had a positive effect of on the abundance of South African citrus Thrips. These results gave rise to recommendations for the agroecological management of mango orchards.

8.7 Sustainable Changes in Agricultural Practices with Positive Socio-Economic Consequences

8.7.1 Implementation of Results and Satisfaction Survey

Overall, the results are very encouraging. It shows that the pilot farms were willing to participate in the design of agroecological production methods and are satisfied with the progress of the project and its achievements. The project marked a break with traditional farming techniques. A survey was carried out with all partners and growers. It emerges that they consider that this agroecological experience is a milestone in the development of agricultural practices in Réunion (Gloanec 2015).

The survey showed that growers are generally satisfied with their experience and the majority of them are willing to continue the agroecological practices on part or all of their orchards. Growers have gained some experience, even self-sufficiency, in vegetal cover management in their orchards (Gloanec et al. 2017). Some of these growers were asked by their neighbours to demonstrate the techniques used. Nine of the 11 growers surveyed rated 7/10 or more, reflecting their satisfaction with improved orchard health. However, the impact of agroecological practices on fruit yield and quality was difficult to assess due to cyclones during two harvest seasons.

These growers, however, consider that agroecological practices are economically viable over the long term.

8.7.2 Production, Production Costs and Treatment Frequency Index (TFI)

Due to two cyclones during the Biophyto project, one in 2013 and the other in 2014, large differences in production were observed between the different plots, whether conventional or agroecological. Overall, equivalent yields were recorded in conventional and Biophyto plots (Gloanec and Guignard 2015; Gloanec et al. 2017). Smaller yield losses, (impossible to state if significant) were only found in 6 out of 24 comparisons, mainly in areas sensitive to the gall midge, especially in plots with high production potential (Cogshall variety, allowing significant intensification). Certain fungicides, although compatible with organic farming, were not systematically used against oidium, as growers can underestimate the incidence of this disease, and this can contribute to production losses. In these few cases of production loss, the decrease in pesticide outlay helped compensate for the decline in gross margin, and sometimes even managed to maintain it.

Analysis of input and labour costs shows that agroecological farming has significantly lower costs than conventional farming: phytosanitary treatments are limited to one or two for oidium during flowering, herbicide treatments are eliminated and mowing is less frequent. The cost of production for plots in agroecological management mode is 35% lower (Gloanec et al. 2017) (Table 8.3).

The Treatment Frequency Index (TFI) decreased from 22.4 before the Biophyto project to 0.3 at the end of the project (Fig. 8.15). In this calculation, biocontrol products such as Ceratipack are not taken into account. Moreover, these orchard farming techniques are Organic, as the few fungicides that were used (for treatment of oidium) are organic approved. Some growers are aware of the negative impact of treatments on human health and the environment and are trying to find ways to reduce or eliminate treatments to make their farms entirely agroecological.

8.7.3 A New Perception of Agroecosystems for Growers

All Biophyto project partners have adopted new methods of assessing biocoenosis in orchards. Previously, focus was almost exclusively on the mango tree production and pests (without being systematically observed) because of their potential damage. Preventive or regular insecticide treatments were used without considering risk thresholds. Similarly, systematic herbicide treatments were designed to keep the plot free from weeds (Fig. 8.15).

Table 8.3 Economic data (time spent and cost in Euros per hectare) between conventional farming and agroecological farming (in Gloanec et al. 2016)

Data	Conventional farming	Agroecological farming
Inputs		
Pesticides	1614 €	39 €
Mass traps (80 traps/ha)	480 €	480 €
Irrigation (for 2,990 m^3)	299 €	299 €
Total inputs	2393 €	818 €
Labour		
Phyto monitoring	473 €	473 €
(time spent)	(50 h)	(50 h)
Phyto treatments	1 031 €	258 €
(time spent)	(16 h)	(4 h)
Surveillance of mass trapping	525 €	525 €
(time spent)	(55 h)	(55 h)
Rotary slashing after cutting	516 €	516 €
(time spent on tractor)	(8 h)	(8 h)
Chemical weed control	645 €	0
(time spent)	(10 h)	
Mowing beween rows	645 €	322 €
(time spent on tractor)	(10 h)	(5 h)
Mowing rows	0 €	151 €
(time spent on strimmer)		(16 h)
Harvesting	1106 €	1106 €
(time spent)	(117 h)	(117 h)
Mowing	908 €	908 €
(time spent)	(96 h)	(96 h)
Total labour	5849 €	4259 €
Production cost	8242 €	5077 €

Today, growers know that agroecosystem ecology is based around interactions between the different trophic groups that constitute functional biodiversity such as mango trees, weeds, non-harmful and harmful herbivores, detritivores, predators, parasitoids or pollinators. They know that insecticide treatments have negative impacts on auxiliaries, including predators and parasitoids, and that stopping herbicide treatments allows weeds to play a positive role in the agroecosystem. Figure 8.16 gives a schematic representation of this evolution in growers' perception of the agroecosystem.

Agricultural professionals (growers, agricultural advisors and technicians) learned to recognize natural enemies of pests during training sessions (see Sect. 8.2.). They are now often able to recognize them in the field and respect the parasitism and predation services of these natural enemies, such as the action of certain parasitoids on fruit flies (Fig. 8.17).

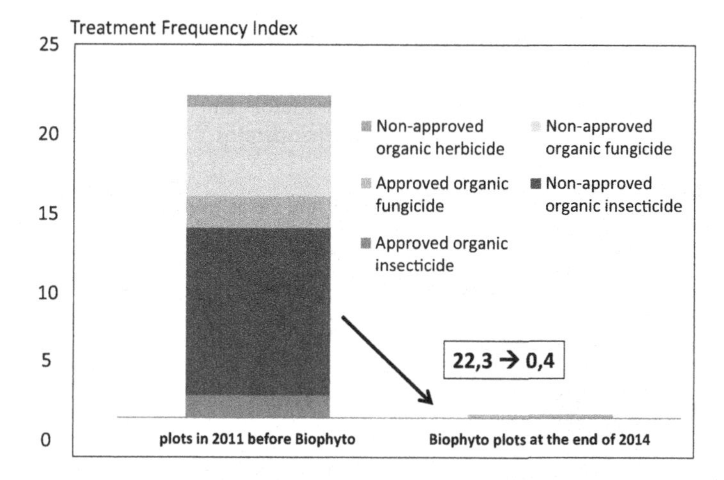

Fig. 8.15 Treatment Frequency index (TFI) before the start and at the end of the Biophyto project. Averages for five Biophyto farms of the Dephy Ferme Mangue Ecophyto network (Réunion Chamber of Agriculture 2015) (Adapted from Gloanec et al. 2017)

8.7.4 Different Marketing Methods

Aside from the technical aspects of the Biophyto project, an important part is a "marketing toolbox" which can help growers to market insecticide-free mangos. This marketing method was analysed via a market study involving 400 consumers and via stakeholder consultations in the grower—distributor chain. These surveys highlighted the commercial potential of insecticide-free mangoes, as well as a better understanding of the most suitable marketing routes (Técher et al. 2015).

An inter-stakeholder group composed of some twenty people representing different sectors of production, distribution, consumption, water, environment, territorial development and food recovery, was set up to seek the best agronomic, economic and marketing solutions for the production of Biophyto mangos depending on the markets (short circuit, supermarket chains, processing, export). This group was consulted throughout the project in meetings and by e-mail in order to take part in the various discussions on the market study and ways marketing techniques.

Consumers judge mangos by the visual appearance of the fruit (its freshness, its calibre, its colour, its shape). Sight is the first sense that is used when eating: when choosing the product, the consumer will prefer a visually appealing fruit compared to a damaged and aged fruit. The second decisive factor is taste. The dozen varieties that can be found in Réunion are all very different. Each consumer will have a different taste preference, which will influence purchase decisions. Price is also a factor in the final decision, marking acceptance or the refusal of the product. Consumers considered an acceptable price to be between 1.7 and 3.5 euros/kilo.

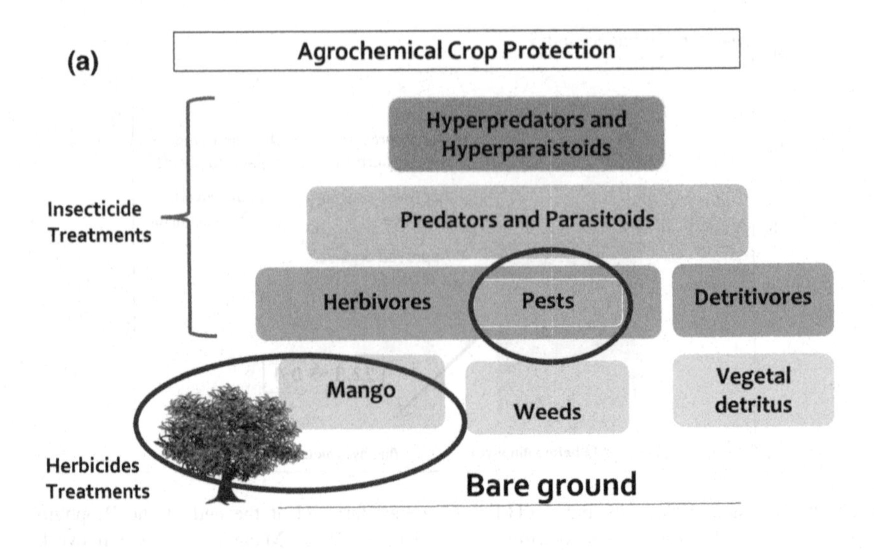

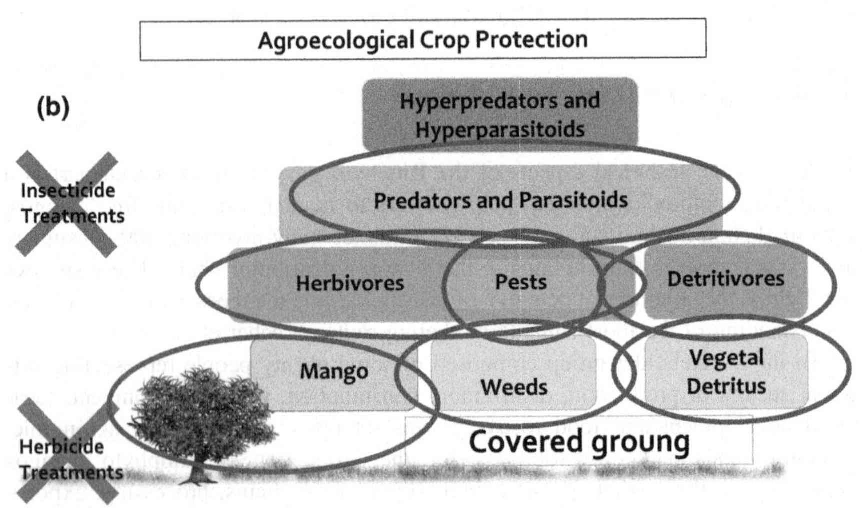

Fig. 8.16 Evolution of growers' perception of biocenosis in mango orchards between the agrochemical protection period (**a**) and the agroecological protection period (**b**)

Various operators involved in the mango sector (production, marketing, processing and institutions) were interviewed as part of the Biophyto project in 2013 and 2014. Scenarios were analysed and different marketing channels identified. For each of them, information about context, product characterization, type of customer, methodology and potential were collected during these interviews. It is clear from this study that a Biophyto mango has marketing potential, but with different communication strategies depending on the different marketing channels (Técher et al. 2015).

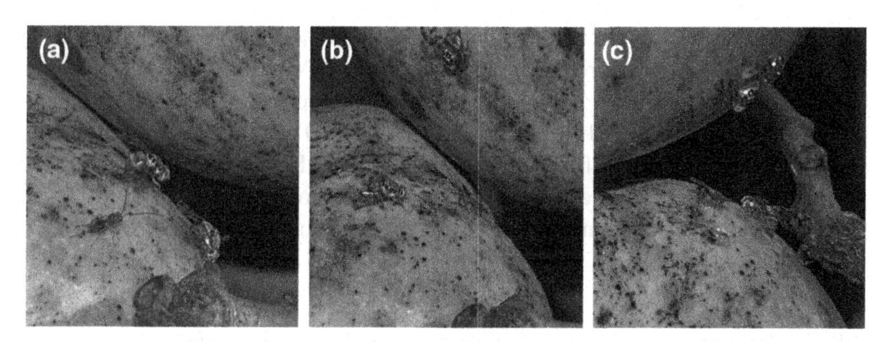

Fig. 8.17 Parasitism of *Ceratitis rosa* eggs by *Fopius arisanus*. **a** *C. rosa* females laying eggs; **b** *F. arisanus* female looking for *C. rosa* eggs; **c** *F. arisanus* female laying in *C. rosa* eggs. (Photos: a, b and c: A. Franck)

As a conclusion, insecticides and herbicides have been significantly reduced or eliminated. The treatment frequency index (TFI) has decreased from 22.4 before the project to 0.3 after the project. Production costs have been reduced by 35% without any loss of yield, except in a few specific circumstances.

8.8 Development of Collaborative Tools for Transfer

8.8.1 A Wide Range of Tools Available to Agricultural Professionals

The creation of a training course for professionals (growers, technicians and agricultural advisers) was an original and innovative idea issued from the Biophyto project. A University Certificate of Professional Qualification (UCPQ) entitled "Agroecological Crop Protection" is now available at the University Institute of Technology (University of Réunion) in partnership with CIRAD, The Chamber of Agriculture, Armeflhor, Fdgdon and Octroi (Laurent et al. 2015). The course took place in 2013–2018 and the UCPQ/ACP has already been awarded to some seventy candidates. This vocational training is now available at the IUT (University Institute of Technology) and the Réunion Chamber of Agriculture.

Information on the project is available through the website http://www.agriculture-biodiversite-oi.org managed by CIRAD as part of the Indian Ocean Regional Plant Protection Program.

A Newsletter was sent to more than 500 recipients to keep them informed of the project. Various communication events took place at local agricultural events and national seminars (e.g. EcophytoDom seminar, November 2013, Biophyto restitution seminar, October 2014, Casdar project restitution seminar, January 2017).

Various educational documents illustrating the project have been published: a guide to agroecological protection of mango trees (Vincenot et al. 2015); The UCPQ/ACP training pack (Laurent et al. 2015); Proceedings of the Biophyto seminar (Deguine et al. 2015b); A film, available on DVD and entitled "Biophyto, agroecological protection of mango trees in Réunion"; Eight posters illustrating the different stages of the project; Mirid identification cards (Atiama 2016). Most of these are available online at http://www.agriculture-biodiversite-oi.org/Biophyto.

8.8.2 Focus on a Diploma Course for Professionals in Agriculture

An innovative and original training module has been developed to assist knowledge transfer: the University Certificate of Professional Qualification in Agroecological Crop Protection (UCPQ/ACP). This is the first diploma course in France aimed at agricultural professionals who wish to acquire operational skills in this field. The objectives reflect those of an agroecological transition project taking into account the constraints of the different actors. The format, programme and organization of this course are the result of collaboration by a multidisciplinary team. It not only teaches knowledge and methods, but it is also a network for exchange and knowledge transfer for various and diverse audiences sometimes with different objectives and projects.

The teaching team consists of researchers, teacher-researchers, engineers, technicians and growers who have worked on the Biophyto project. The 36-hour program mixes theoretical and practical phases over 5 days. The content is initially disciplinary (ecology, entomology, agronomy) and progressive. Starting with these concepts, the trainees are given a transdisciplinary approach to ACP. The course also uses results of projects in Réunion. The field visits give trainees an opportunity to see the techniques for themselves and are an opportunity to meet growers using ACP on a daily basis (Fig. 8.18). At the end of the course, incentives and marketing issues are presented and discussed. The qualification is awarded if the candidate passes two assessments (written and oral). The course ends with an assessment of the classes. Technical points are tackled and the trainers and the trainees provide personalized advice. Finally, the trainees evaluate the quality and relevance of the course by individually and anonymously completing a questionnaire, the results of which will help the University to improve future sessions.

To date, 4 training sessions have been organized, with 50 trainees out of 51 graduating, mostly growers, technicians, agricultural advisers and engineers. The participants represent the diverse professions in the agricultural sectors. This mix is one of the conditions necessary for the success of this course because it allows the participants to enrich themselves through the experiences of others and to discuss crop protection issues, agroecology and its implementation. Growers represented the majority of participants (56.9%), followed by agricultural advisors and

Fig. 8.18 The field visits give trainees an opportunity to see the techniques for themselves and are an opportunity to meet growers using Agroecological Crop Protection on a daily basis. (Photo: D. Vincenot)

technicians (33.3%) and engineers (9.8%). It can be noted that production is represented in roughly equivalent proportion to research, advice and transfer. Men (62.7%) are more represented than women (37.3%). The average age of growers participating in the course (42.8 years ± 8.8) indicates a willingness to change production models after years of conventional agrochemical practices and an awareness of the situation these practices have put them in. They come from a variety of backgrounds: organic farming, rational farming, conventional agriculture, shelter crops, fruit crops, vegetables and cane, and even livestock breeding, often with a mix of crops. The surveys carried out at the end of the course show very high overall satisfaction rates (>90% for all sessions). From 2017, a refresher day was offered to allow all the previous trainees to learn about recent innovations in ACP and to discuss their experiences since completing the course. The University Certificate of Professional Qualification in Agroecological Crop Protection (UCPQ/ACP) will continue to be offered in Réunion until demands are satisfied and will soon be extended to the South-western region of the Indian Ocean in order to meet the needs of regional partners.

In conclusion, the modular UCPQ/ACP is a powerful training tool adapted to current needs in Agroecological Crop Protection. It is also a transfer tool because it offers a recognized operational qualification in innovative practices resulting from research and fieldwork. Finally, it is a powerful awareness and marketing tool because it brings various issues directly to the attention of professionals thanks to the diverse participants. These participants become valuable tools in spreading knowledge of ACP to their peers on the island. This type of flexible course can be adapted to and delivered in many different contexts.

8.8.3 Relays by Transfer Agencies

Since 2015, development agencies, notably the Déphy Ferme Mangue network, have taken over and have continued to train mango farms in new techniques. The Treatment Frequency Index decreased by 43% from 2012 to 2015 on all the farms in the Dephy Ferme Mango network (Réunion Chamber of Agriculture data, 2016). In 2016, this network consisted of 14 farms.

Also in 2015, a 'Biophyto Agri-Environment and Climate Measure' was launched and has been offered to growers, and is now open to other perennial fruit growers (e.g. citrus). In the same year, an association of growers supported by the Chamber of Agriculture applied for Economic and Environmental Interest Group status for its project to implement agroecological techniques in mango orchards. This initiative, by nine growers (some of whom participated in the Biophyto project) covers an area of 80 ha and demonstrates how the profession has adopted these new techniques. The Biophyto project, like any agroecological approach, opened up multiple avenues to be further explored by growers.

Since 2015, training days and one-off hands-on training sessions have been organized in the field. The Biophyto project has triggered a change in the sector's technicians as regards their individual and collective recommendations.

Finally, starting in the 2015–2016 academic year, the Réunion Chamber of Agriculture, in concertation with the University of Réunion, took over the coordination of the "UCPQ—Agroecological Crop Protection" course.

To sum up, the project also gave rise to a large number of transfer assistance tools, particularly in the field of professional training. Trained growers now have a basic knowledge of agroecosystem ecology, especially of mango orchards in Réunion.

8.9 Generic Keys for Agroecological Transition

It is worth pointing out how agroecological protection strategies for mango orchards in Réunion were developed. Although specific to Réunion, the following five keys are generic and are essential for the success of agroecological transition (Fig. 8.19). These five keys may overlap, but are fundamental to agroecological transition.

The transition from conventional agriculture to agroecology requires a significant change in techniques from agricultural professionals, particularly growers. In conventional agriculture, cropping systems generally aim to maximize yield and

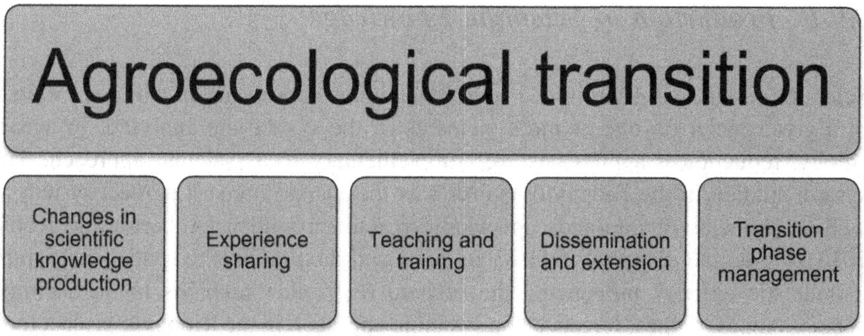

Agroecological transition

Changes in scientific knowledge production

Experience sharing

Teaching and training

Dissemination and extension

Transition phase management

Fig. 8.19 Agroecological transition cornerstones

economic returns for the farmer. Pesticides control pathogens, weeds and pests in a systematic way. In agroecology, from the outset, a prophylactic approach is necessary to manage biotic stresses. Once in place, Agroecological Crop Protection ensures an almost ontological resilience of agroecosystems, which simplifies crop management and especially pest control. One of the essential strategies is the diversification of crops in time and in space. This implies managing a higher level of complexity than specialized agroecosystems. Innovation therefore has a central role to play in agroecological transition (Ricci and Méssean 2015). It should be pointed out here that the term "innovation" refers to a set of practices rarely implemented in agroecosystems. They are sources of improvement and progress and integrate old or new technical solutions with generic characteristics, while presenting specificities related to the production in hand. Nethertheless, agroecology leads to difficulties in management of interstitial spaces, more frequent plot surveillance and high requirements in botanical, epidemiological and ecological knowledge. Moreover, the transition to agroecological farming can lead to doubts and uncertainties, and the link between practices and profits are not always clear.

The preparation and management of an agroecological project requires a collective design. This step is critical. Words and good intentions must be translated into actions. First, key players should be identified and federated. It would be wrong to develop innovative programs without involving, from the outset, actors such as growers, consumers, stakeholders in the sector and environmental associations as well as those in local development, research, agriculture and academia, as well as public authorities. The range of points of view makes it possible to enlarge the scope of the program and ensures solid and coherent implementation of the program, with favourable conditions for knowledge dissemination and appropriation. A collective formalization of objectives, as well as of the methods and means to be used to achieve the objectives, guarantees support between the various partners involved, which will remove many difficulties. In the Biophyto project, the design phase lasted one and a half years (mid-2010–2011) and knowledge sharing took place up to and beyond the end of the Biophyto project.

8.9.1 Production of Scientific Knowledge

Scientific research is often issued from reductionist approaches in which the effects of a given factor on one or more elements of the system are analysed. In agroecology, reductionist approaches are complementary to a systemic approach. The system studied in the Biophyto project was the mango orchard agroecosystem as well as elements of the landscape likely to interact with it (Aubertot and Robin 2013). Given the dispersal capacity of pests and auxiliaries, the spatial scale must include the territory influencing the orchard (e.g. other orchards in the vicinity, hedges, slopes, gullies, forests). The systemic approach in the Biophyto project took place via in situ experiments on agricultural land. The challenge was to embrace an ad hoc level of complexity to analyse the mechanisms underlying biotic regulation in mango orchards. Thus, all arthropod communities were inventoried, along with farming practices, with particular focus on the identification of plant communities present in orchards, as well as climate and the landscape. The collected datasets thus give a systemic representation of the agroecosystems under study.

Biophyto studies revealed the need for more knowledge of functional biodiversity in agroecosystems, including the impacts of agroecological practices and landscape characteristics (Jacquot et al. 2013). Catalogues of arthropods (spiders, parasitoids) and identification cards to aid in recognition of certain families of Miridae were produced during the Biophyto project (Atiama et al. 2016). Versions for growers are also available. The advent of agroecology will require research and development approaches to mobilize new knowledge. New biodiversity indicators will have to be taken into account in experiments and in agricultural field diagnoses. Consequently, agroecosystem function should be characterized by increasing the number of variables identified, through adapted observations and protocols or methodological innovations making it possible to quickly identify the different components of agroecosystems. Special attention should be paid to the impact of agricultural practices on biocenosis dynamics, since all practices, even those not directly related to crop protection, can impact biotic health (Zadoks 1993). Moreover, modeling must also be adapted to deal with higher levels of complexity.

8.9.2 Empirical Knowledge Sharing

In addition to scientific knowledge, empirical knowledge plays a major role in agroecological transition. Contrary to the established pattern where R&D would provide vertical references, tools and methods, agroecology feeds on horizontal transfers, i.e. knowledge acquired by growers without a scientific approach. Experience in a farming environment is essential. To do this, different tools or approaches can be mobilized: meetings between actors with discussions on individual experiences; Companion modeling (D'Aquino et al. 2002; Papazian et al. 2017); Qualitative modeling formalizing experts' proposals (Aubertot and Robin 2013); Elicitation of expertise (Knol et al. 2010). Meetings between growers and their advisors may take the form of technical guidance, if possible in the field, thematic meetings and transfer training. It is important to formalize growers' knowledge so that it can be shared and capitalized upon. Growers observe phenomena and infer hypotheses. The knowledge thus acquired is the object of various horizontal transfers, generally standardized, cumulative and shared. Growers' knowledge helped reconstruct bocage in western France. In addition, digital technology is a powerful tool for sharing and capitalizing on empirical knowledge (e.g. http://www.savoirfairepaysans.fr).

8.9.3 Training and Education

It is important to train those involved in agroecological transition including farmers, advisors and development technicians. Training is not only through involvement in participatory knowledge production and validation but also by spreading research and development work in the widest and most accessible way possible (Levidow et al. 2014). This knowledge was presented in concrete quantified form through various training courses, information and communication campaigns, and made it possible to transfer the results directly to the growers. Various transfer tools have been designed for field campaigns by development agencies (Réunion Chamber of Agriculture, other professional organizations) to promote organic farming techniques acquired during the Biophyto project.

In addition to further education, it is also important that CEAP concepts, methods and knowledge be taught at university. Agricultural education plays a key role in the success of the Réunion Plan for Sustainable Agriculture and Agri-Food (PRAAD), and more broadly, in the France Agroecology Project, as a mechanism for training those involved in farming. Agricultural education in Réunion, with its network of public and private agricultural establishments, facilitates knowledge transfer to current and future agricultural professionals in partnership with research and development institutes.

Training current and future growers in new practices is a major challenge: it involves activities like practical work, tutorials, mini-courses and hands-on that puts learners in close proximity to new knowledge. Demonstrations and information and exchange workshops in 2013 and 2015 in partnership with the agricultural sector made it possible to reach more people and create better links between stakeholders. In addition, a book entitled "Agroecological Crop Protection" (Deguine et al. 2017), a chapter of which is devoted to the Biophyto project (Gloanec et al. 2017), has been published.

The training program was co-designed between regional education and agricultural professionals to define the role of agricultural establishments in agroecological transition. Between 2014 and 2018, this program offers an action plan based on strengthened links between training, farming and the professional world to achieve ambitious technical objectives and to disseminate agroecological innovation to future and current agricultural professionals. This program has been assigned a Rita (innovation and agricultural transfer networks) status in order to guarantee strong links between training and research and development for more effective sharing of agroecological solutions to present and future agricultural professionals.

Thus, a new marketing and transfer program has been created for Réunion, in the form of a professional training module: the University Certificate of Professional Qualification in Agroecological Crop Protection (UCPQ/ACP). The module has a very broad scope. In addition to the diffusion of knowledge and methods, its raison d'être is to form the nucleus of a network, a place where projects are exchanged and mature and where professionals from different backgrounds and sometimes with different objectives may meet. The aim of the module is to support agricultural professionals who wish to adopt Agroecological Crop Protection. For the grower or agricultural technician, this transition requires learning afresh how to manage pest populations and preserve and promote beneficials. This implies a change in techniques and letting ecological processes do their job.

Agroecological transitions go hand-in-hand with digital transitions, and this mango pest management experience will feed the GASCON digital module (Agroecological Management of Crops and Harmful Organisms, Virtual Agroecology University (UVAE, http://www.ea.inra.fr/uvae). Developing digital training in agroecology should not be an exhaustive inventory of agroecological views or techniques. The aim is to build an educational resources for learners to acquire knowledge and autonomy to analyse the diverse knowledge available. Far from replacing face-to-face teaching and training, it is a tool that they can share, evaluate and embrace. Finally, working in partnership and collaboration in digital training projects where distance work is preferred will assist the progress of an integrated vision combining research, development and training. Although it has some benefits, distance self-training alone is not a substitute for face-to-face training, because of weak learner-trainer and learner-learner interaction.

8.9.4 Dissemination and Knowledge Transfer

In this agroecological mango experience, several tools were useful in transferring techniques to growers. To recap, the two main tools are the Dephy Ferme network at the national level and a socio-economic partnership, or GIEE at the local level. These facilitate cooperation for the transfer of innovations. In addition, Rita networks, created in 2011 at the initiative of the Ministry of Agriculture, Food and Forestry, are networks formed between major stakeholders in agricultural development in overseas departments, at the service of professionals (each intervening in its core business in a complementary fashion). The general aim of Rita is to assist in the socio-economic development of local production. In Réunion, three networks have been set up in sugarcane, fruit horticulture and livestock sectors. The challenge is to understand all the facets of a technical problem, and operational technical groups have been set up within Rita composed of researchers, experimenters, technicians and farmers working collaboratively. Within an agroecological framework, different thematic groups have been set up. This approach is innovative, voluntary and exemplary in terms of optimizing public and professional funds, to meet the needs in innovation and transfer for professionals and overseas agricultural and agri-food sectors. Rita has several objectives in Réunion: to promote the link between agriculture and science, taking market logic and economics into account; To promote the transfer of innovations via pilot farm networks and through education and professional training; To ensure that results are effectively converted into technical and economic itineraries; Finally, to strengthen co-operation within sectors and between sectors.

In order to turn innovation into transfer, a specific tool has been designed. In 2014, four research, training, experimentation and regulatory partners (CIRAD, Armeflhor, Anses, Fdgdon) joined forces in a Mixed Technology Unit called "Plant Health and Agroecological Production in the Tropical Environment", to pool their resources and activities in order to give added value and coherence to convert innovation to application. This vocation of this unit within the Rita movement is to better respond to the demands of professionals using an agroecological approach.

It should be emphasized that this is a broad community sharing the same program design. The campaigns need to be effectively coordinated. Of the many different partners involved in an agroecology project, transfer agencies are certainly the best placed to do this. Their relationship with growers, research, education and public authorities facilitate day-to-day program coordination. This is necessary for the proper management of campaigns on the ground.

8.9.5 Management of the Transition Period

The first few years after adapting new practices (elimination of pesticides, establishment of habitat management practices and conservation biological control) are the most difficult financially, given the time required to establish or re-establish ecological balances in orchards. This transition period should be accompanied by financial support or incentives.

During the agroecological mango experience, aside from the schemes promoted at operational level by transfer agencies and supported by public authorities (GIEE, the Dephy Ferme network, etc.), there are the following assistance schemes:

A regulation scheme to encourage transition to agroecological protection techniques. A Biophyto agri-environment and climate scheme was set up to encourage producers to become involved in Agroecological Crop Protection. These measures encourage farmers to protect and enhance the environment by providing remuneration for the provision of environmental services. Farmers undertake, for a minimum period of five years, to adopt environmentally friendly agricultural techniques that go beyond legal obligations. In exchange, they receive financial assistance to offset any additional costs and income losses resulting from the adoption of these practices.

- help for becoming organic. This agroecological experience on mango trees also contributed to the development of organic farming.
- support can be envisaged downstream for marketing a pesticide-free mango.

The adoption in France of "la loi d'avenir" on 13 October 2014 for agriculture, food and forestry now asserts that French agriculture and the agri-food and forestry sectors must rise to the challenge of competitiveness in order to remain competitive internationally and contribute to the development of production in France. This endeavour is part of the challenge of ecological transition. The agroecological project for France aims to put economic and environmental success at the heart of innovative agricultural practices. Public authorities therefore have an important role to play in encouraging the adoption of innovative and environmentally friendly agricultural practices, particularly in the early years of transition.

The government supports growers during the transition phase, but there are countless press articles and television reports highlighting the risks associated with residues in agricultural products. Consumers and citizens more generally are increasingly aware of the health issues associated with crop protection products. Consumers, through their purchasing choices, can therefore also support agroecological transition by choosing organic products or products issued from agroecosystems and sometimes, though not always, buy products with certain appearance defects. Organisations such as the Association for the maintenance of agriculture (Amap, http://www.reseau-amap.org) play an important role in getting consumers involved in agroecological transition. The Amap charter requires producers to respect the principles of agroecology without actually requiring organic certification. Similarly, the digital revolution has brought consumers and local producers together. Different initiatives already exist (e.g. http://laruchequiditoui.fr), allowing consumers to order products from local family-run farms or local artisans. Nevertheless, there are not always rules on the agricultural practices used. The consumer has to be cautious if he or she supports agroecological transition.

Finally, communication campaigns are essential and should not only be aimed at those involved in agriculture but also the general public. Different media can provide support during agroecological transition. Labels, certificates, or simply notices, compatible with logos already used on packaging (e.g. Organic) can help to give added value to new agroecosystems developed in certain sectors.

To summarize, after the Biophyto project and the follow-up studies, agricultural agencies supervised knowledge transfer. The government supported the development of agroecological farming during the transition from agrochemical to agroecological practices. Development tools such as demonstration plots (Dephy Ferme Mango), a Biophyto Agri-Environment and Climate Measure, and an Economic and Environmental Interest Group, have been set up.

8.10 Conclusion

This research into agroecological protection of mango orchards in Réunion, has yielded significant results. Growers have adopted agroecological practices such as vegetal cover in orchards and the elimination or near-elimination of insecticides and herbicides. The treatment frequency index (TFI) decreased from 22.4 before the project to 0.3 after the project and production costs were reduced by 35% without any loss of production except in a few specific situations. Tools and knowledge have been acquired on the ecological changes following the changes in practices, in particular in functional biodiversity.

These results show, for mango orchards in Réunion, that transition from agrochemical crop protection (with its limits and disadvantages) to agroecological protection, is possible. To our knowledge, this is the first large scale experience of its kind (i.e. in a farming environment, over several years) and represents a significant step towards demonstrating that Agroecological Crop Protection is viable for farming environments. It is now possible, based on the knowledge acquired on the role of biodiversity in agroecosystems, to propose generic management strategies which can then be adapted to local contexts and growers' strategies. The training tools developed assist in the transfer of innovative techniques and the socio-economic performance of the farms is not affected by the environmental, health and ecological added value of Agroecological Crop Protection.

Following on from the Gamour experience, the Réunion Biophyto project has opened up a new dynamic in agroecological attitudes towards mango and fruit production, not least with the growers themselves. There is a real interest in these new techniques, as well as a collective awareness of the limits of conventional agriculture and a desire to develop agroecological solutions for the production of other fruit and vegetables using the knowledge from the Biophyto transition experiments.

Acknowledgements The authors of the project wish to thank all mango growers who contributed to this agroecological experience. They were instrumental in informing the general public about the effectiveness of new techniques and the success of the project. Acknowledgments are also extended to all partners involved in the project, especially C. Ajaguin Soleyen, M. Atiama, A. Bailly, J. Brun-Vitelli, C. Cresson, B. Derepas, S. Dinnoo, P. Ferron, S. Gasnier, V. Gazzo, the Gros, K. Le Roux, E. Lucas, R. Michellon, M.-L. Moutoussamy, D. Muru, T. Nurbel, S. Plessix, T. Ramage, A. Reteau, M. Rousse, C. Schmitt, T. Schmitt, W. Suzanne, E. Tarnus, K. Técher and M. Tenailleau. In addition, our gratitude goes to the Ministry of Agriculture, Food and Forestry of the French Government, which enabled us to carry out the Biophyto project through the Trust Account for Agricultural and Rural Development (CASDAR-1138, CASDAR 2011) and the Biophytomang[2] project within the framework of Ecophyto 1. The Biophyto project has been approved by Qualitropic. This agroecological experience was co-financed by the European Union (European Regional Development Fund (ERDF) and Agricultural Fund for Rural Development (EAFRD)), the Conseil Régional de La Réunion, the Conseil Départemental de La Réunion and the Centre de Coopération internationale en Recherche Agronomique pour le Développement (CIRAD) to whom we extend our thanks.

References

Altieri M (1999) The ecological role of biodiversity in agroecosystem. Agric Ecosyst Environ 74:19–31

Amouroux P, Normand F (2013) Survey of mango pests in reunion island, with a focus on pests affecting flowering. Acta Hort 992:459–466

Atiama M (2016) Biéocologie et diversité génétique d'Orthops palus (Heteroptera, Miridae), ravageur du manguier à La Réunion. Université de La Réunion, Thèse de doctorat

Atiama M, Ramage T, Deguine J-P, Jacquot M (2016). Fiche de reconnaissance des punaises Mirides dans les vergers de manguiers à La Réunion. Chambre d'agriculture de La Réunion

Aubertot J-N, Robin M-H (2013) Injury profile SIMulator, a qualitative aggregative modelling framework to predict crop injury profile as a function of cropping practices, and the abiotic and biotic environment I. Conceptual Bases. PLoS ONE 8(9):e73202

Bellon S, Penvern S (2014) Organic Farming. Prototype for Sustainable Agricultures, Springer, Netherlands

Bianchi FJJA, Booij CJH, Tscharntke T (2006) Sustainable pest regulation in agricultural landscapes: a review on landscape composition, biodiversity and natural pest control. Proc Biol Sci 273:1715–1727. https://doi.org/10.1098/rspb.2006.3530

Bourguet D, Guillemaud T (2016) The hidden and external costs of pesticide use. Sustain Agric Rev 19:35–120. https://doi.org/10.1007/978-3-319-26777-7_2

Bouriga J, Graindorge R, (2016) Gestion des couverts végétaux en vergers de manguiers en conduite agroécologique. Rapport d'activité BIOPHYTOMANG[2], Armeflhor, Saint-Pierre

Brym ZT, Reeve JR (2016) Agroecological principles from a bibliographic analysis of the term agroecology. Sustain Agric Rev 19:203–231. https://doi.org/10.1007/978-3-319-26777-7_5

Brose U, Hillebrand H (2016) Biodiversity and ecosystem functioning in dynamic landscapes. Philos Trans R Soc B Biol Sci 371:20150267. https://doi.org/10.1098/rstb.2015.0267

Brun-Vitelli J (2015) Analyse des données du projet Biophyto à La Réunion. Mémoire de fin d'études, Master 2 Méthodes Statistiques des Industries Agronomiques Agro-Alimentaires et Pharmaceutiques, Université de Montpellier II

Cardinale B, Duffy E, Srivastava D et al (2009) Biodiversity, ecosystem functioning and Human wellbeing. In: Naeem S, Bunker DE, Hector A et al (eds) Biodiversity and human impacts, Oxford Bio. Oxford University Press, Oxford, pp 105–120

Cardinale BJ, Duffy JE, Gonzalez A, Hooper DU, Perrings C, Venail P, Narwani A, Mace GM, Tilman D, Wardle DA, Kinzig AP, Daily GC, Loreau M, Grace JB, Larigauderie A, Srivastava DS, Naeem S (2012) Biodiversity loss and its impact on humanity. Nature 489:326. https://doi.org/10.1038/nature11373

Carvalheiro LG, Seymour CL, Nicolson SW, Veldtman R (2012) Creating patches of native flowers facilitates crop pollination in large agricultural fields: Mango as a case study. J Appl Ecol 49:1373–1383. https://doi.org/10.1111/j.1365-2664.2012.02217.x

Carvalheiro LG, Seymour CL, Veldtman R, Nicolson SW (2010) Pollination services decline with distance from natural habitat even in biodiversity-rich areas. J Appl Ecol 47:810–820. https://doi.org/10.1111/j.1365-2664.2010.01829.x

Castagneyrol B, Jactel H, Vacher C, Brockerhoff EG, Koricheva J (2014) Effects of plant phylogenetic diversity on herbivory depend on herbivore specialization. J Appl Ecol 51:134–141. https://doi.org/10.1111/1365-2664.12175

Chaplin-Kramer R, O'Rourke ME, Blitzer EJ, Kremen C (2011) A meta-analysis of crop pest and natural enemy response to landscape complexity. Ecol Lett 14:922–932. https://doi.org/10.1111/j.1461-0248.2011.01642.x

Christensen NL, Bartuska AM, Brown JH, Carpenter S, D'Antonio C, Francis R, Peterson CH (1996) The report of the ecological society of America committee on the scientific basis for ecosystem management. Ecol Appl 6:665–691

Coll M, Wajnberg E (2017) Environmental Pest Management. Wiley, Challenges for Agronomists, Ecologists, Economists and Policy makers

Costanza R, Arge R, de Groot R, Faber S, Grasso M, Hannon B, Limburg K, Naeem S, O'Neill RV, Paruelo J, Raskin RG, Sutton P, van den Belt M (1997) The value of the world's ecosystem services and natural capital. Nature 387:253–260. https://doi.org/10.1038/387253a0

Crowder DW, Northfield TD, Strand MR, Snyder WE (2010) Organic agriculture promotes evenness and natural pest control. Nature 466:109–112. https://doi.org/10.1038/nature09183

D'Aquino P, Le Page C, Bousquet F, Bah A (2002) A novel mediating participatory modelling: the "self-design" process to accompany collective decision making. Int. J. Agricultural Resources Gov Ecol 12:59–74

Darnhofer I, Lindenthal T, Bartel-Kratochvil R, Zollitsch W (2010) Conventionalisation of organic farming practices: from structural criteria towards an assessment based on organic principles. A review. Agron Sustain Dev 30:67–81

Dassou AG, Tixier P (2016) Response of pest control by generalist predators to local-scale plant diversity: a meta-analysis. Ecol Evol 6:1143–1153. https://doi.org/10.1002/ece3.1917

David C, Wezel A (2012) Agroecology and the food system. In: Lichtfouse E (ed) Agroecology and strategies for climate change, vol 8. Sustain Agr Rev. Springer, Dordrecht, pp 17–34

Deguine J-P, Atiama-Nurbel T, Aubertot J-N, Augusseau X, Atiama M, Jacquot M, Reynaud B (2015a) Agroecological cucurbit-infesting fruit fly management: a review. Agr Sust Dev 35:937–965. https://doi.org/10.1007/s13593-015-0290-5

Deguine J-P, Ferron P, Russell D (2009) Crop Protection: from Agrochemistry to Agroecology. Science Publishers, Enfield, NH, USA

Deguine J-P, Gloanec C, Laurent P, Ratnadass A, Aubertot J-N (2016) Protection agroécologique des cultures. Quae, Versailles

Deguine J-P, Gloanec C, Laurent P, Ratnadass A, Aubertot J-N (2017) Agroecological Crop Protection. Springer, ISBN 978-94-024-1184-3, https://doi.org/10.1007/978-94-024-1185-0

Deguine J-P, Gloanec C, Schmitt T (2015b) Biodiversité et protection agroécologique des cultures. Actes du Séminaire Biophyto, Saint-Pierre, La Réunion. 21–24 octobre 2014. Saint-Denis: Chambre d'agriculture de La Réunion, ISBN: 978-2-87614-704-1

Diaz S, Fargione J, Chapin FS, Tilman D (2006) Biodiversity loss threatens human well-being. PLoS Biol 4:1300–1305. https://doi.org/10.1371/journal.pbio.0040277

Doré T, Makowski D, Malézieux E, Munier-Jolain N, Tchamitchian M, Tittonell P (2011) Facing up to the paradigm of ecological intensification in agronomy: revisiting methods, concepts and knowledge. Eur J Agron 34(4):197–210

Duffy JE, Cardinale BJ, France KE, McIntypre PB, Thébault E, Loreau M (2007) The functional role of biodiversity in ecosystems: incorporating trophic complexity. Ecol Lett 10:522–538. https://doi.org/10.1111/j.1461-0248.2007.01037.x

Eveleigh ES, McCann KS, McCarthy PC, Pollock SJ, Lucarotti CJ, Morin B, McDougall GA, Strongman DB, Huber JT, Umbanhowar J, Faria DB (2007) Fluctuations in density of an outbreak species drive diversity cascades in food webs. Proc Natl Acad Sci USA 104:16976–16981. https://doi.org/10.1073/pnas.0704301104

Ferron P, Deguine J-P (2005) Crop protection, biological control, habitat management and integrated farming. Agr. Sust. Dev. 25:17–24

Foley JA, Defries R, Asner GP, BarfordC Bonan G, Carpenter SR, Chapin FS, Coe MT, Daily GC, Gibbs HK, Helkowski JH, Holloway T, Howard EA, Kucharik CJ, Monfreda C, Patz JA,

Prentice IC, Ramankutty N, Snyder PK (2005) Global consequences of land use. Science 309:570–574. https://doi.org/10.1126/science.1111772
Food and Agriculture Organization (2015) FAOSTAT http://www.fao.org/faostat/en/#data/QC
Gaba S, Bretagnolle F, Rigaud T, Philippot L (2014) Managing biotic interactions for ecological intensification of agroecosystems. Front Ecol Evol 2:1–9. https://doi.org/10.3389/fevo.2014.00029
Geertsema W, Rossing WAH, Landis DA, Bianchi FJJA, van Rijn PCJ, Schaminée JHJ, Tscharntke T, van der Werf W (2016) Actionable knowledge for ecological intensification of agriculture. Front Ecol Environ 14:209–216. https://doi.org/10.1002/fee.1258
Gloanec C (2015) Outils et enjeux de la coordination d'un projet partenarial. Outils d'évaluation et observatoire des impacts. In: Deguine J-P, Gloanec C, Schmitt T (eds) Biodiversité et protection agroécologique des cultures. Actes du Séminaire Biophyto, Saint-Pierre, La Réunion. 21–24 octobre 2014. Saint-Denis: Chambre d'agriculture de La Réunion, 216 p. ISBN: 978-2-87614-704-1, pp 12–21
Gloanec C, Deguine J-P, Vincenot D, Jacquot M, Graindorge R (2017) Application in fruit crops: the biophyto experience. In: Deguine J-P, Gloanec C, Laurent P, Ratnadass A, Aubertot J-N (eds) Agroecological crop protection. Springer, ISBN 978-94-024-1184-3, https://doi.org/10.1007/978-94-024-1185-0
Gloanec C, Guignard I (2015) Impact des pratiques phytosanitaires et culturales. In: Deguine J-P, Gloanec C, Schmitt T (eds). Biodiversité et protection agroécologique des cultures. Actes du Séminaire Biophyto, Saint-Pierre, La Réunion. 21–24 octobre 2014. Saint-Denis: Chambre d'agriculture de La Réunion, 216 p. ISBN: 978-2-87614-704-1, pp 60-65
Gurr GM, Scarratt SL, Wratten SD, Berndt L, Irvin N (2004) Ecological engineering, habitat manipulation and pest management. In: Gurr GM, Wratten SD, Altieri MA (eds) Ecological Engineering for Pest Management. CSIRA Publishing, CABI Publishing, Advances in Habitat Manipulation for Arthropods, pp 1–12
Haddad NM, Crutsinger GM, Gross K, Haarstad J, Knops JMH, Tilman D (2009) Plant species loss decreases arthropod diversity and shifts trophic structure. Ecol Lett 12:1029–1039. https://doi.org/10.1111/j.1461-0248.2009.01356.x
Hairston NG, Smith FE, Slobodkin LB (1960) Community structure, population control, and competition. Am Nat 94:421–425
Henri DC, Jones O, Tsiattalos A, Thebault E, Seymour CL, van Veen FJF (2015) Natural vegetation benefits synergistic control of the three main insect and pathogen pests of a fruit crop in southern Africa. J Appl Ecol 52:1092–1101. https://doi.org/10.1111/1365-2664.12465
Hertzog LR, Meyer ST, Weisser WW, Ebeling A (2016) Experimental Manipulation of grassland plant diversity induces complex shifts in aboveground arthropod diversity. PLoS ONE 11: e0148768. https://doi.org/10.1371/journal.pone.0148768
Hogg BN, Daane KM (2011) Diversity and invasion within a predator community: impacts on herbivore suppression. J Appl Ecol 48:453–461. https://doi.org/10.1111/j.1365-2664.2010.01940.x
Holway DA, Lach L, Suarez AV, Tsutsui ND, Caseal TJ (2002) The causes and consequences of ant invasions. Annu Rev Ecol Syst 33:181–233. https://doi.org/10.1146/annurev.ecolsys.33.010802.150444
Hooper DU, Adair EC, Cardinale BJ, Byrnes JEK, Hungate BA, Matulich KL, Gonzalez A, Duffy JE, Gamfeldt L, O'Connor MI (2012) A global synthesis reveals biodiversity loss as a major driver of ecosystem change. Nature 486:105–108. https://doi.org/10.1038/nature11118
Hooper DU, Chapin FS, Ewell JJ, Hector A, Inchausti P, Lavorel S, Lawton JH, Lodge DM, Loreau M, Naeem S, Schmid B, Seta H, Symstad A, Andermeer JV, Werdle DA (2005) Effects of biodiversity on ecosystem functioning: a consensus of current knowledge. Ecol Monogr 75:3–35. https://doi.org/10.1890/04-0922
Iverson AL, Marin LE, Ennis KK, Gonthier DJ, Connor-Barrie BT, Remfert JL, Cardinale BJ, Perfecto I (2014) Do polycultures promote win-wins or trade-offs in agricultural ecosystem services? A meta-analysis. J Appl Ecol 51:1593–1602. https://doi.org/10.1111/1365-2664.12334

Jacquot M (2016) Biodiversité et fonctionnement écologique des agroécosystèmes à base de manguiers à La Réunion. Université de La Réunion, Thèse de doctorat

Jacquot M, Derepas B, Deguine J-P (2016) Seven newly recorded species and families of spiders from Réunion (Malagasy region) (Araneae, Araneomorphae). Bull Soc Entomol Fran 121:421–430

Jacquot M, Massol F, Muru D, Derepas B, Tixier P, Deguine J-P (in prep) Arthropod diversity is governed by bottom-up and top-down forces in a tropical agroecosystem

Jacquot M, Tenailleau M, Deguine J-P (2013) La biodiversité fonctionnelle dans les vergers de manguiers à La Réunion. Effets de facteurs écosystémiques et paysagers sur les arthropodes prédateurs terrestres. Inn Agr 32:365–376

Jacquot M, Tixier P, Flores O, Muru D, Massol F, Derepas B, Chiroleu F, Deguine J-P (2017) Contrasting predation services of predator and omnivore diversity mediated by invasive ants in a tropical agroecosystem. Basic Appl Ecol. https://doi.org/10.1016/j.baae.2016.09.005

Kenis M, Auger-Rozenberg M-A, Roques A, Timms L, Péré C, Cock MJW, Settele J, Augustin S, Lopez-Vaamonde C (2009) Ecological effects of invasive alien insects. Biol Invasions 11:21–45. https://doi.org/10.1007/s10530-008-9318-y

Knol AB, Slottje P, van der Sluijs JP, Lebret E (2010) The use of expert elicitation in environmental health impact assessment: a seven step procedure. Environmental Health 9-19

Laurent P, Deguine J-P, Graindorge R, Jacquot M, Roux E, Rossolin G, Técher K, Vincenot D, Gloanec C (2015) Le Certificat Universitaire de Qualification Professionnell, une formation diplômante adaptable et adaptée aux enjeux de la protection agroécologique des cultures. In: Deguine J-P, Gloanec C, Schmitt T (eds) Biodiversité et protection agroécologique des cultures. Actes du Séminaire Biophyto, Saint-Pierre, La Réunion. 21–24 octobre 2014. Saint-Denis, Chambre d'agriculture de La Réunion, 216 p. ISBN: 978-2-87614-704-1, pp 203-205

Levidow L, Pimbert G, Vanlocqueren G (2014) Agroecological research: conforming or transforming the dominant agro-food regime? Agroecology and sustainable food systems 38:1127–1155

L'Hermier Des Plantes H, (1976) Structuration des tableaux à trois indices de la Statistique: Théorie et Application D'une Méthode D'analyse Conjointe. Université des sciences et techniques du Languedoc

Loreau M, Naeem S, Inchausti P, Bengtsson J, Grime JP, Hector A, Hooper DU, Huston MA, Raffaelli D, Schmid B, Tilman D, Wardle DA (2001) Biodiversity and ecosystem functioning: current knowledge and future challenges. Science 294:804–808. https://doi.org/10.1126/science.1064088

Lowe S, Browne M, Boudjelas S, De Poorter M (2000) 100 of the world's worst invasive alien species. A selection from the global invasive species database. Invasive Species Spec Gr a Spec Gr Species Surviv Comm World Conserv Union 12. https://doi.org/10.1614/wt-04-126.1

Mace GM, Norris K, Fitter AH (2012) Biodiversity and ecosystem services: a multilayered relationship. Trends Ecol Evol 27:19–25. https://doi.org/10.1016/j.tree.2011.08.006

Magurran AE (2004) Introduction: Measurement of (Biological) Diversity. Blackwell Publishing, Oxford

Morin PJ (2011) Community Ecology. Wiley-Blackwell, Chichester

Murphy GEP, Romanuk TN (2014) A meta-analysis of declines in local species richness from human disturbances. Ecol Evol 4:91–103. https://doi.org/10.1002/ece3.909

Muru D, Madl M, Jacquot M, Deguine J-P (2017) A literature-based review of hymenopteran Parasitica from Reunion Island. ZooKeys 652: 55–128. https://doi.org/10.3897/zookeys.652.10729

Myers N, Mittermeier RA, Mittermeier CG, Da Fonseca GA, Kent J (2000) Biodiversity hotspots for conservation priorities. Nature 403:853–858

Normand F, Jessu D, Sinatamby M, Carissimo L, Champarvier K (2015) Changement de pratique d'irrigation lié à une conduite agroécologique du manguier: effets sur le bilan hydrique et la production. Biodiversité et protection agroécologique des cultures. In: Deguine J.-P, Gloanec C, Schmitt T (eds) Biodiversité et protection agroécologique des cultures. Actes du

Séminaire Biophyto, Saint-Pierre, La Réunion. 21–24 octobre 2014. Saint-Denis: Chambre d'agriculture de La Réunion, 216 p. ISBN: 978-2-87614-704-1, pp 39-43

Normand F, Michels T, Lechaudel M, Joas J, Vincenot D, Hoarau I, Desmulier X, Barc G (2011) Approche intégrée de la filière mangue à La Réunion. Inn Agr 17:67–81

Noss RF (1990) Indicators for monitoring biodiversity: a hierarchical approach. Conserv Biol 4:355–364. https://doi.org/10.1111/j.1523-1739.1990.tb00309.x

Nuñez-Elisea R, Davenport TL (1994) Flowering of mango trees in containers as influenced by seasonal temperature and water stress. Scientia Hort 126:65–72

Offenberg J (2015) Ants as tools in sustainable agriculture. J Appl Ecol 52:1197–1205. https://doi.org/10.1111/1365-2664.12496

Oksanen L, Fretwell SD, Arruda J, Niemela P (1981) Exploitation ecosystems in gradients of primary productivity. Amer Nat 118:240–261. https://doi.org/10.1086/283817

Papazian H, Bousquet F, Antona M, d'Aquino P (2017) A stakeholder-oriented framework to consider the plurality of land policy integration in Sahel. Ecol Econ 132:155–168

Pimentel D, Stachow U, Takacs DA, Brubaker HW, Dumas AR, Meaney JJ, O'Neil JAS, Onsi DE, Corzilius DB (1992) Conserving biological diversity in agricultural forestry systems - most biological diversity exists in human-managed ecosystems. Bioscience 42:354–362

Ramankutty N, Evan AT, Monfreda C, Foley JA (2008) Farming the planet: 1. Geographic distribution of global agricultural lands in the year 2000. Global Biogeochem Cycles 22:1–19. https://doi.org/10.1029/2007GB002952

Reddy PP (2017) Agro-ecological Approaches to Pest Management for Sustainable Agriculture. Springer Nature Singapore, ISBN 978-981-10-4324-6

Ricci P, Méssean A (2015) Stratégies intégratives et innovations systémiques: sortir du cadre. Inn Agr 46:147–155

Root RB (1973) Organization of a plant-arthropod association in simple and diverse habitats: the fauna of collards (Brassica Oleracea). Ecol Monogr 43:95–124

Rusch A, Valantin-Morison M, Sarthou JP, Roger-Estrade J (2010) Biological control of insect pests in agroecosystems. Effects of crop management, farming systems, and seminatural habitats at the landscape scale: a review. Elsevier Ltd

Sabatier R, Vivien M, Reynès C (2013) Une nouvelle proposition, l'Analyse Discriminante Multitableaux: STATIS-LDA. J Soc Fr Stat 154:31–43

Scherber C, Eisenhauer N, Weisser WW, Schmid Voigt W, Fischer M, Schulze ED, Roscher C, Weigelt A, Allan E, Beßler H, Bonkowski M, Buchmann N, Buscot F, Clement LW, Ebeling A, Engels C, Halle S, Kertscher I, Klein AM, Koller R, König S, Kowalski E, Kummer V, Kuu A (2010) Bottom-up effects of plant diversity on multitrophic interactions in a biodiversity experiment. Nature 468:553–556. https://doi.org/10.1038/nature09492

Simon S, Bouvier JC, Debras JF, Sauphanor B (2010) Biodiversity and pest management in orchard systems. Agron Sustain Dev 30:139–152. https://doi.org/10.1007/978-94-007-0394-0_30

Simon S, Lesueur-Jannoyer M, Plénet D, Lauri P-E, Le Bellec F (2016) Methodology to design agroecological orchards: learnings from on-station and on-farm experiences. Eur. J. Agr. 82:320–330. https://doi.org/10.1016/j.eja.2016.09.004

Snyder WE, Snyder GB, Finke DL, Straub CS (2006) Predator biodiversity strengthens herbivore suppression. Ecol Lett 9:789–796. https://doi.org/10.1111/j.1461-0248.2006.00922.x

Suarez AV, McGlynn TP, Tsutsui ND (2010) Biogeographic and taxonomic patterns of introduced ants. In: Lach L, Parr CL, Abbott KL (eds) Ant Ecology. Oxford University Press, Oxford, pp 233–244

Técher K, Dijoux A, Gloanec C, Danflous J-P, Guignard I (2015) La question de la valorisation commerciale pour une culture de mangue en protection agroécologique des cultures. In: Deguine J-P, Gloanec C, Schmitt T (eds) Biodiversité et protection agroécologique des cultures. Actes du Séminaire Biophyto, Saint-Pierre, La Réunion. 21–24 octobre 2014. Saint-Denis: Chambre d'agriculture de La Réunion, 216 p. ISBN: 978-2-87614-704-1, p 25

Tittonell P, Klerkx L, Baudron F, Félix GF, Ruggia A, van Apeldoorn D, Dogliotti D, Mapfumo P, Rossing WAH (2016) Ecological intensification: local innovation to address global challenges. Sustain Agric Rev 19:1–34. https://doi.org/10.1007/978-3-319-26777-7_1

Tscharntke T, Klein AM, Kruess A, Steffan-Dewenter I, Thies C (2005) Landscape perspectives on agricultural intensification and biodiversity - Ecosystem service management. Ecol Lett 8:857–874. https://doi.org/10.1111/j.1461-0248.2005.00782.x

Varmuza K, Filzmoser P (2009) Introduction to Multivariate Statistical Analysis in Chemometrics. CRC press

Vincenot D, Amouroux P, Hoarau I, Joas J, Léchaudel M, Michels T, Normand F (2009) Guide de Production intégrée de mangues à la Réunion. Chambre d'agriculture de La Réunion, Saint-Denis

Vincenot D, Deguine J-P, Gloanec C, Dijoux A, Graindorge R (2015) Initiation à la protection agroécologique du manguier à La Réunion. Retour d'expérience - Projet Biophyto 2012–2014. Chambre d'agriculture de La Réunion, Saint-Denis, 56 p. ISBN: 978-2-87614-705-8

Vivien M, Sabatier R (2001) Une extension multi-tableaux de la régression PLS. Rev Stat Appl 49:31–54

Veres A, Petit S, Conord C, Lavigne C (2013) Does landscape composition affect pest abundance and their control by natural enemies? A review. Agric Ecosyst Environ 166:110–117. https://doi.org/10.1016/j.agee.2011.05.027

Wetterer JK (2010) Worldwide spread of the tropical fire ant, Solenopsis geminata (Hymenoptera: Formicidae). Myrmecological News 14:21–35

Wezel A, Bellon S, Doré T, Francis C, Vallod D, David C (2009) Agroecology as a science, a movement, and a practice: a review. Agron Sustain Dev 29:503–515

Zadoks JC (1993) Cultural methods. In: Zadoks JC (ed) Modern crop protection: development and perspectives. Wageningen Press Wageningen, The Netherland, pp 161–170

Chapter 9
Drought and Agricultural Ecosystem Services in Developing Countries

Marzieh Keshavarz and Ezatollah Karami

Abstract Agricultural system serves as an important source of provisioning, regulating, supporting and cultural ecosystem services. However, increased occurrence of drought has reduced ecosystem services provided by agriculture. Climate change is also projected to reduce essential ecosystem services, especially in developing countries. In order to mitigate the negative impacts of climate change on farming systems, it is necessary to improve our understanding about ecosystem services and disservices of agriculture and clarify the physical and human factors that drive the flow of ecosystem services, in developing countries. Since the flows of ecosystem services and disservices rely on how agricultural ecosystems are managed, it is also crucial to gain insight into transition of agricultural systems from conventional to multifunctional production systems. Here, since drought is one of the main drivers of change in ecosystem services, droughts incidence in developing world and their impacts on provisioning and non-provisioning services of agriculture are presented. We explain that the capacity of agricultural systems to provide ecosystem services under drought is only one part of the framing for services equation. The other parts are farmers' willingness to provide additional non-provisioning ecosystem services and their ecosystem-based adaptation to drought. Furthermore, various strategies that are already used to protect soil and water resources or deliver environmental flows during drought are outlined. However, there are some key limitations regarding ecosystem-based adaptation practices, which can hamper their adoption by smallholder farmers, especially in developing countries. Thus, appropriate proactive drought management policies are imperative to facilitate adoption of drought resilient ecosystem based agriculture.

M. Keshavarz
Department of Agriculture, Payame Noor University, P.O. Box 19395-3697, Tehran, Iran

E. Karami (✉)
Department of Agricultural Extension and Education, College of Agriculture, Shiraz University, Shiraz, Iran
e-mail: ekarami@shirazu.ac.ir

S. Gaba et al. (eds.), *Sustainable Agriculture Reviews 28*, Ecology for Agriculture 28,
https://doi.org/10.1007/978-3-319-90309-5_9

Keywords Agriculture · Ecosystem services · Drought · Climate change
Developing countries · Multifunctional agriculture · Smallholder farmers
Adaptation · Adaptive governance · Resilience

9.1 Introduction

While agriculture is inherently sensitive to the vagaries of weather, it is now exposed to unprecedented changes due to climate change, particularly in arid and semi-arid regions. Climate change is projected to increase the frequency, intensity and duration of droughts in arid and semi-arid areas (IPCC 2014b). Therefore, nearly all arid and semi-arid regions are expected to experience significant water stress. Moreover, extreme weather events are expected to put further pressure on natural resources and severely reduce productivity of crops and livestock. Climate change is identified as one of the most important drivers of change in ecosystems and their services to agriculture (MEA 2005). The most serious impacts such as watershed changes, pollution and water resource depletion have been observed in regions under intensive agriculture and areas with water scarcity (Davis et al. 2015; Hrudey and Hrudey 2004; Keshavarz et al. 2013; Maleksaeidi and Karami 2013). The intensification of agriculture has put pressure on natural ecosystems through increased demand for water, food, fiber, minerals and energy (Davis et al. 2015). These demands are the result of rising world population that is predicted to reach 9.7 billion people by 2050 (UNESA 2015), promoted income, greater global focus on economic growth and changing dietary preferences (MEA 2005). While global food production is required to increase by 70% by 2050 (Steduto et al. 2012), reduction of global water availability along with climate change is likely to constrain this production.

Agriculture is the main source of livelihood for more than 2.5 billion rural residents of the developing world (World Bank 2016). A significant portion of the developing countries' farmlands are small, the proportion of which ranges from 62% of Africa to 85% of Asia lands (Vignola et al. 2015). Though most small-scale landholders face various environmental constraints and rapid economic changes (Godfray et al. 2010), they provide more than 80% of the food consumed in the developing countries (IFAD 2013). Water is important in production and an essential factor to sustain livelihoods (Sullivan 2011). Therefore, direct and indirect competition for water resources to irrigate intensifies the impact of climate variability (Davis et al. 2015) and increases vulnerability to climate change. As smallholder farmers have poor access to natural, financial, physical and institutional resources (Harvey et al. 2014; Keshavarz et al. 2013, 2014), they are disproportionately vulnerable to climate change (Vignola et al. 2015).

To reduce the adverse impacts of extreme weather events, effective management of smallholding agricultural systems and natural ecosystems are imperative (Keshavarz et al. 2014; Robertson et al. 2014). However, smallholder farmers in developing countries have limited capacity to adapt to climate change. Although many smallholder farmers have experience in coping with a certain degree of uncertainty (Keshavarz and Karami 2014), most farmers are ill-prepared for the challenge of adapting to climate change induced hazards (Vignola et al. 2015). This low level of adaptation to climate change will have major socio-economic and ecological implications for smallholder farmers and agricultural systems in developing countries.

In this chapter, first ecosystem services and disservices of agriculture are described. Since the flows of ecosystem services and disservices rely on how agricultural ecosystems are managed (Zhang et al. 2007), transition of agricultural systems from conventional to multifunctional production systems is explained. Also, the main ecosystem services and disservices of different forms of agriculture are introduced. After that, the direct and indirect drivers of ecosystem services of agriculture are expressed. These initial sections provide an understanding of the importance of agricultural systems in ecosystem services flow and attempt to identify the physical and human factors that drive generation and evolution of ecosystem services. Since drought is one of the main drivers of change in ecosystem services, in the next step, drought incidence in developing world and its impacts on ecosystem services of agriculture are described. The focus then shifts to explaining farmers' willingness to provide ecosystem services and their ecosystem-based adaptations to drought. Finally, several issues in regard to institutional management of ecosystem services under drought are discussed and some recommendations are offered to make ecosystem based agricultural systems resilient to drought.

9.2 Ecosystem Services and Disservices from Agriculture

Ecosystem services are defined as the benefits that are provided by nature for human populations and society (Costanza et al. 1997). While ecosystem services depend on context, time and space (Zagonari 2016), the Millennium Ecosystem Assessment (2005) recognized four categories of services, all of which directly or indirectly support human survival, health and well-being, through the world (Costanza et al. 1997; Zhang et al. 2017). These categories include provisioning, supporting, regulating and cultural services.

Provisioning services include the work of nature that supplies food, fresh water, fiber, fuel and other raw materials. Supporting services comprise the ecological processes necessary for production of other ecosystem services. Soil retention and formation, biodiversity protection and disturbance regimes are recognized as supporting services. Regulating services consist of eco-physiological functions and ecosystem processes that are imperative for maintenance of the functioning of ecosystems and they regulate the production of provisioning services. These processes regulate climate, flood and disease, provide clean water and maintain biodiversity. Cultural services, represent the non-material benefits and non-consumptive use values of nature, such as aesthetic landscapes, recreation and spiritual reflection (MEA 2005).

Ecosystem services intersect with agriculture in three critical ways (Dale and Polasky 2007). First, agricultural systems provide many provisioning, regulating, supporting and cultural services for people and communities (Fig. 9.1). Second, agriculture needs various ecosystem services as inputs for production, especially soil fertility and pest control (Zhang et al. 2007). Finally, agriculture influences the quality and quantity of ecosystem services that other ecosystems, i.e. forests and estuaries, can provide (Dale and Polasky 2007). The following subsections focus on describing the ecosystem services and disservices of agricultural production systems.

9.2.1 Provisioning Services

Undoubtedly, the key ecosystem service of agriculture is provision of food, fiber and fuel and agricultural ecosystems are generally managed to optimize provision of grain, livestock, fuel, forage and other products which are used to meet subsistence needs. For instance, over the past 55 years, production of cereals- the major energy component of human diets- and meat has been increased by 369.7% and

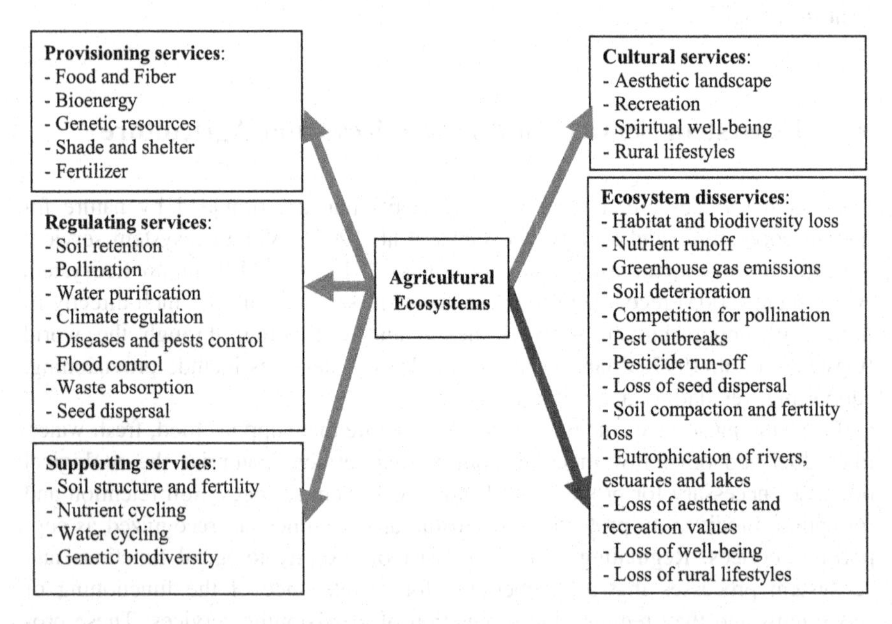

Fig. 9.1 Relationships between agricultural ecosystems and selected ecosystem services; note that green and red arrows indicate ecosystem services, i.e. provisioning, regulating, supporting and cultural services and disservices of agricultural systems, respectively (Dale and Polasky 2007; Kragt and Robertson 2014; Swinton et al. 2007; Tilman et al. 2002; Zhang et al. 2007)

994.6%, respectively, in developing countries (Fig. 9.2). During this period, intensification of cultivated systems, e.g. greater inputs of fertilizer, water and pesticides, new crop strains and other technologies of the 'Green Revolution' (Tilman et al. 2002), has been the primary source of increased yield. However, some developing countries, predominantly found in sub-Saharan Africa, have continuously relied on expansion of harvested land. For example, over the period of 1961–1999, expansion of cultivated area contributed only 29% to growth in crop production of developing countries, while 71% of increased yields in sub-Saharan Africa attributed to land area expansion (MEA 2005).

In addition, a wide variety of crops are used for fiber production such as cotton, flax, hemp and jute. Among these, cotton production has increased more than fourfold in developing countries, yet the land area on which cotton is cultivated has stayed virtually the same (FAOSTAT 2016) indicating that there has been substantial intensification in the crops production. Furthermore, as indicated in Fig. 9.2, the developing countries' timber harvest has increased by 137.3% since 1960 and wood pulp production has increased more than threefold during this period (FAOSTAT 2016). Also over the past five decades, production of charcoal has more than doubled in developing countries (Fig. 9.2). In the process of

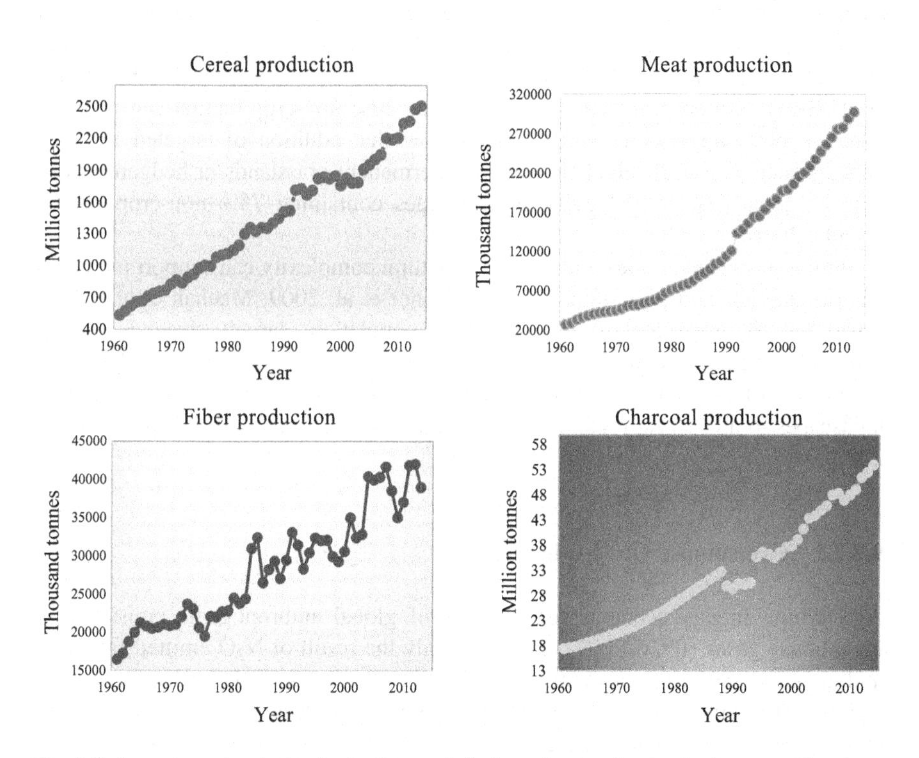

Fig. 9.2 Long-term trends in food, fiber and fuel production in developing countries (own representation based on FAOSTAT 2016)

optimization, agricultural ecosystems depend on a wide variety of regulating and supporting services that determine their biophysical capacity to provide food, fiber and fuel (Zhang et al. 2007).

9.2.2 Regulating Services

Regulating services are the most diverse group of services provided by agriculture (Fig. 9.1). Agricultural ecosystems can regulate the population dynamics of pests, pathogens and pollinators, in addition to fluctuations in levels of soil loss, water quantity and quality, greenhouse emissions and carbon sequestration (Swinton et al. 2007).

9.2.2.1 Pest Control

Multiple studies have examined the relationship between landscape structure and natural enemy populations as well as between predator communities and pest suppression (Armbrecht and Perfecto 2003; Chaplin-Kramer et al. 2011; Perfecto et al. 2004; Philpott et al. 2009; Shackelford et al. 2013; Van Bael et al. 2008; Veres et al. 2013). Estimates indicate that as little as 20% non-crop habitat can preserve effective pest suppression, while others show that addition of targeted resource habitats such as woodlands, field margins, permanent grasslands or hedgerows can improve local pest control even in landscapes containing 75% non-crop habitat (Landis 2016).

Furthermore, landscapes with higher structural complexity can support increased natural enemies and pest suppression (Gardiner et al. 2009; Meehan et al. 2011). Therefore, the presence of natural enemy populations largely depends on the availability of appropriate habitat and prey during periods of the year when crop pests are not available. In order to deliver this service, managing agricultural landscapes to allow pest regulation is imperative (Rusch et al. 2016; Swinton et al. 2007).

9.2.2.2 Greenhouse Gas Mitigation

Agriculture directly accounts for 12–14% of global anthropogenic emissions of greenhouse gases (IPCC 2007). This is mainly the result of N_2O emitted from soil and manure and from methane emitted by ruminant animals and burned crop residues (Robertson et al. 2014). Considering the greenhouse gas costs of agricultural expansion, agronomic inputs, e.g. fertilizers and pesticides, and post-harvest activities, e.g. food processing, transport and refrigeration, agriculture accounts for 26–36% of all anthropogenic greenhouse gas emissions (Barker 2007). However, agricultural practices can effectively reduce or offset greenhouse gas emissions through a variety of processes (Smith et al. 2008). For example, replacing synthetic nitrogen fertilizers with biological nitrogen fixation by legumes can significantly reduce CO_2 emissions from agricultural productions. Effective manure management can also reduce emissions from animal waste.

Various agricultural management practices have different effects on greenhouse gas mitigation may result in trade-offs, e.g. no-till energy savings versus the carbon cost of additional herbicides. They may also be synergistic, e.g. leguminous cover crops in the biologically based systems not only increase soil carbon storage but also reduce the CO_2 emissions of manufactured fertilizer nitrogen (Robertson et al. 2014). Designing optimal agricultural systems is not difficult. For instance, soil conservation practices such as no-till cultivation and conservation tillage can save soil carbon. Crop rotations and cover crops can also significantly reduce degradation of subsurface carbon (Power 2010).

Moreover, production of cellulosic biofuels can mitigate some portions of agriculture' greenhouse gas emissions (Robertson et al. 2008). Bioenergy has the potential to offset fossil fuel use. Also, replacement of fossil fuel-generated energy with solar energy that is captured by photosynthesis has the potential of reducing CO_2, N_2O and NO_x emissions (Power 2010).

9.2.3 Supporting Services

As indicated in Fig. 9.1, enhancing and maintaining soil structure and fertility, nutrient and water cycling, as well as biodiversity protection are recognized as supporting services of agriculture. Arguably, the most important supporting service of agriculture is the maintenance of long-term productivity of soil. Soil provides multiple and multifaceted supporting services such as food and raw material production, water and nutrients cycle regulation and biodiversity conservation (Calzolari et al. 2016). Soil fertility is related to soil structure- porosity, aggregate stability, water holding capacity and erosivity (Robertson et al. 2014) and all of these factors are potentially affected by agronomic practices. Soil organic matter provides mineral nutrients imperative for crop growth. For instance, in intensively fertilized grain crops, soil organic matter provides about 50% of the crop's nitrogen needs (Swinton et al. 2007). However, conservation tillage practices can increase soil organic matter (Kragt and Robertson 2014).

About 50% of soil organic matter is carbon. Carbon provides the chief source of energy for heterotrophic organisms that form the complex soil food web and plays an important role in soil structure. Soil aggregates are formed by mineral particles held together by decomposition products, such as polysaccharides. Aggregates form the basis for a soil structure which improves infiltration, soil water retention and porosity (Swinton et al. 2007). These qualities, in turn, enhance microbial activity and plant growth and therefore provide a more balanced set of services for the agricultural systems (Landis 2016).

9.2.4 Cultural Services

Other ecosystem services provided by agricultural landscapes are cultural benefits including aesthetic landscape, recreation, spiritual well-being and the cultural heritage of rural lifestyles (Fig. 9.1). Economic valuation of these services can be difficult as many of the impacts are intangible.

9.2.5 Ecosystem Disservices

While agricultural production systems provide many ecosystem services, they affect the quality and quantity of ecosystem services that other ecosystems can provide (Dale and Polasky 2007). If the effects on other ecosystems are negative, they are defined as 'disservices' of agriculture. Some agricultural disservices include loss of biodiversity, nutrient runoff, agrochemical contamination and sedimentation of waterways, greenhouse gas emissions and pesticide poisoning of human and non-target species (Fig. 9.1).

9.3 Transition to Multifunctional Agriculture: Improving Ecosystem Services

Historically, the basic function of agriculture was food production. Later at the dawn of economic development, providing industrial raw material and labor (Fig. 9.3) were also considered (Long et al. 2010). After the Second World War, in order to enhance agricultural production, more intensive practices based on mechanization and high levels of inputs such as energy, fertilizers and pesticides were promoted (Tilman et al. 2002). Intensification incorporates intensive tillage regimes and reduction in crop diversity (Matson et al. 1997).

Conventional agriculture has contributed substantially to increasing food production (Craheix et al. 2016; Montanarella 2007) but it has disrupted many of the regulating and supporting ecosystem services that are provided by agricultural ecosystems (Table 9.1). Some of the detrimental impacts of conventional agriculture are:

(1) Depletion of soil ecosystem services: Conventional agriculture can lead to breakdown of soil aggregates, reduced levels of soil organic carbon and total nitrogen and also decreased diversity and abundance of soil organisms due to the monoculture of high-yielding varieties and increased chemical and mechanical inputs (Mazzoncini et al. 2011; Stoate et al. 2009). Soil degradation further reduces production potential and enhances need for greater quantities of external inputs, e.g. fertilizers. As a result, intensive agricultural systems are

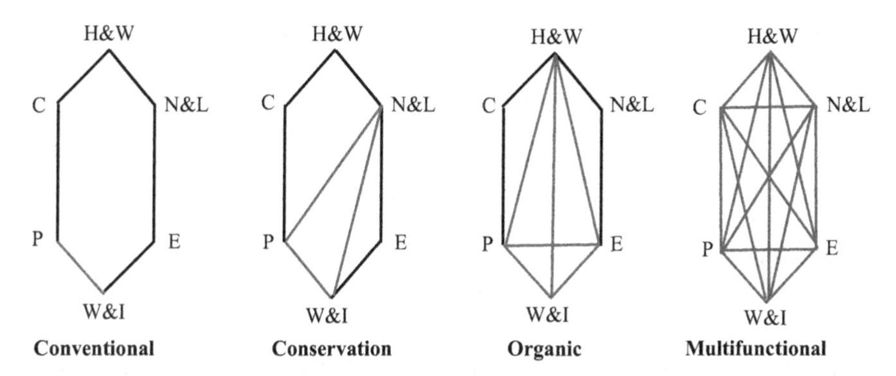

Fig. 9.3 Transition from conventional to multifunctional agriculture; note that prominent functions of each agricultural system are indicated with colored lines. 'H&W', 'C', 'P', 'W&I', 'E' and 'N&L' stand for 'health and well-being', 'climate', 'production', 'work and income', 'environment' and 'nature and landscape', respectively (Vereijken 2002)

Table 9.1 Ecosystem services provided by agricultural systems

Agriculture paradigms	Ecosystem services							Ecosystem services costs			
	Provision of food	Water supply and regulation	disease and natural pest control	biodiversity conservation	greenhouse gas mitigation	soil fertility and erosion control	recreation and aesthetics	food cost	nature and landscape cost	health and well-being cost	payment for ecosystem services
Conventional											
Conservation											
Organic											
Multifunctional											

Cell colors: potential provider of ecosystem services and their costs: high medium low

less able to withstand extreme weather events such as drought (Robertson et al. 2014). Moderately intensified agricultural systems can provide a more balanced set of services. While yields are more modest, other services such as soil retention can be relatively high (Landis 2016).

(2) Greenhouse gas emission: Conventional agriculture contributes to 12–14% of global greenhouse gas emissions through methane and nitrous oxide emissions and also the use of fossil fuel for fertilizer production (IPCC 2007).

(3) Availability and mobility of natural enemies: Reduction in overall complexity of landscapes, fragmentation of natural non-crop habitats and reduction in plant species richness that accompanies agricultural intensification, can lead to the

loss of habitat heterogeneity and significant decline of natural enemy populations (Tscharntke et al. 2005).

(4) Water quality and quantity: In intensified agricultural systems, less than half of the applied nitrogen (N) and phosphorous (P) fertilizer is used by crops. Excess fertilizer is washed away from agricultural fields via surface runoff or leached into the ground water (Jose 2009). Legacy soil N and P contaminate water resources and decline water quality (Horrocks et al. 2014). On the other hand, intensified agriculture has led to increased water withdrawal from surface and ground water resources. It is estimated that almost 70% of water resources are used for irrigation (MEA 2005).

The current trend for increasing ecosystem disservices with conventional agriculture is forecast to continue in developing countries where demand for energy-intensive food is expected to grow (Power 2010; Zhang et al. 2007). The long-term health of agricultural ecosystems, as well as their ability to provide food and other diverse benefits, requires that farmers in developing countries mitigate the negative consequences of conventional farming and expand their management focus to include the provision of ecosystem services (MEA 2005). The three most widely adopted alternative farming practices are conservation agriculture, organic farming and multifunctional agriculture (Fig. 9.3). The following subsections focus on describing these alternative agricultural systems.

9.3.1 Conservation Agriculture

Conservation agriculture is characterized by minimum or no mechanical soil disturbance, permanent organic soil covers by retaining crop residues and crop rotations (Powlson et al. 2016). Conservation agriculture not only preserves high values of landscape and nature and reduces labor costs (Fig. 9.3), but also mitigates climate change through soil carbon sequestration and the regulation and provision of water through soil physical, chemical and biological properties (Haramoto 2014; Palm et al. 2014; Powlson et al. 2016). Conservation agriculture delivers multiple ecosystem services (Table 9.1) including:

(1) Providing soil fertility: Conservation agriculture changes soil properties and processes and reduces soil degradation relative to conventional practices (Palm et al. 2014). For instance, conservation tillage practices can increase soil organic matter which helps water storage and reduces soil erosion (Kragt and Robertson 2014). The presence of cover crops can also increase soil microbial biomass and microbial activity compared to fallow fields (Mendes et al. 1999) and thereby enhance soil ecosystem services through increased plant nutrient uptake and reduced soil erosion (Gianinazzi et al. 2010). Moreover, retaining crop residues can directly reduce soil erosion by minimizing the time that soil is bare and exposed to wind, rainfall and runoff (Palm et al. 2014) and increase

soil carbon sequestration which assists climate change mitigation (Kragt and Robertson 2014).

(2) Providing greenhouse gas mitigation: Conservation agriculture was not initially developed as a practice to reduce greenhouse emission. However, it is now recognized as a potential technology to do so. The net potential of conservation agriculture to serve as a greenhouse gas mitigation strategy depends on the direction and magnitude of changes in soil carbon, nitrous oxide (N_2O) and methane (CH_4) emissions associated with its implementation compared to conventional farming (Palm et al. 2014). Various component practices of conservation agriculture have different effects on soil carbon stock. For instance, leguminous cover crops in the biologically based systems increase soil carbon storage compared to conventional practices (Robertson et al. 2014). Also, no-till cultivation and conservation tillage can save soil carbon (Power 2010). However, changes in soil carbon do not necessarily result in climate regulation if there is not a net transfer of CO_2 from the atmosphere (Palm et al. 2014).

(3) Increasing water quality and quantity: No-till management or minimum tillage methods such as riparian buffers can reduce erosion and runoff (Robertson et al. 2014). Reducing runoff and water erosion also results in lower transport of sediments, nutrients and pesticides and consequently higher water quality in the catchment area (Palm et al. 2014). Moreover, better soil structure, presence of cover crops and low exposure to drying compared to conventional farming, results in more water available for plants in conservation agriculture systems. Water holding capacity of the topsoil is also generally higher due to the increased soil organic matter contents (Palm et al. 2014; Robertson et al. 2014).

(4) Preserving biodiversity: Biodiversity is higher in conservation agriculture relative to conventional farming. This higher diversity can be related to improving regulating services, e.g. pollination (Palm et al. 2014).

Despite these wide-ranging potential ecosystem services, conservation agriculture has been only adopted on large-scale mechanized farms, particularly in Australia, United States and Europe (Craheix et al. 2016; Kertész and Madarász 2014; Basch et al. 2015) and its practical feasibility for African and Asian smallholders in mixed crop/livestock systems, where crop residues are used as animal feed (Lahmar 2010; Valbuena et al. 2012) is questionable. There are also contrasting results regarding the short-term effects of conservation agriculture on crop yield (Aryal et al. 2015; Jat et al. 2012, 2014; Pittelkow et al. 2015; Thierfelder et al. 2015) and pest regulating services (Brainard et al. 2016).

9.3.2 Organic Farming

Organic farming is designed to alleviate the detrimental impacts of chemical-synthetic inputs in conventional agriculture (Duru et al. 2015). The emphasis is on using smart management rather than relying on external chemical inputs. These include crop rotation, biodiversity conservation and managing for biological pest control (Fan et al. 2016). It offers a great potential to develop low input technologies for producing food without causing damage to human health (UN 2008). While crop yields from organic agriculture are often lower than conventional agriculture systems (Fan et al. 2016; Noponen et al. 2012), it can deliver more significant ecosystem services than conventional agriculture (Table 9.1). Some of the environmental benefits of organic farming are:

(1) Providing soil fertility: Organic farming is designed to enhance levels of soil organic carbon and total nitrogen and promote greater microbial biomass and diversity (Reganold et al. 2010; Verbruggen et al. 2010) through various soil fertility management practices including use of organic residues as soil amendments, growing nitrogen fixating crops as the major nitrogen source, use of crop rotation and cultivating species that are diversified in space and time to control erosion and build soil organic matter (Wander 2015). However, there is conflicting evidence on the ability of organic farming to enhance soil fertility above that of conventional agriculture. Gosling and Shepherd (2005) and Kirchmann et al. (2007) found no significant differences in levels of soil organic matter and total nitrogen carbon: nitrogen ratio between conventional and organic farms.
(2) Providing greenhouse gas mitigation: Greenhouse gas emission per hectare in organic agriculture is significantly lower than conventional agriculture, which can be attributed to lower synthetic fertilizer input use in organic farming (Bos et al. 2014; Meng et al. 2017).
(3) Availability and mobility of natural enemies: Pest management agricultural practices often lead to changed food web structure and communities dominated by a few common species, which together contribute to pest outbreaks (Crowder et al. 2010). When enemy evenness is disrupted, pest densities are high and plant biomass is low (Sandhu et al. 2010). Organic farming potentially mitigates this ecological damage by promoting evenness among natural enemy species (Crowder et al. 2010).
(4) Preserving biodiversity: A wide range of birds and mammals, invertebrates and arable flora benefit from organic farming through increasing in abundance and species richness (Holt et al. 2016). Beneficial effects of organic agriculture on fauna and flora can be attributed to prohibition/reduced use of synthetic fertilizers and pesticides and application of environmentally friendly farming practices such as crop rotations (Meng et al. 2017). Planting a variety of crops in rotation increases the diversity of soil microbes underground. This in turn improves soil organic matter and structure and also aids the healthy functioning of the soil (Tiemann et al. 2015).

Despite various environmental benefits, organic farming has not yet succeeded in replacing conventional agriculture (Sandhu et al. 2010). A break-through of organic agriculture will chiefly depend on crop yield enhancement and cost reduction (Table 9.1) so that it competes economically with conventional products (Vereijken 2002).

9.3.3 Multifunctional Agriculture

The concept of multifunctional agriculture was introduced by Japan in the late 1980s (Tao et al. 2014). Multifunctional agriculture aims to maintain sustainable production and minimize the ecological impact of agricultural practices (Marzban et al. 2016), it promises benefits over many other forms of agriculture (Manson et al. 2016) and strengthens the mutual synergies between various ecosystem services (OECD 2001; Palm et al. 2014). However, its potential is unrealized in many places (Jordan and Warner 2010). The notion emerged when it was officially proposed in the Agenda 21 documents of the Rio Earth Summit, in 1992, particularly with regard to food security and sustainable development (UNCED 1992). After being addressed in the Agenda 21, the concept has gradually played an important role in scientific and political debates on the future of agriculture and rural development, particularly the ongoing reform of the Common Agricultural Policy, CAP, of the European Union (Tao et al. 2014).

The meaning attributed to multi-functionality in international debates is ambiguous. Various institutions have adopted the concept with slightly different interpretations and in relation to different policy agendas (Renting et al. 2009). Using the concept of multifunctional agriculture in scientific debates has been most clearly inspired by the work of the Organization for Economic Cooperation and Development, stating that agriculture is multifunctional when it has one or several functions in addition to its primary role of producing food and fiber (Rossing et al. 2007). Another use of the concept can be traced back to the Food and Agriculture Organization, FAO, which focused on the varied nature of agricultural activities, specifically in the developing countries, and its multiple contributions to livelihood strategies of households and rural development. FAO's notion of multiple roles of agriculture covers contributions of agriculture to environmental externalities and development challenges such as food security, poverty alleviation, social welfare and cultural heritage (Renting et al. 2009). A third use of the concept is related to the ongoing reform of CAP. Within this approach, beside producing food and basic materials there is an emphasis on natural resource protection, leisure and recovering space as well as cultural landscape and non-commodity outputs (Wiggering et al. 2003).

Due to the fact that agriculture has different characteristics and functions and human demands differ between countries, adoption of multifunctional agriculture has been very uneven. Some regions such as New Zealand and several European countries, i.e. United Kingdom, Netherlands, France and Norway, consider the multi-dimensional nature of agriculture as a provider of private and public goods and services. However, many developing countries have rarely applied the concept (Manson et al. 2016). The main focus of the developing countries is still on the primary function of agriculture as a supplier of food and basic materials (Tipraqsa et al. 2007) and their support for agriculture does not take into account public goods (Peng et al. 2015).

In developing countries, small-scale farming is widely implemented. Considering the limited resources and small scale, smallholding agriculture cannot fulfil such a broad set of functions at the farm scale. Only larger regions may meet this challenge (Vereijken 2002). Therefore, to achieve multi-functionality, a shift to landscape scale thinking, working across farms and creating a governance that supports food and basic material production whilst ensuring the protection and enhancement of public goods is imperative (Holt et al. 2016). Also, to ensure sufficient work and income for smallholding farmers, ecological functions should be marketed by a wide variety of rural products and services (Vereijken 2002) and be supported by policy.

9.4 Physical and Human Factors Underlying the Ecosystem Services Flow of Agriculture

Rural communities of developing countries are highly dependent on local ecosystems to supply goods and services that support their livelihoods. Therefore, it is imperative to mitigate negative impacts on ecosystem services, e.g. ecosystem disservices from agriculture, through improvement of our understanding about the main drivers of the ecosystem services flow. It is widely acknowledged that multiple direct, mostly physical, and indirect, mostly human, factors drive generation, state and evolution of ecosystem services. In general, at a global and state level, factors including changes in demography, economy, technology, culture and religion, climate conditions, atmospheric carbon dioxide concentrations and land cover are the major drivers of the ecosystem services flow (Jiang et al. 2016; MEA 2005). These factors interact in complex ways to change pressures on ecosystems and use of ecosystem services, in different locations (MEA 2005).

9.4.1 Direct Drivers

Climate variability, i.e. drought, or change is one of the most important direct drivers of changes in ecosystem services. During the last 100 years, the global mean surface temperature has increased about 0.6 °C and precipitation patterns have altered spatially and temporally (MEA 2005). There are strong evidences that most warming can be attributed to human activities such as burning of fossil fuels and changes in land use (IPCC 2013). Current models indicate continued and accelerated climate change, in the future. Based on the Fifth Assessment Report of the Intergovernmental Panel on Climate Change (IPCC 2014a), the global average surface temperature will be very likely to rise in 2100 from 3.7 to 4.8 °C compared to pre-industrial levels, without additional efforts to reduce greenhouse gas emissions. Also, it is very likely that heat waves will occur more often and last longer, heavy precipitation events will become more intense and frequent in tropical regions and drought frequency, intensity and/or duration will increase in arid and semi-arid areas. It should be noted that climate change induced risks are unevenly distributed and are generally greater for developing nations, with rapidly growing human populations (Hernández-Delgado 2015; IPCC 2014a).

Current trends in the changing climate, in combination with mathematical model predictions, have indicated that climate change, especially warmer regional temperatures, has already had or will unequivocally have unprecedented impacts on biophysical processes that underpin ecosystem dynamics (Lavorel and Grigulis 2012; Metzger et al. 2008; Seppelt et al. 2011). There have been changes in species distributions, population sizes and the timing of reproduction or migration events across multiple terrestrial and marine ecosystems (Bellard et al. 2012; Veron et al. 2009) as well as an increase in the frequency of pest and disease outbreaks, especially in forested systems (MEA 2005).

Climate change also drives changes in ecosystem structure and spatial pattern, which would affect key processes such as carbon sequestration (Canadell et al. 2010). The net flux of CO_2 between land and atmosphere is a balance between carbon losses from agricultural practices and carbon gains from plant growth and sequestration of decomposed plant residues in soils (Power 2010). Since 1750–2003, the atmospheric concentration of carbon dioxide has increased by about 32% (from 280 to 376 parts per million), due to combustion of fossil fuels and land use changes, primarily. While this trend is unprecedented in at least the last 800,000 years (IPCC 2014b), it is predicted that the atmospheric concentration of carbon dioxide will exceed 450 ppm CO_2eq by 2030 and will reach CO_2eq concentration levels between 750 and more than 1300 ppm CO_2eq by 2100 (IPCC 2014a). Soil carbon sequestration provides various ecosystem services to agriculture by conserving soil structure and fertility, improving soil quality, increasing use efficiency of agronomic inputs and improving water quality by filtration and denaturing the present pollutants (Smith et al. 2008). However, the switch of some regional ecosystems from carbon sink to carbon sources or vice versa is expected by the end of the 21st century (Tao and Zhang 2010).

Furthermore, ecosystem services are predicted to be directly influenced by land cover change. Land cover is critical in regional climate, biodiversity conservation, provision of ecosystem services and socioeconomic development (Zhao et al. 2015). However, land cover change has increased since 1750 (MEA 2005) and many natural ecosystems have suffered severe degradations from long-term cropland expansion, extensive use and climate change (Liu et al. 2015). Currently, managed lands cover more than 60% of earth's land surface and about 60% of this amount is under agriculture (Foley et al. 2011). Therefore, agriculture is one of the major proximate drivers of land degradation but the extent of degradation is intensified by co-occurrence of the other types of drivers such as increased aridity (Geist and Lambin 2004).

Sandstorms, desertification, deforestation and ecological refugees caused by land cover degradation have threatened ecological security and sustainable development, in many arid and semi-arid countries. The Millennium Ecosystem Assessment (2005) reported that land cover degradation increases surface runoff, topsoil erosion and exposure to rocky surfaces for arid and semi-arid regions. It has also been observed that deforestation increases soil runoff and siltation of water in arid regions (Biswas et al. 2012). Furthermore, large-scale deforestation and desertification in the tropics and sub-tropics lead to reduction of regional precipitation, primarily due to decreased evapotranspiration (MEA 2005).

9.4.2 Indirect Drivers

Collectively, indirect drivers influence the levels of production and consumption of ecosystem services and sustainability of production. From the four major indirect drivers, i.e. changes in demography, economy, technology, culture and religion, unprecedented population and economic growths lead to increased consumption of ecosystem services. However, the negative environmental impacts of increased consumption depend on efficiency of the technologies used in production of ecosystem services (MEA 2005).

Between 1970 and 2015, global population increased by 99.8%, from 3.7 billion to 7.3 billion. Also, total population of developing countries increased from 2.8 to 6.1 billion, i.e. 117.9% growth rate (World Bank 2016). Developing countries have accounted for the most recent population growth in the past 45 years. Changes in population can directly affect the amount of energy use, magnitude of air and water pollution emissions, amount of land required and other direct drivers of ecosystem services (Alcamo and ÓNeill 2005). Furthermore, the global rate of urbanization has increased from 36.5% (1970) to 53.9% (2015) and is projected to reach nearly 60% of human population by 2030 (Alberti 2005; World Bank 2016). While 44 developing countries have populations that are 70–95% urban, some parts of Asia and Africa are still largely rural (World Bank 2016). Income, lifestyle, energy use and the resulting greenhouse gas emissions differ significantly between rural and urban populations (IPCC 2014a). For instance, analyzing the effects of urbanization on energy use over the period of 1975–2005 for 99 countries indicated that the impact of urbanization on energy use is negative in the low-income countries, while this impact is positive and strongly positive in medium and high income countries (IPCC 2014a).

Economic development as one of the main indirect drivers of ecosystem services flow comprises many dimensions such as income levels, economic structure, consumption and income distribution (Alcamo and ÓNeill 2005). Historically, global economic activity has increased by a factor of 26 over the last 45 years. Although there was a strong demographic growth in the same period, per capita GDP growth was about 1.8% per year (World Bank 2016). However, global trends in per capita GDP varied substantially by region. Economic growth was the strongest in Asia, e.g. Hong Kong, South Korea, Singapore and China, averaging 5.0% per year over the 1970–2015 period but was below the global average of 1.8% in many developing countries (World Bank 2016). Rising per capita income leads to greater demand for many ecosystem services. At the same time, the structure of consumption changes by increasing the proportion of fat, meat, fish, fruits and vegetables and rising the proportionate consumption of industrial goods and services (Alcamo and ÓNeill 2005). Also, many taxes and subsidies such as fertilizer taxes dramatically increase rates of resources consumption. For instance, subsidies paid to the agricultural sectors of many developing countries encouraged greater food production that is associated with more water consumption and nutrient and pesticide release, in 2001–2003 (MEA 2005).

The rate of technological change, especially in energy, water and agriculture contexts, is an indirect driver of changes in ecosystem services because it affects the efficiency by which ecosystem services are produced or used (Alcamo and ÓNeill 2005). Technology accounted for more than one third of total GDP growth in United States from 1929 to the early 1980s and for 16–47% of GDP growth in selected OECD countries in 1960–1995 (MEA 2005). Impact of technological change on ecosystem services is evident in the case of food production. With this regard, much of the increase in agricultural outputs have been attributed to the increase in yields per hectare rather than expansion of cultivated area (Burgess and Morris 2009), which has led to reduced need of converting forest or grassland, over the past five decades. Technological change, however, can lead to degradation of ecosystem services (MEA 2005) because technological advancements often require large amounts of goods and materials and can cause new ecological risks (Alcamo and ÓNeill 2005).

Cultural and religious drivers are also important indirect drivers of ecosystem services. These factors can influence the trends of producing energy or consuming food and values related to environmental stewardship. Also, they may be particularly important drivers of environmental change (Alcamo and ÓNeill 2005; MEA 2005). For example, Tengö and von Heland's study in Madagascar (2011) indicated that planting trees serves as a symbol of renewal, purification, agreement and boundary-making culturally, which confirmed the interdependence of cultural beliefs and the generated ecosystem services.

9.5 Drought Incidence and Impacts on Ecosystem Services of Agriculture

Both developing and developed countries are vulnerable to extreme natural disasters. However, developing countries have been more influenced by extreme natural events, on a per capita GDP basis over the past two decades (Munich Re 2013). Furthermore, both the number and severity of natural catastrophes have increased in the world, at the same time (Fig. 9.4). However, the magnitude and duration of

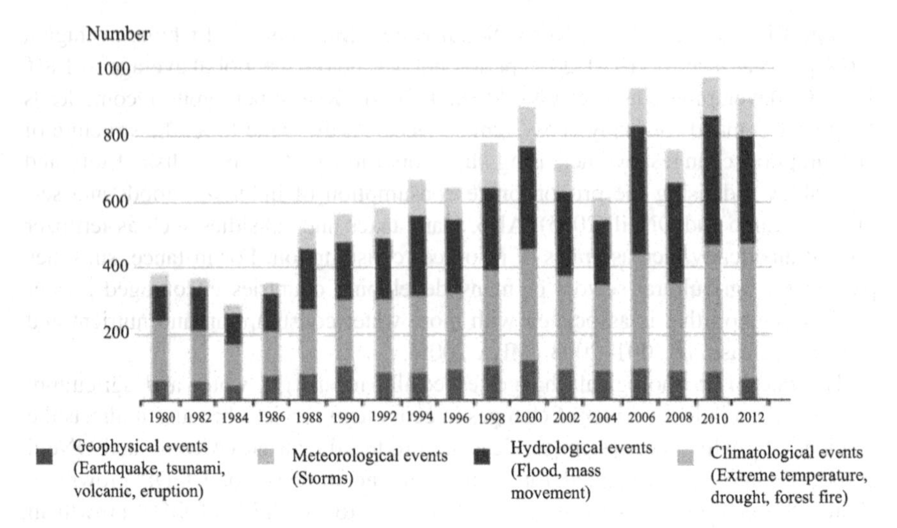

Fig. 9.4 Frequency of natural catastrophes worldwide from 1980 to 2012. (Adapted from Munich Re 2013 with minor changes)

many droughts exceed those captured by instrumental records (Cook et al. 2004) and many regions of Asia, Africa, Australia and America have experienced longer and more intense droughts, in the past four decades (IPCC Special Report 2012).

Also, global and regional studies have projected a significant decrease in mean annual precipitation and higher likelihood of hydrological drought in the twenty-first century, in North and South America, south and central Asia, west and central Australia and central Eurasia (Miyan 2015). Moreover, it is expected that particular hydro-climatic challenges such as seasonal rainfall, intermittent dry spells, recurrent drought years and high evaporative demand make some arid and semi-arid regions, i.e. south Asia and sub-Saharan Africa, more vulnerable to extreme droughts.

9.5.1 Nature of Drought and Its Prevalence in Some Developing Countries

Drought is an insidious, slow-onset and multi-dimensional natural disaster that starts unnoticed and develops cumulatively and its impacts are not immediately observable (Shakya and Yamaguchi 2007). It is commonly classified into four categories. A *meteorological drought* is a period of time, i.e. months to years, with below-normal precipitation over a region (Vrochidou et al. 2013). *Hydrological drought* is a period with insufficient surface and ground water resources for established water usage of given water resources management system (Mishra and Singh 2010). *Agricultural drought* is a period with declining soil moisture and consequent crop failure that results from below-average precipitation, intense but less frequent rain events or above-normal evaporation (Dai 2011). Finally, *socioeconomic drought* is associated with failure of water resources to meet water demands and its effects can be traced to the economic systems (Backerberg and Viljoen 2003; Mishra and Singh 2010).

Drought is pervasive to all continents of the world (Fig. 9.5). While short-term droughts have frequently occurred in various regions of the world, many intense or long-term droughts have been observed in arid and semi-arid regions since 1970s (Dai 2011). For instance, the African Sahel drought that started in 1968 and lasted in 1988 led to the starvation of millions of people (Woods and Woods 2007). Also, very severe droughts occurred in 1972, 1975, 1979, 1981, 1982, 1984, 1989 and 1994–95 in Bangladesh. These events have typically influenced 53% of the population (Adnan 1993). Furthermore, Afghanistan, India, Pakistan and Sri Lanka have reported droughts at least once in every three-year period, in the last 50 years (Miyan 2015). At the same time, in Iran droughts have an unprecedented return frequency of every five years (Amirkhani and Chizari 2010; Habiba et al. 2011) while the international expectation is every 20–30 years (Eskandari 2001).

Since early 2000, high magnitude droughts have affected large areas of Asia and Africa. For example, localized wide-range drought of 2011 affected 14 out of 34 provinces of Afghanistan and approximately three million people (OCHA 2011). Also, Nepal has experienced consecutive annual drought conditions since 2000. However, the 2008–2009 winter drought was unprecedented in both scale and severity and Nepal received less than 50% of average precipitation (Wang et al. 2013). The Iranian drought of 2000–2001 affected 90% of the population and led to serious shortage of safe drinking water in 12 out of 29 provinces (OCHA 2001b). Furthermore, the historic 2011 drought in the Horn of Africa affected over 13

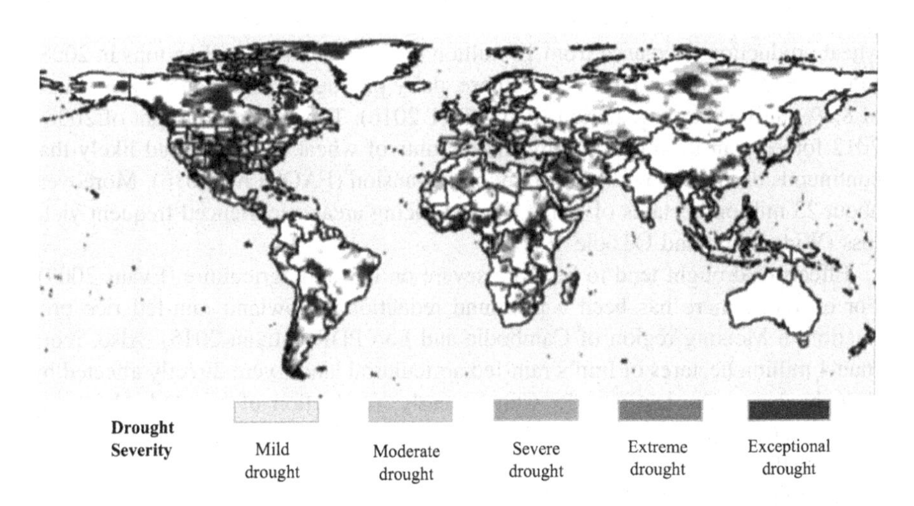

Fig. 9.5 Severity of drought worldwide. (Adapted from University College of London 2016)

million people, caused huge displacement and malnourishment of 30% of the drought affected people (Miyan 2015).

While many arid and semi-arid regions suffered from severe droughts in the recent decades, drought conditions remained constant in South America and intensified in Asia, in 2013 (GDIS 2013). As a result, 258 million people were affected globally by an exceptional drought, in 2012–2015 (University College of London 2016). By 2020s, a large part of Asia is projected to experience increased risk of severe droughts with multiple impacts (Forster et al. 2012).

9.5.2 Drought Impacts on Ecosystem Services of Agriculture

Since drought can negatively affect ecosystem services of agriculture through multiple pathways, investigating the influence of long-term droughts on agro-ecosystems is imperative. Greater understanding of drought impacts on ecosystem services in agriculture can improve efforts to support drought planning and adaptation (Covich 2009).

In largely agrarian countries, drought has widespread and devastating impacts on provisioning services, i.e. food and forage production (Table 9.2), particularly at local levels (Keshavarz et al. 2014). The impacts of drought on agriculture depend on its nature, i.e. chronic and contingent, duration, degree of water/moisture deficiency, periodicity of occurrence within a season and size of the affected area (IPCC 2007). Droughts, in conjunction with socio-economic factors, have significant consequences in terms of reduction/elimination of agricultural production (Chary et al. 2010; Keshavarz et al. 2013; Simelton et al. 2012).

In Asia, drought had considerable impacts on food grain production in Jordan, Iran, Iraq and Syria (Pauw 2005). According to FAO (FAOSTAT 2016), Iran's wheat production decreased from 16 million tons in 2007 to 8 million tons in 2008, causing $775.6 million loss. Furthermore, dairy production experienced a decrease of 8.2% during the same period (FAOSTAT 2016). The ensuing drought of 2008–2012 forced Iran to import significant amounts of wheat and it seemed likely that continuous drought would lead to import expansion (FAOSTAT 2016). Moreover, about 23 million hectares of Asian rice producing areas experienced frequent yield loss (Widawasky and ÓToole 1990).

Effects of drought tend to be more severe on rain-fed agriculture (Evans 2009). For example, there has been a profound reduction in lowland rain-fed rice production in Mekong region of Cambodia and Lao PDR (Miyan 2015). Also, more than 4 million hectares of Iran's rain-fed agricultural lands were directly affected by drought during the last decade. Consequently, production of rain-fed wheat and barley were dropped by 34–75% (OCHA 2001b). Moreover, drought caused serious shortfall of food production in Afghanistan, especially rain-fed wheat was reduced up to 80% (OCHA 2001a).

Table 9.2 Drought impacts on ecosystem services of agriculture

Ecosystem service		Direct impacts of drought	Indirect impacts of drought
Provisioning	Food, fuel and raw material production	Reduction of crop and forage production	• Increased food price • Food insecurity and crisis • Increased livestock and wildlife mortality • Reduced farm families' income
Regulating	Air quality and climate regulation	• Disrupted thermal regulation, e.g. increased daytime temperature and faster evapotranspiration rates • Increased frequency and severity of wildfire	• Increased mortality of less drought and heat stress adapted species • Decrease of genetic diversity • Negative impacts on animal and plant population
	Regulation of water flow	• Reduced water supply and alteration of timing and magnitude of runoff and aquifer recharge • Reduced water quality, e.g. salinity	• Reduced the diversity and abundance of many aquatic taxa • Canopy loss • Conflict about water access and use • Damaged farm infrastructure • Widespread acidification
	Erosion prevention	Increased soil degradation and erosion	• Increased rate of desertification • Permanent decrease in yield potential • Disruption of soil formation
	Soil fertility	• Depletion of soil moisture and productivity • Massive death of drought sensitive microorganisms	• Reduced rate of microbial biomass carbon • Increased wind erosion
	Pollination	Decline in distribution, abundance and effectiveness of pollinators	Reduced seed and fruit production in plants
	Disease and pest control	• Increased disease spread • Increased pest pressure	• Reduced quality of life • Reduced crop yield • Increased food price
Supporting	Nutrient cycling	Reduced nitrogen, phosphorous and carbon uptake	• Reduced plants growth • Reduced crops tolerance to biotic and abiotic stress
	Maintenance of life cycles	• Reduction of vegetative cover • Degraded wildlife habitat	• Negative impacts on temperature and precipitation • Increased soil erosion

(continued)

Table 9.2 (continued)

			• Increased possibility of high intensity wildfires • Increased competition between livestock and wild animals for grazing and water • Decreased reproductive success and survivorship • Increased migration rate in search of new resources
	Maintenance of genetic diversity	Reduced land use and species diversity	Deteriorate nutrient cycling
Cultural	Aesthetics; spiritual and cognitive development	Weakened spiritual benefits people obtain from ecosystems	Reduced quality of life
	Recreation and ecotourism	Decline in quantity and quality of landscapes	Reduced ecotourism income

The impacts of drought are more pronounced on fodder availability compared with that of food grains (Chary et al. 2010). A survey conducted during the 1999–2000 drought in Rajasthan and Gujarat showed that the fodder deficit in arid parts of Rajasthan was 50–75%, which resulted in livestock migration in search of food, fodder and drinking water (Rao and Boken 2005). In extensive grazing lands of Afghanistan, drought also has led to high mortality of livestock. In Samangan, about 30% of animals perished due to drought and 1,400,000 animals were sold at low prices (IRIN 2008). Severe production losses may be tolerated for one year or longer, but when dry conditions persist for several years, rural communities deteriorate particularly for nomads and livestock producers (Karimi et al. 2018; Stehlik et al. 1999). The drought episode of 1979–1984 in Morocco, for example, reduced the small ruminant population by 40–50% (Pauw 2005).

While the impacts of droughts on provisioning services of agriculture have been frequently studied, their effects on other ecosystem services have been less clarified though several types of regulating services of agriculture can be influenced by drought (Table 9.2). Droughts have severe and direct impacts on hydrological ecosystem services such as water supply and quality regulations (Table 9.2). In societies where agriculture is the primary economic activity, deficiencies of precipitation and reduction in annual runoff and aquifer recharge are decreasing the surface and ground water resources with negative impacts on water supply for agricultural and domestic sectors. For instance, consecutive drought years in Yemen (2007–2009), Bangladesh (1993–1995), Nepal (2002–2006 and 2008–2009), Afghanistan (2008–2011) and Iran (2007–2011) caused serious water supply deficit and drying up of some rivers and wet bodies (Keshavarz and Karami 2016; Miyan 2015; OCHA 2011; Wang et al. 2013).

In the early stages of 2007–2011 drought in Iran, some temporary water bodies dried up and in all others, the water levels reduced to a critical level (IRNA 2014). Also, with drought progression, many internationally renowned lakes such as Kaftar, Bakhtegan, Arjan, and Tashk completely dried up (Jam-e-jam 2014) and most of the traditional groundwater irrigation systems (qanats) experienced reduced discharge (Bostani et al. 2009). Changes in water balances greatly reduced the diversity and abundance of many aquatic taxa, from small invertebrates to fish, and led to localized extinctions in some areas such as Bakhtegan and Parishan Lakes (IRNA 2010), which led to decline of farm families' resilience (Maleksaeidi et al. 2016, 2017).

Low flows in rivers and reduced lake levels also increased levels of nutrients and salinity in many drought-stricken areas of developing countries. There is also some evidence that drought induced forest mortality can decrease water quality by increasing nutrient leaching (Beudert et al. 2007). Moreover, reduced flows affected safe drinking water standards in Nepal, Yemen and Iran (Miyan 2015; OCHA 2001b).

For regulating services, drought can in some cases affect thermal regulation (Table 9.2). With this regard, extensive canopy loss has large and spatially complex effects on land surface microclimate and near-ground solar radiation (Royer et al. 2011). Even, reduction of tree cover from 40 to 25% can produce dramatic changes in the amount of solar radiation incident on the land surface that influences microclimates for humans and other biota and can result in increased loss of water via soil evaporation (Royer et al. 2010). Furthermore, there is some evidence that drought can increase frequency and severity of wildfire by setting up ideal fire conditions (Table 9.2). The increasing number of wildfires have negative impacts on animal and plant population (Taninepasargad 2015).

Frequent droughts can reduce soil fertility and increase soil erosion (Table 9.2). While winter rains play an important role in replenishing soil moisture and recharging groundwater aquifers, decreased levels of precipitation during winter season, i.e. winter drought, increased sunlight and warmer temperature caused moisture to evaporate from the soil. Combination of reduced soil moisture and associated crops mortality make the soil tend more to wind and water erosion (Owen 2008). Moreover, increased frequency of droughts can disrupt soil formation by changing erosion rates. If high soil erosion rates are triggered in drought prone areas, then it is expected to result in impacts on soil microbial activity, too. Disruption of soil microbial activity is attributed to massive death of microorganisms that cannot adapt to drought stress (Hueso et al. 2012).

Widespread and severe-sustained droughts have also negative effects on distribution, abundance and effectiveness of pollinators (Table 9.2). Pollinators are vital for transferring genes within and among populations of wild and cultivated plant species (Kearns et al. 1998). However, in addition to drought, pesticide use, land use changes and agricultural intensification have negatively affected plant-pollinator interactions (Kremen et al. 2002; Aguilar et al. 2006; Ricketts et al. 2008). Declined abundance of pollinators and the associated pollen limitation lead to a reduction in seed and fruit production in plants but they have rarely resulted in complete failure to produce seed or fruit (MEA 2005).

Drought can also increase disease spread and pest pressure (Table 9.2). Drought increases the risk of water-related disease, e.g. E coli, cholera and algal bloom, airborne and dust-related disease, e.g. coccidioidomycosis, vector borne disease, e.g. malaria, dengue and West Nile Virus and mental health effects, e.g. depression and anxiety (Stanke et al. 2013). Moreover, long-term drought conditions have negative influences on natural enemy species and thereby increase pest pressure. Water is critical for pests and they congregate where they can find sustaining water and keeping their numbers in balance. In this way, crops are being destroyed.

Supporting services of agriculture can also be modified under drought (Table 9.2). Drought can alter ecosystem nitrogen, phosphorous and carbon cycles (Table 9.2). It can reduce plant growth by decreasing nitrogen and phosphorous uptake, transportation and redistribution (Rouphael et al. 2012; Sardans and Penuelas 2012). Various studies have indicated that plants reduce nitrogen and phosphorous uptake upon reduced soil moisture (Cramer et al. 2009; Waraich et al. 2011) and it usually occurs in the short term, i.e. less than 90 days (He and Dijkstra 2014). Moreover, severe extended droughts can reduce carbon uptake in drought prone areas (Xiao et al. 2009).

Microclimate differences in temperature and soil moisture, due to changes in precipitation interception, can influence understory species composition (Kane et al. 2011), which can in turn affect nutrient cycling (Table 9.2; Royer et al. 2011). Also, the increasing number and severity of droughts can reduce habitat quality for wildlife (Table 9.2). Lack of water, food and habitat protection may decrease reproductive success and increase mortality rates for the most vulnerable animal species, especially endangered species (Breshears et al. 2011; Owen 2008). Moreover, drought may cause existing animal habitats to become patchy by isolating some populations. Other populations will be forced to migrate in search of new resources (Owen 2008). During the 2007–2011 drought event in Iran, loss of a dominant plant species had many effects on ecosystem processes and functions and some animals and birds, e.g. Dalmatian Pelican, responded to the changing conditions by migrating to other areas (Tebna 2013). Also, drought can reduce supply of cultural services (Table 9.2) due to low inflows, poor water quality and people's perceptions about the negative impacts of drought on cultural services (Banerjee et al. 2013).

9.6 Valuing Ecosystem Services of Agriculture Under Drought: The Social Component

The capacity of agricultural systems to provide ecosystem services under drought is only one part of framing for services equation. The other parts are farmers' willingness to implement practices that deliver additional services and the extent at which adoption requires social, economic and institutional supports (Robertson et al. 2014). Farmers are important stewards of ecosystem goods and services (Greenland-Smith et al. 2016). With their selected land use system and drought management strategies, farmers considerably influence provisioning, regulating and supporting services of agro-ecosystems. However, there is often a certain level of trade-off between primary production and provision of ecosystem services to society for farmers (Gordon et al. 2010). Therefore, determining the attitudes and perceptions of farmers regarding how their farming practices affect diverse ecosystem services of agriculture is important and will provide an insight into which ecosystem-oriented management practices would most likely be adopted by farmers (Smith and Sullivan 2014).

Despite increasing abundance of literature on the social values of ecosystem goods and services in developed countries, relatively few environmental valuations have been conducted in developing countries, where the majority of the world's biodiversity is located (Kenter et al. 2011). The limited research has also been directed towards understanding the relevant agricultural practices that mediate between agro-ecosystems and delivery of ecosystem goods and services (Bernués et al. 2016). There is often a focus on particular ecosystem types such as woodlands, forests, grasslands and coastal zones. Furthermore, few studies have solely focused on farmers' willingness to adopt new management practices that provide additional ecosystem services. However, previous studies have revealed that willingness of farmers to provide various provision and non-provision ecosystem goods and services depends on awareness, attitudes, available resources and incentives (Robertson et al. 2014).

9.6.1 Factors Affecting the Perception of Ecosystem Services of Agriculture

Although environmental stewardship is a factor influencing many farmers' decisions, sustained profitability is usually the overriding concern. Particularly for those services related to reducing the environmental impacts of agriculture, farmers are more likely to adopt practices that save labor and inputs or improves farmland water quality without reducing the expected crop revenue (Poppenborg and Koellner 2013; Robertson et al. 2014). Moreover, previous studies have highlighted the importance of financial incentives in securing a change of land use from productive agriculture to provision of ecosystem services (Ma et al. 2012; Robertson et al. 2014; Sullivan et al. 2005; Suter et al. 2008; Yu and Belcher 2011). Recently, there has been a trend towards policies that focus on rewarding farmers for supplying specific ecosystem services (Smith and Sullivan 2014), i.e. water, carbon, biodiversity, cultural heritage and landscape beauty (Scherr et al. 2006; Simelton and Viet Dam 2014). However, only few attempts have been made for implementing Payment for Ecosystem Services (PES) programs in developing countries and most farmers do not get any specific economic compensation for ecosystem services that they deliver. Therefore, they are assigned to less importance compared with these functions (Bernués et al. 2015).

Previous studies have revealed that farm, farmer and institutional factors are important determinants of participation in PES programs. Farm and farmer factors include farm size and location, farm income, farm type, farmer's age, gender and education and farmer's perceptions towards environment. Also, institutional factors include length of scheme and planning horizon, potential development value, bureaucratic load, requirements associated with the scheme, flexibility of conditions, confidence in efficacy of recommended practices and funding certainty (Bremer et al. 2014; Christensen et al. 2011; Kwayu et al. 2014; Ma et al. 2012; Patrick and Barclay 2009; Shultz 2005; Yu and Belcher 2011; Zanella et al. 2014).

While financial rewards are an important reason for farmers to preserve ecosystem services in agriculture, they can also be profoundly driven by non-financial motivations when the issue is providing ecosystem public goods (Greiner and Gregg 2011). Farmers' pro-environmental behavior can be guided by personal or ethical motivations around nature stewardship (Sheidaei et al. 2016; Thomas and Blakemore 2007). Factors such as pursuing personal and family well-being, personal norms and self-identity, care based ethics and very strong stewardship ethics have high association with the importance of ecosystem services on farm (Broch et al. 2013; Fielding et al. 2008; Page and Bellotti 2015).

In some studies, complex sets of farm and socio-demographic factors have been found to be influential on farmers' provision of environmental public goods. These studies include farm size, agricultural land ownership, household assets, enterprise mix, productivity per hectare, age, experience, gender, formal education, off farm employment, experience of agri-environment schemes and availability of suitable equipment (Campos et al. 2012; Hartter 2010; Johnston and Duke 2009; Ma et al. 2012; Pagiola et al. 2010; Poppenborg and Koellner 2013; Shultz 2005; Winter et al. 2007; Yu and Belcher 2011).

Access to adequate information about the benefits of ecosystem services also emerged as an extremely relevant factor driving farmers' participation in nature conservation programs (Opdam et al. 2015; Zanella et al. 2014). But the relationship between information and participation is not straightforward (Zanella et al. 2014). Receivers interpret and disentangle framed information based on their values and beliefs related to nature (Opdam et al. 2015). Still, farmers decline to participate in nature conservation programs when they are informed that the programs may lead to negative consequences for themselves (Zanella et al. 2014).

Interests, social norms and beliefs also determine how receivers respond to information. Rational choice theory suggests that information implying losses to the receiver is likely to be disregarded. For example, when information calls for preserving ecosystem services which cost time or investment, the probability that the receiver will respond positively to the information is low (Opdam et al. 2015). Also, peer influence has been identified as influential in receiver's responses to the information (Yu and Belcher 2011). In certain cases, such as community engagement workshops, it would be strongly appreciated if all farmers commit to protect the ecosystem services of agriculture and the receiver wills to choose pro-environmental behaviors. In addition to the above-mentioned factors that determine if the receiver is willing to understand the message, controlling beliefs may hamper actual behaviors when participants doubt whether they have the resources and ability to respond. For instance, the farmer may doubt whether he or she has sufficient knowledge and understanding to carry out ecological pest control (Opdam et al. 2015).

9.6.2 Determinants of Farmers' Pro-environmental Behavior Under Drought

Few studies focus on valuing ecosystem services of agriculture in the context of drought, particularly. Results of a survey conducted in Australia indicated that farmers value the importance of on-farm regulating and supporting services for maintaining productivity and sustainability of farming enterprise under drought, specifically (Page and Bellotti 2015). Findings of a research in Iran (Keshavarz and Karami 2016) also revealed that some farmers postpone protection measures when encountering a severe sustained drought. Low adoption of conservation practices by farmers suggests that they have expected limited productivity gains from conservation.

Based on the findings of Keshavarz and Karami (2016), there are two major barriers to pro-environmental behavior under drought: (1) the time required to implement the conservation practices; that means farmers with greater dependence on farm income adopt strategies that could be implemented in the short run and reduce the short-term negative impacts of drought, and (2) perception of costs of pro-environmental behavior; that states while the on-farm incomes are extremely reduced because of drought, irrigation and energy costs continue to increase. Therefore, financial constraints confined farmers to adjust to low-cost nature conservation measures. Moreover, debt-to-asset ratio affected the protection behavior. High debt-to-asset ratios prevented farmers from adoption of rather expensive conservation practices. This potentially degrading cycle could be broken through the application of schemes paying for ecosystem services (Smith and Sullivan 2014).

Also, findings of Keshavarz and Karami (2016) indicated that farmers' pro-environmental behavior was significantly influenced by perceived vulnerability. Perceived higher vulnerability to drought threats provisioning, regulating and supporting services of ecosystem and results in more positive evaluation of protection behavior and higher acceptance of conservation practices. Furthermore, results suggested that farmers' trust in their own capabilities to actually implement the conservation measures, i.e. self-efficacy, has a significant influence on their pro-environmental behavior under drought. While renters, who seek profit in the short term, were less likely to adopt conservation practices, owners adopted some protective measures in order to reduce the environmental impacts of drought.

Moreover, operators of large farms were more likely to adopt conservation measures with the aim of reducing pressure on drought-affected lands. Their findings illustrated that income is related to pro-environmental behavior, too. Farmers who attempted to develop off-farm income sources and those who earned more money from agriculture were less likely to adopt pro-environmental measures (Keshavarz and Karami 2016).

9.7 Ecosystem-Based Adaptation of Farmers to Drought

Historically, communities in semi-arid dryland ecosystems have adapted to extreme weather conditions, such as drought (Egoh et al. 2012; Keshavarz and Karami 2016; Shackleton and Shackleton 2006). However, climate change, coupled with socioeconomic, cultural and political changes, have increased their vulnerabilities and undermined the supply, utilization and management of ecosystem services, over the last few decades (Boafo et al. 2016). In order to reduce the adverse effects of drought and ensure farm-based sustainable livelihoods, adaptation of agricultural sector is imperative. Adaptation to drought is recognized as a process that moderate drought-induced risks and supports system transformation through an understanding of the impacts on the system as well as the attributes of the exposed system that drives vulnerability (Schoon et al. 2011). There is a wide array of farm management strategies that can help farmers improve their agricultural production systems and increase the resiliency of their systems to drought (Wezel et al. 2014). However, adaptation is location specific and depends on many socio-economic and agro-ecological factors and also climate conditions. Therefore, there is no single solution which can fit for all regions or countries (Alam 2015).

Research results show that in Ethiopia, Jamaica and South Africa, using different crop varieties, sacrificing a portion of the crops under cultivation, crop diversification, changing planting dates, switching from on-farm to off-farm activities, seasonal migration, borrowing money to buy water, increased use of irrigation and developed soil and water conservation practices have been the most common adaptation strategies to drought (Bryan et al. 2009; Campbell et al. 2011; Deressa et al. 2009; Ifejika Speranza and Scholz 2013; Nhemachena and Hassan 2007). Moreover, various ex post adaptation strategies, e.g. crop diversification, changing cropping intensity, crop mix, crop type and location, and ex ante strategies, e.g. crop insurance, extension services, income diversification, food reservation and storage, migration, land-use change and adoption of new technologies, have been considered by Indian farmers in order to reduce drought impacts (Mwinjaka et al. 2010). Meantime, in Europe, farmers have inclined to redesign their cropping systems to cope with drought by introducing new crops with higher water use efficiency and reducing water consumption through crop rotation (Huntjens et al. 2010; Willaume et al. 2014).

Findings of researches conducted in China have revealed that implementing comprehensive adaptive measures, such as changing varieties, adjusting seeding and/or harvesting dates, enhancing irrigation intensity, implementing precise irrigation approaches and diversifying both on- and off-farm activities are essential to mitigate negative impacts of drought (Sun et al. 2012; Wang et al. 2015; Zhang et al. 2011). Also, results of the studies in Iran have indicated that growing new crop varieties, reducing cultivation area, increasing fertilizer use, using conservation or minimizing tillage and leveling land, water harvesting, re-excavation of rivers and canal, improving water transfer systems, constructing water reservoir, use of treated wastewater, water resource exploitation, digging irrigation well and deepening well, crop intensification, crop diversification, changing cropping pattern, income diversification and expenditure management adaptation strategies have been used by farmers in response to drought (Keshavarz et al. 2014; Keshavarz and Karami 2014, 2016; Sheidaei et al. 2016).

Despite the established nexus between drought adaptation and sustainable agricultural practices, adaptation research does not take into account sustainability issues in designing and implementing adaptation policies and strategies (Wall and Smith 2005). Whereas, many smallholder farmers are already implementing adaptation measures that maintain complex agro-biodiversity that results in higher capacity of their production units to resist, cope with and/or recover from drought (Lin 2007). Moreover, several authors and international organizations have highlighted the general importance and benefits of ecosystem-based strategies for adaptation to climate change (Vignola et al. 2015). However, few studies have defined what ecosystem-based adaptation means in the context of agriculture and they have used this definition to identify which practices are already in place.

Identifying ecosystem-based management practices in agricultural systems and landscapes can help farmers adapt to drought and also conserve the agro-ecosystems' capacity to provide ecosystem goods and services (Tilman et al. 2002). Ecosystem-based adaptation is defined as agricultural management practices which use or take advantage of biodiversity or ecosystem services or processes to help increase the ability of crops or livestock to adapt to climate change, in agricultural systems (SCBD 2009; Vignola et al. 2015). Under this definition, ecosystem-based agricultural practices must be based on conservation, restoration and sustainable management of biodiversity and ecosystem services, such as nutrient cycling and water regulation (Vignola et al. 2015). Ecosystem-based adaptation practices can be considered appropriate or useful for smallholder farmers if they help farmers to improve their livelihoods, increase or diversify their sources of income generation, take advantage of local knowledge, rely on local inputs and have low implementation and labor costs (Vignola et al. 2015). For example, many empirical studies have asserted that diverse agroforestry systems can ensure a broader source of crop resistance-capacity against the negative effects of drought and also enhance food security of smallholder farmers and increase or diversify their sources of income generation (Altieri and Nicholls 2008; Jackson et al. 2010).

Various drought management strategies are already used in developing countries to protect soil and water resources or deliver environmental flows during drought (Campbell et al. 2011; Keshavarz et al. 2010; Keshavarz and Karami 2016; Robertson et al. 2014; Swinton et al. 2014; Vignola et al. 2015). Use of shade trees, as one of the adaptation measures, can help to ensure continued provision of key ecosystem services, e.g. pollination, natural pest control and conservation of water and soil. Also, use of shade trees assures more stable production under climate-related stresses and provides clear socioeconomic benefits to smallholder farmers (Vignola et al. 2015). However, this practice often results in lower yields in normal years and requires significant knowledge, technical skills and labor for site-specific management of shade (Avelino et al. 2011; Lopez-Bravo et al. 2012). Furthermore, use of mulching or local species, as winter cover crops, can help conserving soil structure, humidity and nutrients. Utilizing minimum or no-till practices and also crop rotation strategies can help restoring soil fertility without reducing expected crop revenue and also ensuring continued provision of nutrient regulation services on which farming depends (Lavorel et al. 2015; Robertson et al. 2014).

On the other hand, some adaptation strategies are clearly not ecosystem-based including converting natural habitats to cultivated land, construction of dams for irrigation water, use of inorganic fertilizers and pesticides, application of genetically modified varieties and relocation of crops (Alam 2015; Keshavarz and Karami 2016; Davis et al. 2015; Vignola et al. 2015). These adaptation strategies substitute the role of agricultural systems in providing ecosystem functions and services. However, the last decades have seen an intensifying use of these coping strategies, in many drought-prone areas.

It is worth noting that, there are some key limitations to ecosystem-based adaptation practices which can hamper their adoption by smallholder farmers, especially in developing countries. Some of these practices, such as using cover crops, require farmers to make difficult trade-offs between long-term adaptation benefits and short-term costs, e.g. labor investment needed for their establishment and maintenance (Jha et al. 2014).

9.8 Adaptive Governance of Ecosystem Services Under Drought

Increasing frequency of droughts makes farm level adaptation difficult. Appropriate drought management policies are necessary to facilitate successful forms of adaptive responses and enhance resiliency of individuals and communities, in changing socio-ecological systems (O'Brien et al. 2006; Shinn 2016). While pre-drought planning can anticipate the spread of drought risk through communities and devise actions that will limit the negative impacts, many developing countries are still applying a traditional, reactive approach that relies on crisis management (Karami 2009; Wilhite and Svoboda 2000).

For many years, policymakers and administrators of many developing countries assumed drought as a rare occurrence that did not require preparation (Kartez and Kelly 1988). However, managing drought as a crisis has disrupted many provisioning, regulating and supporting ecosystem services (Dey et al. 2011; Keshavarz and Karami 2016). Meantime, recent changes in the natural ecosystems and complexity of socio-ecological systems have increased the need for a comprehensive understanding of drought, which differs significantly from general institutional assumptions (Keshavarz and Karami 2013). The dynamic patterns, complexity and nonstructural nature of droughts emphasize the importance of adaptive institutions and governance strategies which focus on 'learning to live with droughts', rather than 'controlling them' (Farhad et al. 2015; Mann et al. 2015).

Although the ecosystem service approach often takes into account a range of issues about socio-ecological systems, they do not provide direct solutions to ecosystem services' governance. The limited attention to policy implementation and governance in the ecosystem service is due to the fact that ecosystem service approaches do not take existing administrative and governance structures and practices as a starting point (Primmer and Furman 2012). While existing adaptive management studies have proposed new governance strategies that are able to respond to changing biophysical dynamics effectively (Shinn 2016), these theoretical constructs often pay poor attention to the feasibility of implementation and issues of power within multi-scalar environmental governance systems (Eriksen et al. 2015). In this way, those designing polices and making institutional adaptation decisions are struggling with integrating vulnerability of ecosystem services and the underlying ecosystem functions of agriculture into sustainable use and management (Primmer and Furman 2012).

The design and implementation become complicated in the context of drought due to the fact that interventions can contradict or conflict with one another (Loft et al. 2015). Besides primary environmental objectives, drought management policies often target non-environmental objectives, e.g. rural development and sustainable livelihood management, whereas achieving one policy objective, like significant reduction of chemical fertilizers and pesticides and also irrigation water use, may have negative impacts on the livelihoods of small scale farmers. Thus, inevitable trade-off should be considered between achieving environmental goals and satisfying other policy objectives, in many cases (Loft et al. 2015). Policies that focus on providing few ecosystem services such as food may erode the very structures and functions that maintain the state required to produce the services of interest and eventually they may lead to declined or collapsed ecosystem services (Gunderson et al. 2016). Consequently, adaptive governance is required to identify essential structures and functions and their responses to the environmental crisis, i.e. drought, to avoid critical thresholds (McFadden et al. 2011; Birge et al. 2016).

Adaptive governance is a set of institutions and frameworks that facilitate and foster adaptive management (Gunderson and Light 2006). Adaptive management is based on the idea that acknowledging uncertainty is imperative (Green et al. 2015) and environmental governance systems should have the ability to change activities based on new insights and experiences (Pahl-Wostl 2009). Adaptive governance assists farmers to respond uncertainties effectively and make necessary changes before negative impacts occur (Ostrom 2007). Adaptive governance can be contrasted with other forms of governance in key attributes of: (i) engaging formal and informal institutions, (ii) cross-scale interactions and poly-centricity, and (iii) focus on production and dissemination of new social and ecological knowledge (Chaffin et al. 2014b; Pahl-Wostl et al. 2007). Based on the adaptive governance approach, engaging those formal and informal actors who understand, manage and benefit from the ecosystem services (Primmer and Furman 2012) leads to higher adaptive capacity for resource management and governance systems (Pahl-Wostl 2009). Poly-centricity, i.e. smaller, more local units of governance existing within large, more general ones, assists effective response to change and uncertainty (Bakker and Morinville 2013; Chaffin et al. 2014a).

There are some potential problems with adaptive governance systems. First, the feasibility of implementing adaptive strategies in places with limited resources is questionable (Bakker and Morinville 2013). In examining governance barriers in Mozambique (Sietz et al. 2011), Australia (Mukheibir et al. 2013) and Iran (Keshavarz and Karami 2013), resource limitations including human, technical and financial constrains have been reported as a key challenge for adaptive management which comprises competing priorities due to limited operational resourcing. Second, poor communication and coordination between communities and government (Mukheibir et al. 2013) might lead to local or community needs being overlooked in adaptation efforts. Limited engagement between formal national adaptation efforts and communities is considered as the most common governance barrier in many developing countries such as Kiribati (Kuruppu 2009), Vanuatu and Solomon Islands (Wickham et al. 2009), Malaysia (Mann et al. 2015), Iran (Karami 2009; Karami and Keshavarz 2016) and Haiti (Pelling 2011). Third, a focus on adaptive governance might increase attention to adaptation at the expense of important mitigation efforts (Bakker and Morinville 2013). Fourth, adaptive governance systems have the potential to prevent power dynamics between the actors involved in governance (Bakker and Morinville 2013).

The key message from the body of literature on this topic is that adaptive governance requires joint actions of multiple parties (Farhad et al. 2015). To achieve this goal, empowering local people through participatory processes is imperative (Ronneberg et al. 2013). However, most adaptation efforts aim to strengthen four main types of activities that are deemed to improve the capacity of farmers to manage ecosystem services of agriculture under drought including: (i) developing new technologies such as early warning systems, (ii) facilitating government support, subsidies, insurance, technical assistance, etc., (iii) assisting farmers in accessing credit, capital and risk-insurance, and (iv) adapting farm management practices (Smit and Skinner 2002; Vignola et al. 2015; Wang et al. 2015; Etemadi and Karami 2016).

A survey conducted in southwest Iran (Keshavarz and Karami 2013) indicated that government does not have the capacity to provide the first three activities in the short-term (Table 9.3) due to the time lag and external support required to put in place the necessary enabling conditions, i.e. appropriate policy, governance structure and process, resource and infrastructure. According to the survey, the government lacks comprehensive drought management plan; including an effective early warning system and preparedness schemes; an efficient system for continuous

Table 9.3 Macro-level drought management strategies in Iran (adapted from Keshavarz and Karami 2013 with minor changes)

Strategies	Extent of use†	Strategies	Extent of use†
Conserving water resources	2.63	Developing emergency response systems	2.68
Conserving soil resources	2.17	Allocating drought relief funds	2.53
Encouraging reduced public water consumption	1.37	Providing technical support	2.86
Introducing proper farm management technologies	2.41	Implementing and improving early warning systems	1.51
Establishing crop improvement programs	2.47	Setting up a system of information management	1.97
Introducing drought resistant varieties	2.44	Improving drought management information through extension systems	2.66
Promoting land consolidation and farm development programs	2.22	Establishing awareness-raising campaign to promote household economy	1.83
Promoting land leveling programs	2.71	Leading education and awareness campaigns using media (TV and radio)	2.36
Promoting optimal crop patterns	1.98	Encouraging local participation	2.19

†The mean values relate to a five-point scale from 1 = less than 30% to 5 = 91–100%

assessment of drought impacts and dissemination of accurate information to reduce vulnerability. Also, while different drought assistance schemes are implemented for affected rural communities, they are not sustained through awareness-raising approaches (Table 9.3). Therefore, an immediate and direct way to help farmers providing agricultural ecosystem services is to focus on helping them to use drought management practices based on agro-biodiversity and ecosystem services that provide adaptation benefits (van Noordwijk et al. 2011).

9.9 The Way Forward: Designing Drought Resilient Socio-Ecological Agricultural Systems

Agricultural management practices are keys to realizing the benefits of ecosystem services (Power 2010). However, rising human population, income growth and changing dietary preferences (MEA 2005) have driven agriculture to higher intensities. Intensification of crop production, i.e. conventional agriculture, can increase ecosystem disservices, particularly in developing countries (Zhang et al. 2007). Therefore, directing intensified agriculture in a way that it enhances delivery of non-provisioning ecosystem services, i.e. multifunctional agriculture, is imperative (Robertson et al. 2014). However, in many developing countries, conventional agriculture is still dominant and transition to multifunctional agriculture requires revisiting policies that are recently undermining adoption of this agricultural paradigm and also designing new policy instruments to improve our understanding about multifunctional agriculture and its non-provisioning ecosystem services.

An increasing occurrence of drought has threatened the crops' natural resilience and has reduced provisioning services of conventional agriculture (Lereboullet et al. 2013). Moreover, drought has directly influenced hydrological ecosystem services, i.e. water quality regulations and support of natural habitats (Jiang et al. 2016). These challenges will be magnified in the face of climate change. Climate change is expected to significantly influence the supply of ecosystem services across all socio-ecological agricultural systems of the world (Boafo et al. 2016). However, developing nations are predicted to be among the most vulnerable countries to climate change because they depend heavily on agriculture and lack the required infrastructures to respond effectively to increased variability, i.e. drought, in addition to having limited capitals to invest on innovative adaptations (Lybbert and Sumner 2012).

In order to reduce the negative impacts of drought and preserve ecosystem services of agriculture, policy-makers and resource managers need to find effective ways of building resilience of socio-ecological agricultural systems. Delaying actions will result in further environmental degradation and will detach agriculture more from its biological roots. It seems that building resilient socio-ecological agricultural systems through management of ecosystem processes under drought has the potential to increase food and fuel production, i.e. provisioning services, while it can also minimize some of the negative impacts of agricultural intensification on non-provisioning ecosystem services (Power 2010). Resilience of an ecosystem relates to the functioning of the system, rather than the stability of its components or the ability to maintain a steady ecological state (Adger and O'Riordan 2000). Therefore, resilience of socio-ecological agricultural systems can be achieved when a high level of accessibility to food, health care, credit, public programs, markets, alternative technologies and education is provided (McCune et al. 2012).

In many developing countries, designing socio-ecological agricultural systems that are productive, resilient and able to deliver a rich suite of ecosystem services (Robertson et al. 2014) in the context of drought is ignored. For most developing countries, responses to droughts are reactive, reflecting what is commonly referred to as the crisis management approach. However, this approach is frequently ineffective, poorly coordinated and untimely, and more importantly, it does little to reduce the risks associated with drought (Karami 2009; Keshavarz and Karami 2013; Miyan 2015). Therefore, governments should play more proactive roles than hitherto. With this regard, clear governance arrangements, appropriate policy and legislation should be considered to reduce drought related risks and improve ecosystem services of agriculture. Adoption of drought resilient ecosystem based agriculture (DREbA) in appropriate farming systems and encouraging continued use of this approach in areas where farmers are already applying DREbA can increase adaptation of the agriculture sector to drought. Also, use of drought resilient ecosystem based adaptation strategies offers opportunities to help farmers adapt to climate change while providing sustainable livelihood and environmental co-benefits (Vignola et al. 2015). Many drought coping strategies can help maintain the ability of agricultural systems to provide ecosystem services, e.g. soil and water conservation, reduced tillage or no-till management, use of fallows, crop rotation and crop and farm diversification. However, the current natural, structural, financial, technical and human constraints limit widespread adoption of these strategies among smallholder farmers. In order to increase adoption of drought resilient ecosystem based practices, policy makers and resource managers should promote the use of appropriate adaptive strategies and support their widespread adoption.

Three ways are proposed in which the use of drought resilient ecosystem based adaptation practices can be promoted by farmers of developing countries. First, it is important to improve our understanding and collate scientific evidence of the long-term effectiveness of different drought coping strategies and ecosystem based adaptation practices in order to enhance the resilience of crops, livestock and farming systems in the face of climate change (FAO 2013). There are still few studies that explicitly compare the relative performance and cost-effectiveness of ecosystem based drought coping strategies versus non-ecosystem based drought coping options under different agricultural systems (Vignola et al. 2015). Also, the long-term impacts of many changes that have occurred in natural ecosystems should be recognized. It seems likely that developing countries have greatly underestimated the impact of past land use intensification. As an outcome, they are ill-prepared for the negative consequences of any future intensification (Davis et al. 2015).

In addition, more information is needed about which drought coping ecosystem based adaptation strategies are the most appropriate for farmers living in different socioeconomic and agro-ecological contexts (Vignola et al. 2015), by considering the ecosystem services, especially provisioning ecosystem services, as parts of a wider livelihood strategy portfolio of natural resource dependent communities (Boafo et al. 2016). This is due to the fact that the relative effectiveness of coping practices is likely to be highly context-dependent and no one-size-fit adaptation strategy can enhance the resilience of all households and communities. Ecosystem based adaptation is strongly connected to the social elements of communities and involves local knowledge, values and practices, so joint actions of multiple governmental and local parties are required. Policy tools based on internal motivations, such as participatory policies for determining appropriate adaptation strategies, often entail considerable time and risk. However, their effects on building resilience of socio-ecological systems is more evident (Dedeurwaerdere et al. 2016).

Second, better articulation of agricultural and climate change policies is needed to promote incentives or actions that achieve production targets, maintain the ecosystem services of agriculture and improve farmer livelihoods under drought, simultaneously (Vignola et al. 2015). Indeed, agricultural producers, especially smallholder farmers, are likely to adopt drought resilient ecosystem based adaptation strategies if they perceive that these practices will help them achieve their production goals (Page and Bellotti 2015) or if they receive direct incentives for their implementation, e.g. payment for ecosystem services (Robertson et al. 2014). This means that various ecosystem based adaptation practices would need to be included in payment schemes for ecosystem services. Also, recognition of farmers' values of ecosystem services is imperative to establish which regulating and supporting services are the most important services to them, as well as the perceived drivers of drought and climate change (Smith and Sullivan 2014). Through explicit financial and non-financial supports, i.e. creating a better information and evidence basis, capacity building and technical support for improved protection of agro-ecosystems and the services they provide, farmers will be in a much better position to secure and enhance their important societal role in the future (Mann et al. 2015; Smith and Sullivan 2014). At the same time, governments should revisit policies that promote simplification of agro-ecosystems and use of agrochemicals and fossil fuels (Altieri and Koohafkan 2008).

Third, it is recommended that governments strengthen and provide evolving support for agricultural extension programs, technical assistance services and universities to assure that their curricula and outreach activities include the promotion of drought resilient ecosystem based adaptation practices (Vignola et al. 2015). Many developing countries have reduced their extension programs noticeably while the need for such support, especially for smallholder farmers facing severe sustained droughts, is greater than any other time (Chang 2009; Porter et al. 2014). Effective extension services are imperative to promote information exchange on drought management and ecosystem based adaptation activities from research centers to agricultural producers and vice versa (Vignola et al. 2010). Therefore, greater investment in research, training and transfer of technology services is required to ensure that farmers have enough access to the best available resources and information related to drought adaptation strategies and they can make informed decisions about their farming systems (Keshavarz and Karami 2014; Vignola et al. 2010).

References

Adger WN, O'Riordan T (2000) Population, adaptation and resilience. In: O'Riordan T (ed) Environmental science for environmental management. Longman, Harlow

Adnan S (1993) Living without floods: lessons from the drought of 1992. Research and Advisory Services, Dhaka

Aguilar R, Ashworth L, Galetto L, Aizen MA (2006) Plant reproductive susceptibility to habitat fragmentation: review and synthesis through a meta-analysis. Ecol Lett 9:968–980. https://doi.org/10.1111/j.1461-0248.2006.00927.x

Alam K (2015) Farmers' adaptation to water scarcity in drought-prone environments: a case study of Rajshahi District, Bangladesh. Agric Water Manag 148:196–206. https://doi.org/10.1016/j.agwat.2014.10.011

Alberti M (2005) The effects of urban patterns on ecosystem function. Int Reg Sci Rev 28(2):168–192. https://doi.org/10.1177/0160017605275160

Alcamo J, ÓNeill BC (2005) Changes in ecosystem services and their drivers across the scenarios. In: Millennium Ecosystem Assessment (ed) Ecosystem and human well-being, vol 2. Scenarios. Island Press, Washington DC

Altieri MA, Koohafkan P (2008) Enduring farms: climate change, smallholders and traditional farming communities. Third World Network (TWN), Penang, Malasya

Altieri MA, Nicholls CI (2008) Scaling up agroecological approaches for food sovereignty in Latin America. Development 51:472–480. https://doi.org/10.1057/dev.2008.68

Amirkhani S, Chizari M (2010) Factors influencing drought management in Varamin Township. In: Third congress of agricultural extension and natural resources, Mashhad, Iran

Armbrecht I, Perfecto I (2003) Litter-twig dwelling ant species richness and predation potential within a forest fragment and neighboring coffee plantations of contrasting habitat quality in Mexico. Agric Ecosyst Environ 97:107–115. https://doi.org/10.1016/S0167-8809(03)00128-2

Aryal JP, Sapkota TB, Jat ML, Bishno ID (2015) On-farm economic and environmental impact of zero-tillage wheat: a case of north–west India. Exp Agric 51:1–16. https://doi.org/10.1017/S001447971400012X

Avelino J, Ten Hoopen G, deClerck FAJ (2011) Ecological mechanisms for pest and disease control in coffee and cacao agroecosystems of the Neotropics. In: Rapidel B, deClerck FAJ, Le Coq J, Beer J (eds) Ecosystem services from agriculture and agroforestry: measurement and payment. Earthscan Publications, London, UK

Backeberg R, Viljoen F (2003) Drought management in South Africa. Paper presented at a workshop of the ICID working group on irrigation under drought and water scarcity, Tehran, Iran

Bakker K, Morinville C (2013) The governance dimensions of water security: a review. Philos Trans R Soc A 371. http://dx.doi.org/10.1098/rsta.2013.0116

Banerjee O, Bark R, Connor J, Crossman ND (2013) An ecosystem services approach to estimating economic losses associated with drought. Ecol Econ 91:19–27. https://doi.org/10.1016/j.ecolecon.2013.03.022

Barker T (2007) Technical summary. In: Metz B, Davidson OR, Bosch P, Dave R, Meyer L (eds) Climate change 2007: mitigation. Intergovernmental panel on climate change. Cambridge University Press, Cambridge

Basch G, Friedrich T, Kassam A, Gonzalez-Sanchez E (2015) Conservation agriculture in Europe. In: Farooq M, Siddique K (eds) Conservation agriculture. Springer, Cham. https://doi.org/10.1007/978-3-319-11620-4_15

Bellard C, Bertelsmeier C, Leadley P, Thuiller W, Courchamp F (2012) Impacts of climate change on the future of biodiversity. Ecol Lett 15(4):365–377. https://doi.org/10.1111/j.1461-0248.2011.01736.x

Bernués A, Rorrígez-Ortega T, Alfnes F, Clemetsen M, Eik LO (2015) Quantifying the multifunctionality of fjord and mountain agriculture by means of sociocultural and economic valuation of ecosystem services. Land Use Policy 48:170–178. https://doi.org/10.1016/j.landusepol.2015.05.022

Bernués A, Tello-García E, Rorrígez-Ortega T, Ripoll-Bosch R, Casasũs I (2016) Agricultural practices, ecosystem services and sustainability in High Nature Value farmland: unraveling the perceptions of farmers and nonfarmers. Land Use Policy 59:130–142. https://doi.org/10.1016/j.landusepol.2016.08.033

Beudert B, Klocking B, Schwartze R (2007) Selected case studies of forest- hydrological research in German low mountain ranges. In: Puhmann H, Schwarze R, Federov SF, Marunich SV (eds) Forest hydrology: results of research in Germany and Russia. German International Hydrological Program/Hydrology and Water Resources Program

Birge HE, Allen CR, Garmestani AS, Pope KL (2016) Adaptive management for ecosystem services. J Environ Manag 183:343–352. https://doi.org/10.1016/j.jenvman.2016.07.054

Biswas S, Swanson ME, Vacik H (2012) Natural resources depletion in hill areas of Bangladesh: a review. J Mt Sci 9(2):147–156. https://doi.org/10.1007/s11629-012-2028-z

Boafo YA, Saito O, Jasaw GS, Otsuki K, Takeuchi K (2016) Provisioning ecosystem services-sharing as a coping and adaptation strategy among rural communities in Ghana's semi-arid ecosystem. Ecosyst Serv 19:92–102. https://doi.org/10.1016/j.ecoser.2016.05.002

Bos JFFP, de Haan J, Sukkel W, Schils RLM (2014) Energy use and greenhouse gas emissions in organic and conventional farming systems in the Netherlands. NJAS- Wagening J Life Sci 68:61–70. https://doi.org/10.1016/j.njas.2013.12.003

Bostani A, Salari Sardari F, Adeli J (2009) Impact assessment of climate variability on water resources (a case of Darab County). In: The national conference on water crisis management. Marvdasht Branch of Azad Islamic University, Iran

Brainard DC, Bryant A, Noyes DC, Haramoto ER, Szendrei Z (2016) Evaluating pest-regulating services under conservation agriculture: a case study in snap beans. Agric Ecosyst Environ 235:142–154. https://doi.org/10.1016/j.agee.2016.09.032

Bremer LL, Farley KA, Lopez-Carr D (2014) What factors influence participation in payment for ecosystem services programs? An evaluation of Ecuador's SocioParamo program. Land Use Policy 36:122–133. https://doi.org/10.1016/j.landusepol.2013.08.002

Breshears DD, Hoffman LL, Graumlich LJ (2011) When ecosystem services crash: preparing for big, fast, patchy climate change. AMBIO: A J Hum Environ 40:256–263. https://doi.org/10.1007/s13280-010-0106-4

Broch SW, Strange N, Jacobsen JB, Wilson KA (2013) Farmers' willingness to provide ecosystem services and effects of their spatial distribution. Ecol Econ 92:78–86. https://doi.org/10.1016/j.ecolecon.2011.12.017

Bryan E, Deressa TT, Gbetibouo GA, Ringler C (2009) Adaptation to climate change in Ethiopia and South Africa: Options and constraints. Environ Sci Policy 12:413–426. https://doi.org/10.1016/j.envsci.2008.11.002

Burgess PJ, Morris J (2009) Agricultural technology and land use features: The UK case. Land Use Policy 26S:S222–S229. https://doi.org/10.1016/j.landusepol.2009.08.029

Calzolari C, Ungaro F, Filippi N, Guermandi M, Malucelli F, Marchi N, Staffilani F, Tarocco P (2016) A methodological framework to assess the multiple contributions of soils to ecosystem services delivery at regional scale. Geoderma 261:190–203. https://doi.org/10.1016/j.geoderma.2015.07.013

Campbell D, Barker D, McGregor D (2011) Dealing with drought: Small farmers and environmental hazards in southern St. Elizabeth, Jamaica. Appl Geogr 31:146–158. https://doi.org/10.1016/j.apgeog.2010.03.007

Campos M, Velázquez A, Verdinelli GB, Priego-Santander ÁG, McCall M, Boada M (2012) Rural people's knowledge and perception of landscape: a case study from the Mexican Pacific Coast. Soc Nat Res 25:759–774. https://doi.org/10.1080/08941920.2011.606458

Canadell JG, Ciais P, Dhakal S, Dolman H, Friedlingstein P, Gurney KR, Held A, Jackson RB, Le Que´re´ C, Malone EL (2010) Interactions of the carbon cycle, human activity, and the climate system: a research portfolio. Curr Opin Environ Sustain 2:301–311. https://doi.org/10.1016/j.cosust.2010.08.003

Chaffin BC, Craig RK, Gosnell H (2014a) Resilience, adaptation and transformation in the Klamath River basin social-ecological system. Idaho Law Rev 51:157–193

Chaffin BC, Gosnell H, Cosens BA (2014b) A decade of adaptive governance scholarship: Synthesis and future directions. Ecol Soc 19(3):56. https://doi.org/10.5751/ES-06824-190356

Chang HJ (2009) Rethinking public policy in agriculture: Lessons from history, distant and recent. J Peasant Stud 36:477–515. https://doi.org/10.1080/03066150903142741

Chaplin-Kramer R, O'Rourke ME, Blitzer EJ, Kremen C (2011) A meta-analysis of crop pest and natural enemy response to landscape complexity. Ecol Lett 14:922–932. https://doi.org/10.1111/j.1461-0248.2011.01642.x

Chary GR, Vittal KPR, Venkateswarlu B, Mishra PK, Rao GGSN, Pratibha G, Rao KV, Sharma KL, Rajeshwara Rao G (2010) Drought hazards and mitigation measures. In: Jha MK (ed) Natural and anthropogenic disasters: vulnerability, preparedness and mitigation. Springer, Dordrecht, Netherlands

Christensen T, Pedersen AB, Nielsen HO, Mørkbak MR, Hasler B, Denver S (2011) Determinants of farmers' willingness to participate in subsidy schemes for pesticide free buffer zones-a choice experiment study. Ecol Econ 70:1558–1564. https://doi.org/10.1016/j.ecolecon.2011.03.021

Cook ER, Woodhouse C, Eakin CM, Meko DM, Stahle DW (2004) Long-term aridity changes in the western United States. Science 306:1015–1018. https://doi.org/10.1126/science.1102586

Costanza R, d'Arge R, De Groot R, Farber S, Grasso M, Hannon B, Limburg K, Naeem S, O'neill RV, Paruelo J (1997) The value of the world's ecosystem services and natural capital. Nature 387:253–260

Covich AP (2009) Emerging climate change impacts on freshwater resources: a perspective on transformed watersheds. http://rff.org/rff/documents/RFF-Rpt-Adaptation-Covich.pdf. Accessed 9 Nov 2016

Craheix D, Angevin F, Dore T, de Tourdonnet S (2016) Using a multicriteria assessment model to evaluate the sustainability of conservation agriculture at the cropping system level in France. Eur J Agron 76:75–86. https://doi.org/10.1016/j.eja.2016.02.002

Cramer MD, Hawkins HJ, Verboom GA (2009) The importance of nutritional regulation of plant water flux. Oecologia 161:15–24. https://doi.org/10.1007/s00442-009-1364-3

Crowder DW, Northfield TD, Strand MR, Snyder WE (2010) Organic agriculture promotes evenness and natural pest control. Nature 466(7302):109–112. https://doi.org/10.1038/nature09183

Dai A (2011) Drought under global warming: a review. WIREs Clim Change 2:45–65. https://doi.org/10.1002/wcc.81

Dale VH, Polasky S (2007) Measures of the effects of agricultural practices on ecosystem services. Ecol Econ 64:286–296. https://doi.org/10.1016/j.ecolecon.2007.05.009

Davis J, O'Grady AP, Dale A, Arthington AH, Gell PA, Driver PD, Bond N, Casanova M, Finlayson M, Watts RJ, Capon SJ, Nagelkerken I, Tingley R, Fry B, Page TJ, Specht A (2015) When trends intersect: the challenge of protecting freshwater ecosystems under multiple land use and hydrological intensification scenarios. Sci Total Environ 534:65–78. https://doi.org/10.1016/j.scitotenv.2015.03.127

Dedeurwaerdere T, Admiraal J, Beringer A, Bonaiuto F, Cicero L, Fernandez-Wulff P, Hagens J, Hiedanpää J, Knights P, Molinario E, Melindi-Ghidi P, Popa F, Šilc U, Soethe N, Soininen T, Vivero JL (2016) Combining internal and external motivations in multi-actor governance arrangements for biodiversity and ecosystem services. Environ Sci Policy 58:1–10. https://doi.org/10.1016/j.envsci.2015.12.003

Deressa TT, Hassan RM, Ringler C, Alemu T, Yusef M (2009) Determinants of farmers' choice of adaptation methods to climate change in the Nile Basin of Ethiopia. Global Environ Change 19(2):248–255. https://doi.org/10.1016/j.gloenvcha.2009.01.002

Dey NC, Alam MS, Sajjan AK, Bhuiyan MA, Ghose L, Ibaraki Y, Karimi F (2011) Assessing environmental and health impact of drought in the northwest Bangladesh. J Environ Sci Nat Res 4(2):89–97. https://doi.org/10.3329/jesnr.v4i2.10141

Duru M, Therond O, Martin G, Martin-Clouaire R, Magne MA, Justes E, Journet EP, Aubertot JN, Savary S, Bergez JE, Sarthou J (2015) How to implement biodiversity-based agriculture to enhance ecosystem services: a review. Agron Sustain Dev 35:1259–1281. https://doi.org/10.1007/s13593-015-0306-1

Egoh BN, O'Farrell PJ, Charef A, Gurney JL, Koellner T, Abi NH, Egoh M, Willemen L (2012) An African account of ecosystem service provision: use, threats and policy option for sustainable development. Ecosyst Serv 2:71–81. https://doi.org/10.1016/j.ecoser.2012.09.004

Eriksen SH, Nightingale AJ, Eakin H (2015) Reframing Adaptation: The political nature of climate change adaptation. Global Environ Change 35:523–533. https://doi.org/10.1016/j.gloenvcha.2015.09.014

Eskandari N (in Farsi, 2001) Review of the past 32 years average annual precipitation of Iran. Jihad-e-Agriculture Ministry of Iran

Etemadi M, Karami E (2016) Organic fig growers' adaptation and vulnerability to drought. J Arid Environ 124:142–149. https://doi.org/10.1016/j.jaridenv.2015.08.003

Evans JP (2009) 21st century climate change in the Middle East. Clim Change 92:417–432. https://doi.org/10.1007/s10584-008-9438-5
Fan F, Henriksen CB, Porter J (2016) Valuation of ecosystem services in organic cereal crop production systems with different management practices in relation to organic matter input. Ecosyst Serv 22:117–127. https://doi.org/10.1016/j.ecoser.2016.10.007
FAO (2013) Submission by the Food and Agriculture Organization of the United Nations (FAO) on the support to least developed and developing countries in the national adaptation plan process regarding the integration of agriculture, fisheries and forestry perspectives. FAO, Rome, Italy
FAOSTAT (2016) FAOSTAT database. http://faostat3.fao.org/download/Q/QA/E. Accessed 11 Oct 2016
Farhad S, Gual MA, Ruiz-Ballesteros E (2015) Linking governance and ecosystem services: the case of Isla Mayor(Andalusia, Spain). Land Use Policy 46:91–102. https://doi.org/10.1016/j.landusepol.2015.01.019
Fielding KS, Terry DJ, Masser BM, Hogg MA (2008) Integrating social identity theory and the theory of planned behaviour to explain decisions to engage in nature conservation practices. Br J Soc Psychol 47:23–48. https://doi.org/10.1348/014466607x206792
Foley JA, Ramankutty N, Brauman KA, Cassidy ES, Gerber JS, Johnston M, Mueller ND, O'Connell C, Ray DK, West PC, Balzer C, Bennett EM, Carpenter SR, Hill J, Monfreda C, Polasky S, Rockstrom J, Sheehan J, Siebert S, Tilman D, Zaks DPM (2011) Solutions for a cultivated planet. Nature 478:337–342. https://doi.org/10.1038/nature10452
Forster P, Jackson L, Lorenz S, Simelton E, Fraser E, Bahadur K (2012) Food security: near future projections of the impact of drought in Asia. Center for Low Carbon Future, UK
Gardiner MM, Landis DA, Gratton C, DiFonzo CD, O'Neal M, Chacon JM (2009) Landscape diversity enhances biological control of an introduced crop pest in the north-central USA. Ecol Appl 19:143–154
GDIS (2013) Global drought information system. http://www.drought.gov/gdm/current-conditions. Accessed 9 Nov 2016
Geist HJ, Lambin EF (2004) Dynamic causal patterns of desertification. Bioscience 54:817–829
Gianinazzi S, Gollotte A, Binet MN, van Tuinen D, Redecker D, Wipf D (2010) Agroecology: the key role of arbuscular mycorrhizas in ecosystem services. Mycorrhiza 20:519–530. https://doi.org/10.1007/s00572-010-0333-3
Godfray HCJ, Beddington JR, Crute IR, Haddad L, Lawrence D, Muir JF, Pretty J, Robinson S, Thomas SM, Toulmin C (2010) Food security: the challenge of feeding 9 billion people. Science 327:812–818. https://doi.org/10.1126/science.1185383
Gordon LJ, Finlayson CM, Falkenmark M (2010) Managing water in agriculture for food production and other ecosystem services. Agric Water Manag 97:512–519. https://doi.org/10.1016/j.agwat.2009.03.017
Gosling P, Shepherd M (2005) Long-term changes in soil fertility in organic arable farming systems in England, with particular reference to phosphorus and potassium. Agr Ecosyst Environ 105:425–432. https://doi.org/10.1016/j.agee.2004.03.007
Green OO, Garmestani AS, Allen CR, Gunderson LH, Ruhl JB, Arnold CA, Graham NAJ, Cosens B, Angeler DG, Chaffin BC, Holling CS (2015) Barriers and bridges to the integration of social-ecological resilience and law. Front Ecol Environ 13:332–337. https://doi.org/10.1890/140294
Greenland-Smith S, Brazner J, Sherren K (2016) Farmer perceptions of wetlands and water bodies: using social metrics as an alternative to ecosystem service valuation. Ecol Econ 126:58–69. https://doi.org/10.1016/j.ecolecon.2016.04.002
Greiner R, Gregg D (2011) Farmers' intrinsic motivations, barriers to the adoption of conservation practices and effectiveness of policy instruments: empirical evidence from northern Australia. Land Use Policy 28(1):257–265. https://doi.org/10.1016/j.landusepol.2010.06.006
Gunderson L, Light SS (2006) Adaptive management and adaptive governance in the everglades ecosystem. Policy Sci 39:323–334. https://doi.org/10.1007/s11077-006-9027-2

Gunderson LH, Cosens B, Garmestani AS (2016) Adaptive governance of riverine and wetland ecosystem goods and services. J Environ Manag 183(P2):353–360. https://doi.org/10.1016/j.jenvman.2016.05.024

Habiba U, Shaw R, Takeuchi Y (2011) Socioeconomic impact of drought in Bangladesh. In: Shaw R, Nguyen H (eds) Droughts in Asian monsoon region, community, environment and disaster risk management, vol 8. Emerald Group Publishing Limited

Haramoto ER (2014) Weed population dynamics, profitability and nitrogen loss in strip-tilled sweet corn and cabbage. PhD Dissertation, Michigan State University

Hartter J (2010) Resource use and ecosystem services in a forest park landscape. Soc Nat Res 23:207–223. https://doi.org/10.1080/08941920903360372

Harvey CA, Rakotobe ZL, Rao NS, Dave R, Razafimahatratra H, Rabarijohn RH, Rajaofara H, Mackinnon JI (2014) Extreme vulnerability of smallholder farmers to agricultural risks and climate change in Madagascar. Philos Trans R Soc B Biol Sci 369. https://doi.org/10.1098/rstb.2013.0089

He M, Dijkstra FA (2014) Drought effect on plant nitrogen and phosphorous: a meta-analysis. New Phytol 204:924–931. https://doi.org/10.1111/nph.12952

Hernández-Delgado (2015) The emerging threats of climate change on tropical coastal ecosystem services, public health, local economies and livelihood sustainability of small islands: cumulative impacts and synergies. Mar Pollut Bull 101:5–28. https://doi.org/10.1016/j.marpolbul.2015.09.018

Holt AR, Alix A, Thompson A, Maltby L (2016) Food production, ecosystem services and biodiversity: we can't have it all everywhere. Sci Total Environ. https://doi.org/10.1016/j.scitotenv.2016.07.139

Horrocks CA, Dungait JAJ, Cardenas LM, Heal KV (2014) Does extensification lead to enhanced provision of ecosystems services from soils in UK agriculture? Land Use Policy 38:123–128. https://doi.org/10.1016/j.landusepol.2013.10.023

Hrudey SE, Hrudey EJ (2004) Safe drinking water: lessons from recent out-breaks in affluent nations. IWA Publishing, London

Hueso S, Garcia C, Hernandez T (2012) Severe drought conditions modify the microbial community structure, size and activity in amended and un-amended soils. Soil Biol Biochem 50:167–173. https://doi.org/10.1016/j.soilbio.2012.03.026

Huntjens P, Pahl-Wostl C, Grin J (2010) Climate change adaptation in European river basins. Reg Environ Change 10:263–284. https://doi.org/10.1007/s10113-009-0108-6

IFAD; International Fund for Agricultural Development (2013) Smallholders, food security and the environment. International Fund for Agricultural Development, UNEP, Rome

Ifejika Speranza C, Scholz I (2013) Special issue adaptation to climate change: analyzing capacities in Africa. Reg Environ Change 13:471–475. https://doi.org/10.1007/s10113-013-0467-x

IPCC (2007) Contribution of working group III to the fourth assessment report of the Intergovernmental Panel on Climate Change. Cambridge University Press, Cambridge, UK

IPCC (2013) Climate change 2013: the physical science basis. In: Stocker TF, Qin D, Plattner GK, Tignor M, Allen SK, Boschung J, Nauels A, Xia Y, Bex V, Midgley PM (eds) Contribution of working group I to the fifth assessment report of the intergovernmental panel on climate change. Cambridge University Press, Cambridge

IPCC (2014a) Climate change 2014: Mitigation of climate change. In: Edenhofer OR, Pichs-Madruga Y, Sokona E, Farahani S, Kadner K, Seyboth A, Adler I, Baum S, Brunner P, Eickemeier B, Kriemann J, Savolainen S, Schlömer C, von Stechow T, Zwickel C, Minx JC (eds) Contribution of working group III to the fifth assessment report of the intergovernmental panel on climate change. Cambridge University Press, Cambridge

IPCC (2014b) Climate change 2014: synthesis report. In: Pachauri RK, Meyer L (eds) Cambridge University Press, Cambridge

IPCC Special Report (2012) Managing the risks of extreme events and disasters to advance climate change adaptation. Special report of the Intergovernmental Panel on Climate Change, IPCC working group II. Cambridge University Press, Cambridge

IRIN (2008) Yemen: Drought displaces thousands in mountainous northwest. http://www.irinnews.org/report/78048/yemen-drought-displaces-thousands-in-mountainous-northwes. Accessed 12 Oct 2016

Islamic Republic News Agency; IRNA (2010) Drought impacts on lakes, lagoons and livelihood of farmers in Fars province. http://www.irna.ir/fa/News. Accessed 02 Jun 2015

Islamic Republic News Agency; IRNA (2014) Drought impacts on agriculture sector of Fars province. http://www.irna.ir/fa/news. Accessed 24 Jan 2015

Jackson L, van Noordwijk M, Bengtsson J, Foster W, Lipper L, Pulleman M, Said M, Snaddon J, Vodouhe R (2010) Biodiversity and agricultural sustainability: from assessment to adaptive management. Curr Opin Environ Sustain 2:80–87. https://doi.org/10.1016/j.cosust.2010.02.007

Jam-e-jam (2014) Six lagoons of Fars province are still dried up. http://www.jamejamonline.ir/newspreview. Accessed 24 Jan 2015

Jat RA, Wani SP, Sahrawat KL (2012) Conservation agriculture in the semi-arid tropics: prospects and problems. Adv Agron 117:191–273. https://doi.org/10.1016/B978-0-12-394278-4.00004

Jat RK, Sapkota TB, Sing RG, Jat ML, Kumar M, Gupta RK (2014) Seven years of conservation agriculture in a rice–wheat rotation of Eastern Gangetic Plains of South Asia: yield trends and economic profitability. Field Crops Res 164:199–210. https://doi.org/10.1016/j.fcr.2014.04.015

Jha S, Bacon CM, Philpott SM, Ernesto Méndez V, Läderach P, Rice RA (2014) Shade coffee: update on a disappearing refuge for biodiversity. Bioscience 64:416–428. https://doi.org/10.1093/biosci/biu038

Jiang C, Li D, Wang D, Zhang L (2016) Quantification and assessment of changes in ecosystem services in the Three-River Headwaters Region, China as a result of climate variability and land cover change. Ecol Ind 66:199–211. https://doi.org/10.1016/j.ecolind.2016.01.051

Johnston RJ, Duke JM (2009) Willingness to pay for land preservation across states and jurisdictional scale: implications for benefit transfer. Land Econ 85(2):217–237

Jordan N, Warner KD (2010) Enhancing the multifunctionality of US Agriculture. Bioscience 60:60–66. https://doi.org/10.1525/bio.2010.60.1.10

Jose S (2009) Agroforestry for ecosystem services and environmental benefits: An overview. Agrofor Syst 76:1–10. https://doi.org/10.1007/s10457-009-9229-7

Kane JM, Meinhardt KA, Chang T, Cardall BL, Michalet R, Whitham TG (2011) Drought-induced mortality of a foundation species (Juniperus monosperma) promotes positive afterlife effects in understory vegetation. Plant Ecol 212:733–741. https://doi.org/10.1007/s11258-010-9859-x

Karami E (2009) Drought management: knowledge and information system. In: National drought seminar: issues and mitigation. Shiraz University, Iran, May 13–14

Karami E, Keshavarz M (2016) Natural resource conservation: the human dimensions. Iran Agric Ext Educ J 11(2):101–120 (in Farsi)

Karimi K, Karami E, Keshavarz M (2018) Vulnerability and adaptation of livestock producers to climate variability and change. Rangeland Ecol Manage 71 (2):175–184

Kartez JD, Kelly WJ (1988) Research based disaster planning: conditions for implementation. In: Comfort LK (ed) Managing disaster: strategies and policy perspectives. Duke University Press, Durham, USA

Kearns CA, Inouye DW, Waser NM (1998) Endangered mutualisms: the conservation of plant-pollinator interactions. Ann Rev Ecol Syst 29:83–112

Kenter JO, Hyde T, Christie M, Fazey I (2011) The importance of deliberation in valuing ecosystem services in developing countries—evidence from the Solomon Islands. Global Environ Change 21:505–521. https://doi.org/10.1016/j.gloenvcha.2011.01.001

Kertész A, Madarász B (2014) Conservation agriculture in Europe. Int Soil Water Conserv Res 2 (1):91–96. https://doi.org/10.1016/S2095-6339(15)30016-2

Keshavarz M, Karami E (2013) Institutional adaptation to drought: the case of Fars Agricultural Organization. J Environ Manag 127:61–68. https://doi.org/10.1016/j.jenvman.2013.04.032

Keshavarz M, Karami E (2014) Farmers' decision making process under drought. J Arid Environ 108:43–56. https://doi.org/10.1016/j.jaridenv.2014.03.006

Keshavarz M, Karami E (2016) Farmers' pro-environmental behavior under drought: an application of protection motivation theory. J Arid Environ 127:128–136. https://doi.org/10.1016/j.jaridenv.2015.11.010

Keshavarz M, Karami E, Kamgare-Haghighi A (2010) A typology of farmers' drought management. Am-Eurasian J Agric Environ Sci 7(4):415–426

Keshavarz M, Karami E, Vanclay F (2013) The social experience of drought in rural Iran. Land Use Policy 30(1):120–129. https://doi.org/10.1016/j.landusepol.2012.03.003

Keshavarz M, Karami E, Zibaie M (2014) Adaptation of Iranian farmers to climate variability and change. Reg Environ Change 14(3):1163–1174. https://doi.org/10.1007/s10113-013-0558-8

Kirchmann H, Bergstrom L, Katterer T, Mattsson L, Gesslein S (2007) Comparison of long-term organic and conventional crop-livestock systems on a previously nutrient-depleted soil in Sweden. Agron J 99:960–972. https://doi.org/10.2134/agronj2006.0061

Kragt ME, Robertson MJ (2014) Quantifying ecosystem services trade-offs from agricultural practices. Ecol Econ 102:147–157. https://doi.org/10.1016/j.ecolecon.2014.04.001

Kremen C, Williams NM, Thorp RW (2002) Crop pollination from native bees at risk from agricultural intensification. Proc Natl Acad Sci USA 99:16812–16816. https://doi.org/10.1073/pnas.262413599

Kuruppu N (2009) Adapting water resources to climate change in Kiribati: the importance of cultural values and meanings. Environ Sci Policy 12(7):799–809. https://doi.org/10.1016/j.envsci.2009.07.005

Kwayu EJ, Sallu SM, Paavola J (2014) Farmer participation in the equitable payments for watershed services in Morogoro, Tanzania. Ecosyst Serv 7:1–9. https://doi.org/10.1016/j.ecoser.2013.12.006

Lahmar R (2010) Adoption of conservation agriculture in Europe: lessons of the KASSA project. Land Use Policy 27:4–10. https://doi.org/10.1016/j.landusepol.2008.02.001

Landis DA (2016) Designing agricultural landscapes for biodiversity-based ecosystem services. Basic Appl Ecol. https://doi.org/10.1016/j.baae.2016.07.005

Lavorel S, Colloff MJ, McIntyre S, Doherty MD, Murphy HT, Metcalfe DJ, Dunlop M, Williams RJ, Wise RM, Williams KJ (2015) Ecological mechanisms underpinning climate adaptation services. Glob Change Biol 21:12–31. https://doi.org/10.1111/gcb.12689

Lavorel S, Grigulis K (2012) How fundamental plant functional trait relationships scale-up to trade-offs and synergies in ecosystem services. J Ecol 100:128–140. https://doi.org/10.1111/j.1365-2745.2011.01914.x

Lereboullet AL, Beltrando G, Bardsley DK (2013) Socio-ecological adaptation to climate change: a comparative case study from the Mediterranean wine industry in France and Australia. Agr Ecosyst Environ 164:273–285. https://doi.org/10.1016/j.agee.2012.10.008

Lin BB (2007) Agroforestry management as an adaptive strategy against potential microclimate extremes in coffee agriculture. Agric For Meteorol 144:85–94. https://doi.org/10.1016/j.agrformet.2006.12.009

Liu SL, Zhao HD, Su XK, Deng L, Dong SK, Zhang X (2015) Spatio-temporal variability in rangeland conditions associated with climate change in the Altun Mountain National Nature Reserve on the Qinghai-Tibet Plateau over the past15 years. Rangel J 37:67–75

Loft L, Mann C, Hansjürgens B (2015) Challenges in ecosystem services governance: Multi-levels, multi-actors, multi-rationalities. Ecosyst Serv 16:150–157. https://doi.org/10.1016/j.ecoser.2015.11.002

Long H, Liu Y, Li X, Chen Y (2010) Building new countryside in China: a geographical perspective. Land Use Policy 27:457–470. https://doi.org/10.1016/j.landusepol.2009.06.006

Lopez-Bravo DF, Virginio ED, Avelino J (2012) Shade is conducive to coffee rust as compared to full sun exposure under standardized fruit load conditions. Crop Prot 38:21–29. https://doi.org/10.1016/j.cropro.2012.03.011

Lybbert TJ, Sumner DA (2012) Agricultural technologies for climate change in developing countries: policy options for innovation and technology diffusion. Food Policy 37:114–123. https://doi.org/10.1016/j.foodpol.2011.11.001

Ma S, Swinton SM, Lupi F, Jolejole-Foreman C (2012) Farmers' willingness to participate in payment-for-environmental-services programmes. J Agric Econ 63(3):604–626. https://doi.org/10.1111/j.1477-9552.2012.00358.x

Maleksaeidi H, Karami E (2013) Social- ecological resilience and sustainable agriculture under water scarcity. Agroecol Sustain Food Syst 37:262–290. https://doi.org/10.1080/10440046.2012.746767

Maleksaeidi H, Karami E, Zamani GH, Rezaei-Moghaddam K, Hayati D, Masoudi M (2016) Discovering and characterizing farm households' resilience under water scarcity. Environ Dev Sustain 18(2):499–525. https://doi.org/10.1007/s10668-015-9661-y

Malek Saeidi H, Keshavarz M, Karami E, Eslamian S (2017) Climate change and drought: Building resilience for unpredictable future. In: Eslamian S, Eslamian F (eds) Handbook of drought and water scarcity: environmental impacts and analysis of drought and water scarcity. Taylor and Francis, CRC Group, Boca Raton, pp 163–186

Mann C, Loft L, Hansjürgens B (2015) Governance of ecosystem services: lessons learned for sustainable institutions. Ecosyst Serv 16:275–281. https://doi.org/10.1016/j.ecoser.2015.11.003

Manson SM, Jordan NR, Nelson KC, Brummel RF (2016) Modeling the effect of social networks on adoption of multifunctional agriculture. Environ Model Softw 75:388–401. https://doi.org/10.1016/j.envsoft.2014.09.015

Marzban S, Allahyari MS, Damalas CA (2016) Exploring farmers' orientation towards multifunctional agriculture: insights from northern Iran. Land Use Policy 59:121–129. https://doi.org/10.1016/j.landusepol.2016.08.020

Matson PA, Parton WJ, Power AG, Swift MJ (1997) Agricultural intensification and ecosystem properties. Science 277:504–509. https://doi.org/10.1126/science.277.5325.504

Mazzoncini M, Sapkota TB, Bárberi P, Antichi D, Risaliti R (2011) Long-term effect of tillage, nitrogen fertilization and cover crops on soil organic carbon and total nitrogen content. Soil Tillage Res 114:165–174. https://doi.org/10.1016/j.still.2011.05.001

McCune MN, Guevara-Hernández F, Nahed-Toral J, Mendoza-Nazar P, Ovando-Cruz J, Ruiz-Sesma B, Medina-Sanson L (2012) Social-ecological resilience and maize farming in Chiapas, Mexico. In: Curkovic S (ed) Sustainable development–authoritative and leading edge content for environmental management. INTECH open Access Publisher, Croatia

McFadden JE, Hiller TL, Tyre AJ (2011) Evaluating the efficacy of adaptive management approaches: is there a formula for success? J Environ Manag 92:1354–1359. https://doi.org/10.1016/j.jenvman.2010.10.038

Meehan TD, Werling BP, Landis DA, Gratton C (2011) Agricultural landscape simplification and insecticide use in the Midwestern United States. Proc Natl Acad Sci USA 108:11500–11505. https://doi.org/10.1073/pnas.1100751108

Mendes IC, Bandick AK, Dick RP, Bottomley PJ (1999) Microbial biomass and activities in soil aggregates affected by winter cover crops. Soil Sci Soc Am J 63:873–881. https://doi.org/10.2136/sssaj1999.634873x

Meng F, Qiao Y, Wu W, Smith P, Scott S (2017) Environmental impacts and production performances of organic agriculture in China: a monetary valuation. J Environ Manag 188:49–57. https://doi.org/10.1016/j.jenvman.2016.11.080

Metzger M, Schröter D, Leemans R, Cramer W (2008) A spatially explicit and quantitative vulnerability assessment of ecosystem service change in Europe. Reg Environ Change 8:91–107. https://doi.org/10.1007/s10113-008-0044-x

Millennium Ecosystem Assessment (2005) Ecosystems and human well-being- synthesis. Island Press, Washington, DC

Mishra AK, Singh VP (2010) A review of drought concepts. J Hydrol 391:202–216. https://doi.org/10.1016/j.jhydrol.2010.07.012

Miyan MA (2015) Droughts in Asian least developed countries: vulnerability and sustainability. Weather Climate Extremes 7:8–23. https://doi.org/10.1016/j.wace.2014.06.003

Montanarella L (2007) Trends in land degradation in Europe. In: Sivakumar MVK, Ndiang'ui N (eds) Climate and land degradation, environmental science and engineering. Springer, Berlin

Mukheibir P, Kuruppu N, Gero A, Herriman J (2013) Overcoming cross-scale challenges to climate change adaptation for local government: a focus on Australia. Clim Change 121 (2):271–283. https://doi.org/10.1007/s10584-013-0880-7

Munich Re (2013) 2012 natural catastrophe year in review. https://www.munichre.com/…/MunichRe_III_NatCat01032013.pdf. Accessed 13 Nov 2016

Mwinjaka O, Gupta J, Bresser T (2010) Adaptation strategies of the poorest farmers of drought-prone Gujarat. Climate Dev 2(4):346–363

Nhemachena C, Hassan R (2007) Micro-level analysis of farmers adaptation to climate change in Southern Africa. International Food Policy Research Institute, Washington DC, Discussion paper 714

Noponen MRA, Edwards-Jones G, Haggar JP, Soto G, Attarzadeh N, Healey JR (2012) Greenhouse gas emissions in coffee grown with differing input levels under conventional and organic management. Agr Ecosyst Environ 151:6–15. https://doi.org/10.1016/j.agee.2012.01.019

O'Brien G, O'Keefe P, Rose J, Wisner B (2006) Climate change and disaster management. Disasters 30(1):64–80. https://doi.org/10.1111/j.1467-9523.2006.00307.x

OCHA (2001a) Afghanistan: three million people are affected by drought. http://www.unocha.org/top-stories/all-stories/afghanistan-three-million-people-are-affected-drought. Accessed 9 Aug 2011

OCHA (2001b) Iran- drought OCHA situation report No. 1. http://reliefweb.int/node/83752. Accessed 9 Aug 2011

OCHA (2011) Afghanistan: Three million people are affected by drought. http://www.unocha.org/top-stories/all-stories/afghanistan-three-million-people-are-affected-drought. Accessed 28 Dec 2016

OECD (2001) Multifunctionality: towards and analytical framework. Organization for Economic Cooperation and Development, Paris Cedex, France

Opdam P, Coninx I, Dewulf A, Steingröver E, Vos C, van der Wal M (2015) Framing ecosystem services: affecting behaviour of actorsin collaborative landscape planning? Land Use Policy 46:223–231. https://doi.org/10.1016/j.landusepol.2015.02.008

Ostrom E (2007) A diagnostic approach for going beyond panaceas. Proc Nat Acad Sci USA 104 (39):15181. https://doi.org/10.1073/pnas.0702288104

Owen G (2008) Drought and the environment. http://www.southwestclimatechange.org/impacts/land/drought. Accessed 09 Jan 2017

Page G, Bellotti B (2015) Farmers value on-farm ecosystem services as important, but what are the impediments to participation in PES schemes? Sci Total Environ 515–516:12–19. https://doi.org/10.1016/j.scitotenv.2015.02.029

Pagiola S, Rios AR, Arcenas A (2010) Poor household participation in payments for environmental services: lessons from the Silvopastoral project in Quindío, Colombia. Environ Res Econ 47:371–394. https://doi.org/10.1007/s10640-010-9383-4

Pahl-Wostl C (2009) A conceptual framework for analyzing adaptive capacity and multi-level learning processes in resource governance regimes. Global Environ Change 19(3):354–365. https://doi.org/10.1016/j.gloenvcha.2009.06.001

Pahl-Wostl C, Sendzimir J, Jeffrey P, Aerts J, Berkamp G, Cross K (2007) Managing change toward adaptive water management through social learning. Ecol Soc 12(2):30, http://www.ecologyandsociety.org/vol12/iss2/art30

Palm C, Blanco-Canqui H, DeClerck F, Gatere L, Grace P (2014) Conservation agriculture and ecosystem services: an overview. Agr Ecosyst Environ 187:87–105. https://doi.org/10.1016/j.agee.2013.10.010

Patrick I, Barclay E (2009) If the price is right: farmer attitudes to producing environmental services. Aust J Environ Manag 16:36–46

Pauw ED (2005) Monitoring agricultural derought in the Near East. In: Boken VK, Cracknell AP, Heathcote RL (eds) Monitoring and predicting agricultural drought: a global study. Oxford University Press, New York

Pelling M (2011) Urban governance and disaster risk reduction in the Caribbean: the experiences of Oxfam GB. Environ Urban 23(2):383–400. https://doi.org/10.1177/0956247811410012

Peng J, Liu Z, Liu Y, Hu X, Wang A (2015) Multifunctionality assessment of urban agriculture in Beijing city, China. Sci Total Environ 537:343–351. https://doi.org/10.1016/j.scitotenv.2015.07.136

Perfecto I, Vandermeer J, Bautista G, Nuñez G, Greenberg R, Bichier P, Langridge S (2004) Greater predation in shaded coffee farms: the role of resident neotropical birds. Ecology 85:2677–2681. https://doi.org/10.1890/03-3145

Philpott S, Soong O, Lowenstein J, Pulido A, Lopez D, Flynn D, DeClerck F (2009) Functional richness and ecosystem services: bird predation on arthropods in tropical agroecosystems. Ecol Appl 19:1858–1867. https://doi.org/10.1890/08-1928.1

Pittelkow CM, Liang X, Linquist BA (2015) Productivity limits and potentials of the principles of conservation agriculture. Nature 517:365–368. https://doi.org/10.1038/nature13809

Poppenborg P, Koellner T (2013) Do attitudes toward ecosystem services determine agricultural land use practice? An analysis of farmers' decision-making in a South Korean watershed. Land Use Policy 31:422–429. https://doi.org/10.1016/j.landusepol.2012.08.007

Porter JR, Xie L, Challinor A, Cochrane K, Howden SM, Iqbal MM, Lobell D, Travasso MI (2014) Food security and food production systems. In: Aggarwal P, Hakala K (eds) Climate change 2014: impacts, adaptation, and vulnerability. Contribution of working group II to the fifth assessment report of the Intergovernmental Panel on Climate Change. IPCC, New York, US

Power AG (2010) Ecosystem services and agriculture: tradeoffs and synergies. Philos Trans R Soc B Biol Sci 365(1554):2959–2971. https://doi.org/10.1098/rstb.2010.0143september2010

Powlson D, Stirling CM, Thierfelder C, White RP, Jat ML (2016) Does conservation agriculture deliver climate change mitigation through soil carbon sequestration in tropical agro-ecosystems? Agr Ecosyst Environ 220:164–174. https://doi.org/10.1016/j.agee.2016.01.005

Primmer E, Furman E (2012) Operationalising ecosystem service approaches for governance: do measuring, mapping and valuing integrate sector-specific knowledge systems? Ecosyst Serv 1:85–92. https://doi.org/10.1016/j.ecoser.2012.07.008

Rao AS, Boken VK (2005) Monitoring and managing agricultural drought in India. In: Boken VK, Cracknell AP, Heathcote RL (eds) Monitoring and predicting agricultural drought: a global study. Oxford University Press, New York

Reganold JP, Andrews PK, Reeve JR, Carpenter-Boggs L, Schadt CW, Alldredge JR, Ross CF, Davies NM, Zhou J (2010) Fruit and soil quality of organic and conventional strawberry agro-ecosystems. PLoS ONE 5:e12346. https://doi.org/10.1371/journal.pone.0012346

Renting H, Rossing WAH, Groot JCJ, Van der Ploeg JD, Laurent C, Perraud D, Stobbelaar DJ, Van Ittersum MK (2009) Exploring multifunctional agriculture: a review of conceptual approaches and prospects for an integrative transitional framework. J Environ Manag 90:S112–S123. https://doi.org/10.1016/j.jenvman.2008.11.014

Ricketts TH, Regetz J, Steffan-Dewenter I, Cunningham SA, Kremen C, Bogdanski A, Gemmill-Herren B, Greenleaf SS, Klein AM, Mayfield MM, Morandin LA, Ochieng A, Viana BF (2008) Landscape effects on crop pollination services: are there general patterns? Ecol Lett 11:1121. https://doi.org/10.1111/j.1461-0248.2008.01157.x

Robertson GP, Dale VH, Doering OC, Hamburg SP, Melillo JM, Wander MM, Parton WJ, Adler PR, Barney JN, Cruse RM, Duke CS, Fearnside PM, Follett RF, Gibbs HK, Goldemberg J, Mladenoff DJ, Ojima D, Palmer MW, Sharpley A, Wallace L, Weathers KC, Wines JA, Wilhelm WW (2008) Sustainable biofuels redux. Science 322:49–50. https://doi.org/10.1126/science.1161525

Robertson GP, Gross KL, Hamilton SK, Landis DA, Schmidt TM, Snapp SS, Swinton SM (2014) Farming for ecosystem services: an ecological approach to production agriculture. Bioscience. https://doi.org/10.1093/biosci/biu037

Ronneberg E, Nakalevu T, Leavai P (2013) Report on adaptation challenges in Pacific Island Countries. Secretariat of the Pacific Regional Environment Programme. https://www.sprep.org/publications/adaptation-challenges-in-pics. Accessed 11 Nov 2016

Rossing WAH, Zander P, Josien E, Groot JCJ, Meyer BC, Knierim A (2007) Integrative modelling approaches for analysis of impact of multifunctional agriculture: a review for France, Germany and the Netherlands. Agr Ecosyst Environ 120:41–57. https://doi.org/10.1016/j.agee.2006.05.031

Rouphael Y, Cardarelli M, Schwarz D, Franken P, Colla G (2012) Effects of drought on nutrient uptake and assimilation in vegetable crops. In: Aroca R (ed) Plant responses to drought stress. Springer, Berlin

Royer PD, Breshears DD, Zou CB, Cobb NS, Kurc SA (2010) Ecohydrological energy inputs for coniferous gradients: Responses to management- and climate-induced tree reductions. For Ecol Manage 260:1646–1655. https://doi.org/10.1016/j.foreco.2010.07.036

Royer PD, Cobb NS, Huang CY, Breshears DD, Adams HD, Villegas JC (2011) Extreme climatic event-triggered overstorey vegetation loss increases understorey solar input regionally: primary and secondary ecological implications. J Ecol 99:714–723. https://doi.org/10.1111/j.1365-2745.2011.01804.x

Rusch A, Chaplin-Kramer R, Gardiner MM, Hawro V, Holland J, Landis D, Thies C, Tscharntke T, Weisser WW, Winqvist C, Woltz M, Bommarco R (2016) Agricultural landscape simplification reduces natural pest control: a quantitative synthesis. Agr Ecosyst Environ 221:198–204. https://doi.org/10.1016/j.agee.2016.01.039

Sandhu HS, Wratten SD, Cullen R (2010) Organic agriculture and ecosystem services. Environ Sci Policy 13:1–7. https://doi.org/10.1016/j.envsci.2009.11.002

Sardans J, Penuelas J (2012) The role of plants in the effects of global change on nutrient availability and stoichiometry in the plant-soil system. Plant Physiol 160:1741–1761. https://doi.org/10.1104/pp.112.208785

SCBD (2009) Connecting biodiversity and climate change: report of the second ad hoc technical expert group on biodiversity and climate change. CBD, UNEP, Montreal, Canada

Scherr SJ, Bennett MT, Loughney M, Canby K (2006) Developing future ecosystem service payments in China: lessons learned from international experience. Forest Trends, Washington, DC

Schoon M, Fabricius C, Anderies JM, Nelson M (2011) Synthesis: vulnerability, traps, and transformations—long-term perspectives from archaeology. Ecol Soc 16(2):24, http://www.ecologyandsociety.org/vol16/iss2/art24

Seppelt R, Dormann CF, Eppink FV, Lautenbach S, Schmidt S (2011) A quantitative review of ecosystem service studies: approaches, shortcomings and the road ahead. J Appl Ecol 48:630–636. https://doi.org/10.1111/j.1365-2664.2010.01952.x

Shackelford G, Steward PR, Benton TG, Kunin WE, Potts SG, Biesmeijer JC (2013) Comparison of pollinators and natural enemies: a meta-analysis of landscape and local effects on abundance and richness in crops. Biol Rev 88:1002–1021. https://doi.org/10.1111/brv.12040

Shackleton CM, Shackleton SE (2006) Household wealth status and natural resource use in the Kat River Valley South Africa. Ecol Econ 57(2):306–317. https://doi.org/10.1016/j.ecolecon.2005.04.011

Shakya N, Yamaguchi Y (2007) Drought monitoring using vegetation and LST indices in Nepal and northeastern India. In: 28th Asian conference on remote sensing, Japan

Sheidaei F, Karami E, Keshavarz M (2016) Farmers' attitude towards wastewater use in Fars province, Iran. Water Policy 18(2):355–367. https://doi.org/10.2166/wp.2015.045

Shinn JE (2016) Adaptive environmental governance of changing social-ecological systems: empirical insights from the Okavango Delta, Botswana. Global Environ Change 40:50–59. https://doi.org/10.1016/j.gloenvcha.2016.06.011

Shultz SD (2005) Evaluating the acceptance of wetland easement conservation offers. Rev Agric Econ 27:259–272. https://doi.org/10.1111/j.1467-9353.2005.00225.x

Sietz D, Boschütz M, Klein RJT (2011) Mainstreaming climate adaptation into development assistance: Rationale, institutional barriers and opportunities in Mozambique. Environ Sci Policy 14(4):493–502. https://doi.org/10.1016/j.envsci.2011.01.001

Simelton E, Fraser E, Termansen M, Benton T, Gosling SN, South A, Arnell NW, Challinor A, Dougill A, Forster P (2012) The socioeconomics of food crop production and climate change vulnerability: a global scale quantitative analysis of how grain crops are sensitive to drought. Food Secur 4(2):163–179. https://doi.org/10.1007/s12571-012-0173-4

Simelton E, Viet Dam B (2014) Farmers in NE Viet Nam rank values of ecosystems from seven land uses. Ecosyst Serv 9:133–138. https://doi.org/10.1016/j.ecoser.2014.04.008

Smit B, Skinner M (2002) Adaptation options in agriculture to climate change: a typology. Mitig Adapt Strat Glob Change 7:85–114. https://doi.org/10.1023/A:1015862228270

Smith HF, Sullivan CA (2014) Ecosystem services within agricultural landscapes—Farmers' perceptions. Ecol Econ 98:72–80. https://doi.org/10.1016/j.ecolecon.2013.12.008

Smith P, Martino D, Cai Z, Gwary D, Janzen H, Kumar P, McCarl B, Ogle S, O'Mara F, Rice C, Scholes B, Sirotenko O, Howden M, McAllister T, Pan G, Romanenkov V, Schneider U, Towprayoon S, Wattenbach M, Smith J (2008) Greenhouse gas mitigation in agriculture. Philos Trans R Soc B Biol Sci 363:789–813. https://doi.org/10.1098/rstb.2007.2184

Stanke C, Kerac M, Prudhomme C, Medlock J, Murray V (2013) Health effects of drought: a systematic review of the evidence. https://www.ncbi.nlm.nih.gov/pubmed/23787891. Accessed 09 Jan 2017

Steduto P, Faures JM, Hoogeveen J, Winpenny J, Burke J (2012) Coping with water scarcity: an action framework for agriculture and food security. Food and Agriculture Organization of the United Nations, Rome

Stehlik D, Gray I, Lawrence G (1999) Drought in the 1990s: Australian farm families' experiences. RIRDC Pub. No. 99/14. Rural Industries Research and Development Corporation, Canberra, Australia

Stoate C, Baldi A, Beja P, Boatman ND, Herzon I, van Doorn A (2009) Ecological impacts of early 21st century agricultural change in Europe- a review. J Environ Manage 91:22–46. https://doi.org/10.1016/j.jenvman.2009.07.005

Sullivan CA (2011) Quantifying water vulnerability: a multi-dimensional approach. Stoch Env Res Risk Assess 25(4):627–640. https://doi.org/10.1007/s00477-010-0426-8

Sullivan J, Amacher GS, Chapman S (2005) Forest banking and forest landowners foregoing management rights for guaranteed financial returns. Forest Policy Econ 7:381–392. https://doi.org/10.1016/j.forpol.2003.07.001

Sun Y, Zhou H, Wang J, Yuan Y (2012) Farmers' response to agricultural drought in paddy field of southern China: a case study of temporal dimensions of resilience. Nat Hazards 60(3):865–877. https://doi.org/10.1007/s11069-011-9873-x

Suter JF, Poe GL, Bills NL (2008) Do landowners respond to land retirement incentives? Evidence from the conservation reserve enhancement program. Land Econ 84:17–30

Swinton SM, Lupi F, Robertson GP, Hamilton SK (2007) Ecosystem services and agriculture: cultivating agricultural ecosystems for diverse benefits. Ecol Econ 64:245–252. https://doi.org/10.1016/j.ecolecon.2007.09.020

Swinton SM, Rector N, Robertson GP, Jolejole-Foreman MCB (2014) Farmers decisions about adopting environmentally beneficial practices. In: Hamilton SK, Doll JE, Robertson GP, Lupi F (eds) The ecology of agricultural ecosystems: long-term research on the path to sustainability. Oxford University Press, New York

Taninepasargad (2015) More than 11000 hectares of Fars province' forests impacted by drought. http://taninepasargad.ir/news/show/3249. Accessed 6 Feb 2015

Tao FL, Zhang Z (2010) Dynamic responses of terrestrial ecosystems structure and function to climate change in China. J Geophys Res Biogeosci 115. https://doi.org/10.1029/2009jg001062

Tao J, Fu M, Sun J, Zheng X, Zhang J, Zhang D (2014) Multifunctional assessment and zoning of crop production system based on set pair analysis- a comparative study of 31 provincial regions

in mainland China. Common Nonlinear Sci Number Simular 19:1400–1416. https://doi.org/10.1016/j.cnsns.2013.09.006
Tebna news agency (2013) Dalmatian pelican migration under drought. http://tebna.ir/fa/news/view/43339. Accessed 10 Feb 2015
Tengö M, von Heland J (2011) Trees and tree-planting in southern Madagascar: sacredness and remembrance. In: Tidball K, Krasny M (eds) Greening in the red zone–disaster, resilience and community greening. Springer, New York
Thierfelder C, Matemba-Mutasa R, Rusinamhodzi L (2015) Yield response of maize (Zea mays L.) to conservation agriculture cropping system in Southern Africa. Soil Tillage Res 146:230–242. https://doi.org/10.1016/j.still.2014.10.015
Thomas RH, Blakemore FB (2007) Elements of a cost-benefit analysis for improving salmonid spawning habitat in the River wye. J Environ Manage 82:471–480. https://doi.org/10.1016/j.jenvman.2006.01.004
Tiemann LK, Grandy AS, Atkinson EE, Marin-Spiotta E, McDaniel MD (2015) Crop rotational diversity enhances belowground communities and functions in an agroecosystem. Ecol Lett. https://doi.org/10.1111/ele.12453
Tilman D, Cassman KG, Matson PA, Naylor R, Polasky S (2002) Agricultural sustainability and intensive production practices. Nature 418:671–677. https://doi.org/10.1038/nature01014
Tipraqsa P, Craswell ET, Noble AD, Schmidt-Vogt D (2007) Resource integration for multiple benefits: Multifunctionality of integrated farming systems in Northeast Thailand. Agric Syst 94:694–703. https://doi.org/10.1016/j.agsy.2007.02.009
Tscharntke T, Klein AM, Kruess A, Steffan-Dewenter I, Thies C (2005) Landscape perspectives on agricultural intensification and biodiversity-ecosystem service management. Ecol Lett 8:857–874. https://doi.org/10.1111/j.1461-0248.2005.00782.x
UN (2008) Organic agriculture and food security in Africa. United Nations Conference on Trade and Development-United Nations Environment Program, Capacity-building Task Force on Trade, Environment and Development. http://www.unep-unctad.org/cbtf/publications/UNCTAD_DITC_TED_2007_15.pdf. Accessed 2 Apr 2011
UNESA (2015) World population projected to reach 9.7 billion by 2050. http://www.un.org/en/development/desa/news/population/2015-report.html. Accessed 25 Sept 2016
United Nations Conference on Environment and Development; UNCED (1992) Agenda 21-An action plan for the next century. Endorsed at the United Nations Conference on Environment and Development, New York
Valbuena D, Erenstein O, Tui SH (2012) Conservation agriculture in mixed crop–livestock systems: scoping crop residue trade-offs in Sub-Saharan Africa and South Asia. Field Crops Res 132:175–184. https://doi.org/10.1016/j.fcr.2012.02.022
Van Bael S, Philpott S, Greenberg R, Bichier P, Barber N, Mooney K, Gruner D (2008) Birds as predators in tropical agroforestry systems. Ecology 89:928–934. https://doi.org/10.1890/06-1976.1
van Noordwijk M, Hoang MH, Neufeldt H, Öborn I, Yatich T (2011) How trees and people can co-adapt to climate change: reducing vulnerability through multifunctional agroforestry landscapes. World Agroforestry Centre (ICRAF), Nairobi, Kenya
Verbruggen E, Roling WFM, Gamper HA, Kowalchuk GA, Verhoef HA, van der Heijden MGA (2010) Positive effects of organic farming on below-ground mutualists: large-scale comparison of mycorrhizal fungal communities in agricultural soils. New Phytol 186:968–979. https://doi.org/10.1111/j.1469-8137.2010.03230.x
Vereijken PH (2002) Transition to multifunctional land use and agriculture. NJAS-Wageningen J Life Sci 50(2):171–179. https://doi.org/10.1016/S1573-5214(03)80005-2
Veres A, Petit S, Conord C, Lavigne C (2013) Does landscape composition affect pest abundance and their control by natural enemies? A review. Agr Ecosyst Environ 166:110–117. https://doi.org/10.1016/j.agee.2011.05.027
Veron JEN, Hoegh-Guldberg O, Lenton TM, Lough JM, Obura DO, Pearce-Kelly P, Sheppard CRC, Spalding M, Stafford-Smith MG, Rogers AD (2009) The coral reef crisis:

the critical importance of b350 ppm CO_2. Mar Pollut Bull 58:1428–1436. https://doi.org/10.1016/j.marpolbul.2009.09.009

Vignola R, Harvey CA, Bautista-Solis P, Avelino J, Rapidel B, Donatti C, Martinez R (2015) Ecosystem-based adaptation for smallholder farmers: definitions, opportunities and constraints. Agr Ecosyst Environ 211:126–132. https://doi.org/10.1016/j.agee.2015.05.013

Vignola R, Koellner T, Scholz RW, McDaniels TL (2010) Decision-making by farmers regarding ecosystem services: Factors affecting soil conservation efforts in Costa Rica. Land Use Policy 27:1132–1142. https://doi.org/10.1016/j.landusepol.2010.03.003

Vrochidou AEK, Tsanis IK, Grillakis MG, Koutroulis AG (2013) The impact of climate change on hydro-meteorological droughts at a basin scale. J Hydrol 476:290–301. https://doi.org/10.1016/j.jhydrol.2012.10.046

Wall E, Smith B (2005) Climate change adaptation in light of sustainable agriculture. J Sustain Agric 27:113–123. https://doi.org/10.1300/J064v27n01_07

Wander M (2015) Soil fertility in organic farming systems: Much more than plant nutrition. http://articles.extension.org/pages/18636/soil-fertility-in-organic-farming-systems:-much-more-than-plant-nutrition. Accessed 1 Jan 2017

Wang J, Yang Y, Huang J, Chen K (2015) Information provision, policy support, and farmers' adaptive responses against drought: an empirical study in the North China Plain. Ecological Modelling 318:275–282. https://doi.org/10.1016/j.ecolmodel.2014.12.013

Wang SY, Yoon H, Cho C (2013) What caused the winter drought in western Nepal during recent years? J Clim 26(21):8241–8256. https://doi.org/10.1175/JCLI-D-12-00800.1

Waraich EA, Ahmad R, Saifullah U, Ashraf MY, Ehsanullah (2011) Role of mineral nutrition in alleviation of drought stress in plants. Aust J Crop Sci 5:764–777

Wezel A, Casagrande M, Celette F, Vian JF, Ferrer A, Peigné J (2014) Agroecological practices for sustainable agriculture. A review. Agronomy for Sustainable Development 34:1–20. https://doi.org/10.1007/s13593-013-0180-7

Wickham F, Kinch J, Lal P (2009) Institutional capacity within Melanesian countries to effectively respond to climate change impacts, with a focus on Vanuatu and the Solomon Islands. Samoa, SPREP, pp 1–84

Widawasky DA, ÓToole JC (1990) Prioritizing the rice biotechnology research agenda for Eastern India. Rockefeller Foundation Press, New York

Wiggering H, Mueller K, Werner A, Helming K (2003) The concept of multifunctionality in sustainable land development. In: Helming K, Wiggering H (eds) Sustainable development of multifunctional landscapes. Springer, Berlin, pp 3–18

Wilhite DA, Svoboda MD (2000) Drought early warning systems in the context of drought preparedness and mitigation. In: Wilhite DA, Sivakumar MVK, Wood DA (eds) Early warning systems for drought preparedness and management. World Meteorological Organization Switzerland, Geneva

Willaume M, Rollin A, Casagrande M (2014) Farmers in southwestern France think that their arable cropping systems are already adapted to face climate change. Reg Environ Change 14(1):333–345. https://doi.org/10.1007/s10113-013-0496-5

Winter SJ, Prozesky H, Esler KJ (2007) A case study of landholder attitudes and behaviour toward the conservation of Renosterveld, a critically endangered vegetation type in Cape Floral Kingdom, South Africa. Environ Manage 40:46–61. https://doi.org/10.1007/s00267-006-0086-0

Woods M, Woods MB (2007) Droughts. Learner Publications Minneapolis, USA

World Bank (2016) World development indicators. http://databank.worldbank.org/data/reports.aspx?source=world-development-indicators. Accessed 25 Sept 2016

Xiao JF, Zhuang Q, Liang E, McGuire AD, Moody A, Kicklighter DW, Melillo JM (2009) Twentieth century droughts and their impacts on terrestrial carbon cycling in China. Earth Interact 13(10):1–31. https://doi.org/10.1175/2009EI275.1

Yu J, Belcher K (2011) An economic analysis of landowners' willingness to adopt wetland and riparian conservation management. Can J Agric Econ 59:207–222. https://doi.org/10.1111/j.1744-7976.2011.01219.x

Zagonari F (2016) Using ecosystem services in decision-making to support sustainable development: critiques, model development, a case study, and perspectives. Sci Total Environ 548–549:25–32. https://doi.org/10.1016/j.scitotenv.2016.01.021

Zanella MA, Schleyer C, Speelman S (2014) Why do farmers join Payments for Ecosystem Services (PES) schemes? An assessment of PES water scheme participation in Brazil. Ecol Econ 105:166–176. https://doi.org/10.1016/j.ecolecon.2014.06.004

Zhang B, Zhang N, Zhang Y (2011) Performance assessment of agricultural adaptive measures to climate change. J Agrotech Econ 7:43–49. https://doi.org/10.1016/j.njas.2010.11.002

Zhang W, Ricketts TH, Kremen C, Carney K, Swinton SM (2007) Ecosystem services and dis-services to agriculture. Ecol Econ 64:253–260. https://doi.org/10.1016/j.ecolecon.2007.02.024

Zhang Z, Gao J, Fan X, Lan Y, Zhao M (2017) Response of ecosystem services to socioeconomic development in the Yangtze River Basin, Chana. Ecol Ind 72:481–493. https://doi.org/10.1016/j.ecolind.2016.08.035

Zhao HD, Liu SL, Dong SK, Su XK, Wang XX, Wu XY, Wu L, Zhang X (2015) Analysis of vegetation change associated with human disturbance using MODIS data on the rangelands of the Qinghai-Tibet Plateau. Rangel J 37:77–87. https://doi.org/10.1071/rj14061

Index

E. Lichtfouse et al. (eds.), *Sustainable Agriculture Reviews 28*, Sustainable Agriculture Reviews 28, https://doi.org/10.1007/978-3-319-90309-5

Printed in the United States
By Bookmasters